The Economics of Industrial Health

HD
7654

The Economics of Industrial Health

History, Theory, Practice

Joseph F. Follmann, Jr.

A Division of American Management Associations

Library of Congress Cataloging in Publication Data

Follmann, Joseph Francis, 1908-
    The economics of industrial health.

    Bibliography: p.
    Includes index.
    1. Industrial hygiene--United States. 2. Labor
and laboring classes--Medical care--United States.
I. Title.
HD7654.F65   331.2'55   77-25077
ISBN 0-8144-5444-5

© 1978 AMACOM
A division of American Management Associations, New York.
All rights reserved. Printed in the United States of America.

This publication may not be reproduced, stored in a retrieval system, or transmitted in whole or in part, in any form or by any means, electronic, mechanical, photocopying, recording, or otherwise, without the prior written permission of AMACOM, 135 West 50th Street, New York, N.Y. 10020.

First Printing

TO KAREN AND JEFFREY

For their very considerable assistance, sincere thanks are extended to

Edward M. Dolinsky
Metropolitan Life Insurance Company

Nixon de Tarnowsky
American National Standards Institute

John J. Walsh, M.D.
Union Carbide Corporation

Contents

Part I INTRODUCTION AND BACKGROUND

1 Man's Quest for Health   3
2 The Long History of Concern   5
   The Scourges of Mankind; Medical Care as It Was; The Forerunners of Industrial Health; Early Industrial Health Efforts; The Role of Voluntary Agencies; The Role of Unions; The Concern of Insurance Companies; The Beginnings of Health Insurance
3 Health Status in the United States   28
   Mortality; Diseases, Mental Illness; Alcoholism; Drug Abuse; Physical Impairments; Health Hazards; Accidents; Incidence of Disability
4 The Economic Costs   48
   Costs of Health Care; Costs of Disability; Costs of Certain Illnesses; Costs of Accidents; Controlling Costs and Expenditures

Part II THE HEALTH OF EMPLOYED PEOPLE

5 Incidence of Diseases and Accidents   67
   Diseases, Accidents; Incidence and Durations of Disability; Mental Illness; Premature Death
6 The Economic Costs   84
   Measurable Costs; Unmeasurable Costs

Part III THE POTENTIAL OF PREVENTIVE MEDICINE

7 The Concept   97
8 The Many Aspects of Preventive Medicine   100
   Public Health Measures; Combating Age-Old Enemies of Health; Immunization and Control of Infectious and Communicable Diseases; Eugenics; Prenatal and Postnatal Care and Pediatric Management; Prevention of Mental Illness; Accident Prevention; Occupational Health and Industrial Medicine; Health Education; Changing Lifestyles; Physical Examinations and Screening; Research; Rehabilitation
9 Those Engaged in Prevention   123
   The Private Sector; The Public Sector
10 Some Deterrents   127
   Public Attitudes and Behaviors; The Supply of Professional Personnel; The Question of Economic Deterrents; The Question of Financial Incentives
11 The Potential   139

Part IV INDUSTRIAL HEALTH PROGRAMS

12 The Concept   155
13 The Aspects of Industrial Health Programs   160
   Occupational Accidents and Diseases; Nonoccupational Diseases and Accidents; The Community at Large; Rehabilitation
14 The Role of Employers   178
   Occupational Safety and Health; Nonoccupational Health and Safety; The Community at Large; Rehabilitation; The Future
15 The Role of Unions   204
16 The Role of Government   211
   OSHA; Environmental Control; Product Safety; Public Health Efforts; Other Legislative Approaches

Part V WHAT IS BEING DONE?

17 Setting Up an Industrial Health Program   233
   Top Management Support; A Written Policy Statement; Applicability of the Program; Union Cooperation; Staffing and Facilities; Confidentiality of Records; The Role of Other Specialists; The Supervisor's Role; Insurance Protection; Reevaluation

18 What Does It Cost? 248
19 Who Is Doing It? 263
20 The Problems of Small Employers 279
21 Where to Receive Help 291
22 Some Special Problems 297
    Mental Illness; Alcoholism; Drug Abuse

Part VI RELATIONSHIP OF FINANCING PROGRAMS FOR HEALTH CARE AND LOSS OF INCOME

23 Publicly Established Programs 305
    Workmen's Compensation; Social Security Disability Benefits; Medicaid; CHAMPUS; Public Assistance Programs; Public Provision of Care; Temporary Disability Benefits; The Question of National Health Insurance
24 Private Insurance Programs and Employee Benefits 321
    Health Insurance; Disability Insurance; Salary-Continuance Programs; Life Insurance; The Role of Insurers in Prevention and Rehabilitation

Part VII THE COST-EFFECTIVENESS OF INDUSTRIAL HEALTH

25 Evaluating the Cost-Effectiveness of Preventive Medicine 347
    The Concept; The Effectiveness of Specific Programs; Effectiveness of Physical Examinations and Screening
26 The Cost-Effectiveness of Industrial Health Programs 368
    The Concept; The Problem; Some Findings; Alcoholism Control Programs; Pollution Control; The Need for Better Information

Part VIII THE FUTURE

27 Some Aspects of the Future of Industrial Health 401

APPENDIXES

A Occupational Safety and Health Act of 1970 415
B Established Industrial Health Programs 446
C Sources of Information, Assistance, and Guidance 450
D Annotated Bibliography 461

INDEX 463

Part I

Introduction and Background

Industrial health is not an isolated subject. It is inextricably interwoven into the entire warp and fabric of man's eternal quest for a healthy existence, a long fruitful life, and a sense of well-being. It is an indivisible part of the concern of a society to reduce to the greatest degree possible the incidence of disease, accidents, disability, and premature death. It evolves, rather logically, from a long history of concern on the part of those who, for whatever reason, feel and assume a sense of responsibility for the welfare of a people. As Thomas Mann has said in *The Magic Mountain:* "All interest in disease and death is only another expression of the interest in life."

The ultimate goal of an industrial health program, then, is the prevention of illnesses and injuries and the reduction of the occurrence of disability and premature death among employed people and within the community at large. Beyond the obvious physical aspects and the effects on personal and family lives, the economic consequences of such hazards are considerable: the inability to work, the cost of treatment and care, the loss of savings and assets, disruption of the family, and reduced education for the children. Employers are affected by absenteeism, labor turnover, reduced production, inefficiency, impaired employee morale, deteriorated customer and public relations, and increased costs for health and welfare benefits and insurance. Society feels the impact in many ways, including the production loss to the nation, the need to support various types of government programs made necessary by such occurrences, the financing of hospitals and other health facilities, and the contributions to the many voluntary agencies and other forms of charity.

Thus the health of people is not just a matter of individual concern. Modern society recognizes a responsibility that stems from humane considerations for fellow creatures as well as from a natural instinct for self-preservation. Today there is general recognition that the good health and economic security of a people are vital assets to a vigorous society. In order that that goal be accomplished, consideration must be given not only to the treatment and cure of disease but also to the preservation of health, education in health, and the prevention of illnesses and accidents to the greatest degree possible. The means available range from improved sanitation, water and air purification, pure food and drug surveillance, and pest control, to immunization, the inculcation of healthful lifestyles, and the adherence to safe methods of functioning. These are vital considerations in the expansion of health perspectives.

Here it cannot be overlooked that while modern industrialization, agriculture, and transportation have been accountable, directly or indirectly, for a variety of health hazards, the products that have accompanied the process of their development have at the same time perhaps been the single most significant factor in improving the health status of people generally. The results have been an increasingly improved standard of living, the general availability of a greater degree of education, adequate nutrition, safer and more wholesome housing, improved sanitation, advancing techniques in medical care, a broad scope of public health efforts, research into many aspects of health, improved working conditions and greater attention to the hazards of the workplace, shorter workdays and workweeks, an increased availability of recreation and vacations, and the presence generally of insurance protection against the costs of medical care and the partial replacement of income. Granted, such gains are not universal nor do they reach every member of society, but the forward movement is constant and continuous.

Thus Arnold Toynbee predicted: "The twentieth century will be remembered chiefly not as an age of political conflicts and technical inventions, but as an age in which human society dared to think of the health of the whole human race as a practical objective."

1

Man's Quest for Health

Mankind has always placed primary value on good health and a long life. From the beginning people must have had a concern about such matters. Clearly any condition that created pain or suffering, that resulted in the inability to follow customary and necessary pursuits, or that brought about early death would be a matter for attention. Man had, then, to gradually acquire the knowledge and the means for overcoming such conditions. Certainly he would observe that sufficient food and adequate clothes and shelter were essential to this end. He would seek cures for those ailments that did develop, and gradually he would seek the means of preventing the onset of such conditions.

Yet, as Victor Fuchs has observed, while universally we give lip service to the idea that health comes first, in the daily performance of our lives we must constantly make choices—choices concerning the use of our time, the expenditure of our resources, and the exertion of effort and willpower—and these choices most usually incline toward the creature comforts, the status symbols, and the easiest way.[1]

To this day there is, within the human species, an imponderable that would appear to defy comprehension. Man, with his sense of reason, is perhaps the only living form that knows that someday he is to die. This he resists with every fiber in his being, and he has had considerable success in devising the means by which the inevitable is postponed. Yet at the same time he quite obviously goes out of his way to bring about his early demise. He disdains safety precautions at work, on the highway, and even in his own home. He overeats and goes on crash diets. He permits himself a totally sedentary existence and then overexercises. He subjects himself to undue stress both on and off the job. He ignores his family history and breeds with infinitely less concern than a farmer raises pigs. He ignores the advice of his physician and neglects immunization. He contracts and spreads venereal disease. He pollutes his environment with his own filth. He owns, carries, and uses firearms, smokes like a chimney, drinks like a fish, and freely consumes pep pills and sleeping pills, not to mention a host of other drugs. He burns Edna St. Vincent Millay's candle at both ends. Knowing full well that he is mortal, he clearly behaves as though, like the gods, he were immortal.

Perhaps to overcome this contradiction man many ages ago postulated and evolved the concept of a life after death. Archaeological findings make clear beyond any doubt that for as far back as there are evidences of human existence there has been a belief in a life hereafter. In fact, it is to this concept, more than any one other fact, that we today are indebted for what we know of human life long before historical records were made and preserved. The burial places of the Mesopotamians, the Egyptians, the Scythians, the Vikings, the Mayan Indians, the Chinese, and many others have bequeathed an enormous treasure displaying the ingenuity, skill, and artistry of early man.

REFERENCES

1. Victor Fuchs, *Who Shall Live?* (New York: Basic Books, 1975).

2

The Long History of Concern

LIFE AND DEATH THROUGH THE AGES

For practically all of the millenia of history, in fact until the present century, a human being's span of life was exceedingly brief. Bronowski has speculated that some five million years ago the cousin of man, Australopithecus, died before the age of 20.[1] History is full of exceptions, but the average span of life in ancient Greece is generally accepted to have been about 20 years. In the Roman Empire this increased slightly to 23 years. In England of the Middle Ages it is thought to have been about 33 years. In America, at the time of the signing of the Declaration of Independence, it was about 35 years. Slow progress indeed, although between ancient Greece and nineteenth-century England the average span of life had doubled. Unfortunately in many nations in the world today life expectancy is hardly better than that in the world of the ancients.

THE SCOURGES OF MANKIND

Scientific examinations of the remains of humans who died several thousands of years ago bear evidence of many of the diseases with which we are familiar today. The spine of Heidelberg Man (7000 B.C.) bears evidence of Pott's disease, as do several of the Egyptian mummies buried later. Other human remains show the presence of tapeworms, roundworms, clubfoot, and schistosomiasis, polio, ruptured spleens, arthritis deformans, leprosy, gout, fever, and heart disease. Evidences have also been found of substantial carbon deposits in the lungs, probably as a result of smoke from cooking fires, particularly in caves.

Written history documents the presence of blindness, deafness, and dumbness, and the records of ancient Rome reveal cancer, liver disorders, spinal deformities, elephantiasis, vertigo, palsy, stones, gout, pleurisy, fever, nausea, epilepsy, rheumatism, stomach weakness, and infectious disease. The records of the Egyptians, the Greeks, and the Israelites contain vague descriptions of chest pains. Pliny the Younger (circa 79 A.D.) tells us of fatal ulcers "in those parts which modesty conceals." Plutarch also notes such causes of death as drowning, drunkenness, lightning, poison, strangulation, and suicide.

Other destroyers of life also tormented human beings. Earthquakes, floods, and great winds were not infrequent. Rains of frogs, serpents, and lizards are told of, as are sheets of heavenly fire such as those that struck the Papal Palace at Avignon. Air pollution from pollens, gases, mists, and volcanic ash and dust must at times have been deadly. And such debilitating occurrences as broken bones, sprains, and toothaches certainly added to the miseries.

Where humans appeared particularly helpless was in the mass onslaughts of death and disease that came with famines and epidemics. For a long time mortals were extremely vulnerable in their presence.

Famine only disappeared in Western Europe toward the end of the eighteenth century although Ireland had severe famines in 1821-22, and 1849, and famine has continued to appear in the twentieth century in China, India, and other nations.

The plague, an ill-defined disease entity, devastated humanity throughout time. In the fourteenth century alone a quarter to a half of the population of Europe, some 25 million persons, died of it. Other deadly epidemic diseases included typhus, cholera, dysentary, diarrhea, diphtheria, influenza, malaria, yellow fever, smallpox, pellagra, measles, and whooping cough.

For a long time these scourges served to keep the population of the world in balance. Not infrequently did death exceed the number of births in a given place. It appears to have been only about the later eighteenth century when life began to gain over death. Thus while the population of the world has been estimated to have been 500 million in 1650, it increased to some 700 million by 1750, to 900 million by 1800, to over one billion by 1850. By 1900 it had grown to one and a half billion, and in the next 50 years another billion were added to the planet's population.² Today the world population approaches 4 billion, and it is estimated that by the turn of the century it will be 6 billion and by 2050, 9 billion.

In all this devastation, people were distinct contributors to their own early demise. Their lack of personal cleanliness, their unsanitary living conditions, their toleration of pests and rodents, the eating of rotten foods, and their general filth all invited a host of disease. For centuries in Europe chamber pots and slop were emptied out of upper story windows, and Hogarth has left us graphic prints of such practices (which is why a gentlemen to this day, walking with his lady, walks on the outside of the sidewalk). Dwellings, in fact, were so constructed as to accommodate the practice. The streets were like sewers, flowing with filth. Drinking water was contaminated with human refuse. Bathrooms and toilets were a rarity in Europe until after the seventeenth century, despite the fact that Sir John Harington invented an English-style water closet in 1596. And even the great and glamorous palace at Versaille was described by a royal visitor as reeking with foul odors.

Violence was common: murder, robbery, and rape. In high places assassination was the rule, even among parents and children, brothers and sisters. And warfare between tribes, political factions, religious sects, and nations appears to have been a constant state of affairs.

People also did themselves in very directly, with no assistance from the outside. The Emperor Hadrian, for example, noted that overeating was a Roman vice and that the people poisoned themselves with spices. Syphilis has also been destroying human beings for a very long time, interred prehistoric skeletons bearing distinct marks of the disease. The abuse of alcoholic beverages of all types also produced deadly results.³

Throughout the ages, then, people have been the victims of all forms of scourges. Of some they were more or less innocent victims. Of others they were direct contributors, even if unwittingly. For some they found cures. For others they overtly continue to pursue a course that leads to their own destruction. Happily, albeit slowly, people discerned the causes of the vast majority of these sourges and in many instances have found the means of preventing their occurrence.

MEDICAL CARE AS IT WAS

When one considers this sorry state of affairs through the ages, it is of interest to take a brief look at the available medical care.

For many millenia reliance for the prevention and cure of human ills was placed in large part in a variety of religious rites, sacrifices, exorcisms, and incantations. Recourse was had to a priest, a witch doctor, or a magician. Faith was rested in gods, signs, symbols, numbers, charms, amulets, and astrology.

At the same time many types of herbs and drugs were used quite effectively. Gradually the physician emerged, at first apparently in ancient Egypt, Greece, and India. Among the famous were Im-hotep in Egypt (2900 B.C.) and Ascelepius, Hippocrates, and Galen in Greece. From the ninth to the twelfth centuries Arabian physicians such as Rhase, Haly Abbas, and Avicenna dominated the field. By the twelfth century Europeans such as Roger of Salerno assumed importance. While interest in anatomy and diagnosis gradually developed, the care provided left much to be desired. Meanwhile surgery, with antecedents in prehistoric man and later in Egypt (2500 B.C.), gradually refined its processes as the necessary instruments were improved.

Hospitals, with a history dating to 800 B.C. in India, gradually came into being. Medical schools, from one in Athens in 522 B.C. to those at Salerno, Montpellier, Paris, and Bologna in twelfth century Europe, became more numerous.

Finally the processes of medical care became more scientific with the knowledge of blood developed by Harvey in 1628, the improvements in anesthesia commencing in 1646, and the concept of immunization in 1796. It remained for the nineteenth century, however, to bring about such developments as the thermometer, the microscope, and the X ray; to recognize the value of antiseptics; and to discover bacteria.

Before leaving the subject of medical care, one final thought is relevant. Despite all their concern about life and health, people have continued to place primary value on the services of the healer. Reliance on the curative processes has taken marked precedence over the less costly and less devastating approach of preventing a disease or accident. Thus the curative process has been more tangible, more gratifying, more exciting, and more rewarding in every way.

For this we have had to pay dearly, albeit complaining all the while. Attempts to rectify the situation run through history. About 2250 B.C., Hammurabi's Code in ancient Babylon included the conditions of medical practice and the establishment of the fees to be charged a man of means, a freeholder, and a slave. In 1214 when the physician Hugh of Lucca was called to Bologna, a statute required that he treat everyone, except those suffering from hernia, at certain fees fixed according to the social status of the patient. Much later, in North America the Assembly of Maryland regulated surgeons' fees in 1638, and the colony of Virginia passed a law in 1646 providing that because of the "immoderate and excessive rates and prices exacted by practitioners of Physick and Chyrygery" a person might bring under arrest the offending physician who must swear to the "true value, worth, and quantity of the drugs and medicines." The court would thereupon determine the proper fee. In 1661 this law was amended to include the value of "care, visits, and attendance." In 1736 a specific fee schedule was established. Meanwhile, in 1685 an ordinance in Prussia regulated all medical fees. And while celebrating the Bicentennial we were struggling with systems of prospective rating for institutions of care, Comprehensive Health Planning, PSROs, and a host of other attempts to contain the costs of our health care.

THE FORERUNNERS OF INDUSTRIAL HEALTH

At this point it might well be asked what the relationship of the foregoing discussion is to industrial health. There is a direct relationship in that those events that killed or disabled human beings through the ages, and the means by which people attempted to cope with them, provided the very beginnings for what eventually came to be known as industrial health. As has been noted, the processes of industrial health do not exist in a vacuum but are responsive to and function in relation to the entire spectrum of health matters.

There is, furthermore, a body of historical evolution that quite directly relates to the concepts of industrial health. It relates in the sense that throughout history religious leaders, political leaders, and military commanders—those who had or assumed responsibility for the health and efficient functioning of their people—took certain measures devoted to that end. The problems they faced in so doing were in essence the same as those that face employers and union leaders today. In either case diseases and accidents, disability, and premature death impair the functioning of a body of people at full capacity and are problems that must be faced as an entity. In either case the goal is to prevent the occurrence of such events to the maximum extent possible.

Certainly the process has always been sporadic in the sense of moving forward, slipping back, totally forgetting what has already been learned, then going forward again. Where laxity has been severe or prolonged enough, whole societies have disappeared, wars been lost, and empires or civilizations fallen from positions of eminence to mediocrity.

One very early act toward the preservation of health probably was the disposal of the dead by burial, cremation, or casting into the waters. In recorded history, religious leaders have been in the forefront of endeavors to improve the public health of their people. In the Old Testament guidelines for the preservation of health are clearly laid down. Personal cleanliness, a sound diet, proper rest, and sanitation measures are called for. Jesus of Nazareth provided excellent advice when he taught temperance in all things. Then Mohammed cautioned against several deleterious health habits, advising a proper diet and demanding fasting. In the twentieth century it was Gandhi, who by personal example taught the people of India such a fundamental precaution against disease as the burying of their own excrement.

In ancient Greece healthful living was encouraged by the political leaders and as a consequence there was more or less general interest in the subject. Public baths, for example, were established and their remains exist today. Important developments grew out of this interest. It was about 500 B.C. when Empedocles suggested as a means for preventing malaria that the swamplands be drained (a precaution subsequently forgotten until this approach was adopted in the nineteenth century, thereby eliminating the transmitter of the disease, the mosquito). Later Hippocrates described the effects of lead poisoning. He also noted that "when more food than is proper has been taken, it occasions disease."

The Romans, too, had a keen interest in the prevention of disease and stressed the importance of personal cleanliness, exercise, and rest. Julius Caesar, knowing full well the devastating effects of such diseases as dysentary, scurvy, pellagra, beriberi, and typhus and other fevers on his legions, enforced sanitary practices and dietary measures. And the Emperor Hadrian demanded that the city streets be sprayed with clear water.

Other aspects of health were also explored. The health hazards of mining were described by Paracelsus, Agricola, and Ramazzini; and in 100 A.D. Pliny developed the idea of using sheep bladders as respiratory masks. Eating habits were, however, a problem. Horace tells us: "We rise from table pale from overeating." Later the practice of using emetics to make possible a greater amount of eating came into vogue; a villa at Pompeii bears evidence of this overindulgence in the form of a small side room adjoining the dining hall that served as a vomitorium.

Water supply pipes, a drainage system, and public baths and toilets were rather general throughout the Roman Empire by 21 A.D., and the Romans originated the practice of using soap for cleansing purposes. Later the Anglo-Saxons were known to have prescribed baths for therapeutic purposes, and circa 800 A.D. the Mayan Indians in Central America had recourse to purifying steam baths for sweating. All this was forgotten, too. Public baths disappeared in Europe (except, apparently, in Finland and Russia), and with them the idea of bathing.

The obtainment of a sufficient supply of pure water was also a nagging problem. In most of the cities there was recourse to public wells, but this presented a problem of contamination. In London in 1236 and later in 1285 water was piped in from outside sources. By the seventeenth century, however, the degree of retrogression is exemplified by the fact that in

Paris the means for supplying the city with water took the form of 20,000 water carriers.

When steam pumps were invented, the carriers protested for fear of mass unemployment. At the same time, incidentally, the matter of an adequate water supply was a problem in China, but was at least partially solved in Peking by storing the winter ice. Small wonder that among the relics of ancient civilizations are the means for establishing public policy for governing water rights.

Housing also had a most harmful effect on health. The walls and floors were damp, heating facilities were inadequate and smoky, lighting came from mere holes in the wall or from lamps that gave off foul oil fumes, and the floors were covered with rush or straw that contained the remains of food, bones, and vermin. These conditions were by no means limited to the poor. The White Tower at the Tower of London, which housed the royal family and their guests, had the same unsanitary conditions. It was not until 1624 that Savot brought about improvements in fireplaces and flues, 1745 until Cooke introduced steam heating, 1783 until gas lighting was introduced in Louvain, 1821 until Duvoir developed hot water heating, and 1878 until Edison developed the electric lamp. It cannot pass unnoted that in 1696, in England, a window tax served no other purpose than to deprive all but the rich of much needed light and fresh air.

Complicating the situation was the severe overcrowding in European cities. The streets were ridden with the filth dumped by mills, forges, stables, and slaughterhouses. Gradually ordinances attempted to cope with such practices, and men known as scavengers were appointed to dispose of rubbish, but without much success. In London, for example, a 1281 ordinance forbade swine in the streets, and a royal order in 1371 prohibited the slaughtering of animals in London. In 1388, at the time of Richard II, the First English Sanitary Act became law. Under Henry VII, in 1486, a law was passed regulating the operation of slaughterhouses, and in 1495 a law attempted to cope with contagion by fomites (bedding).

Water pollution was a persistent problem. In the European towns, from the thirteenth to the fifteenth centuries, brewers, dyers, tanners, butchers, fishmongers, washerwomen, and others contaminated the rivers, with further pollution being provided by latrines, sewage, garbage, and rubbish. In England religious orders attempted to counteract these practices by the construction of conduits to carry water from unpolluted springs to public wells and houses. By 1343, however, it was necessary to call for a report on faulty sanitation in London, and in 1375 a royal order prohibited the throwing of filth into the Thames. By 1532 a Commission of Sewers was established in England. Not until 1855 was the filtration of water made compulsory in London, with similar action being taken in Berlin the following year. In that same year the sewers of Paris were constructed. In 1859 sewage disposal was provided for in London and the next year in Munich. In the New World they had their problems, too. As early as 1647 Massachusetts had to enact a law prohibiting the pollution of Boston Harbor. In 1857 a National Sanitary Convention was held in Philadelphia. By 1800 Philadelphia had established a public sewage and water system. In 1848 Chicago completed a system for sewage disposal and in 1867 developed a system of tunnels for the supply of water. In 1885 New York City built its first conduit to supply water to its growing population.

EARLY INDUSTRIAL HEALTH EFFORTS

It was about the sixteenth century that an awareness of the health hazards arising from the processes of industry began to dawn upon communities, although there were a few antecedents, some in the ancient world having been noted earlier.

In 1546 the concern began to gain momentum when Agricola noted the problem of ventilation in the mines, and when Paracelsus reported on miner's phthisis in 1567. In 1572 the presence of lead poisoning was observed at Poitou and in 1575 Pare described monoxide poisoning. In 1630 arsenic poisoning was found in Paris, and 11 years later it was noted that the waters of the Seine, used by Parisians for drinking, cooking, and bathing, were being polluted by the dyers in the vicinity and some corrective measures were attempted.

In 1700 Ramazzini published his treatise on trade diseases, and that served to open up thought on the matter of industrial health. In 1743 Hales published his treatise on the importance of ventilation and then invented artificial ventilation. In 1775 Pott made the first discovery of occupation-related cancer when he noted the incidence of scrotal cancer among London chimney sweeps. The following year Benjamin Franklin found that the "West Indies gripes" was due to lead colic resulting from the consumption of Jamaica rum, which was manufactured in lead stills. Four years later Morveau stressed the dangers of industrial lead poisoning. In 1785 Watt invented a device for smoke abatement. Three years later the French Academy of Sciences expressed grave concern over the defective emptying of cesspools in Paris. At the end of the century von Humboldt invented the gas mask and a safety lamp for use in the mines (a refinement in the safety lamp being brought about by Davy in 1815).

The Industrial Revolution was now gaining momentum in Europe and in America. It brought with it many benefits, not the least of which was employment for the underemployed who had left the rural areas for the towns and cities. It provided many means for improvement in the health of people. Such basics as soap, cotton underwear, the iron coal range, glass for windows, and iron bedsteads all served to better the health conditions of ordinary people.

By the nineteenth century an awareness began to take hold, however, that industrialization was bringing in its wake a series of health hazards of consequence: long hours of work, child labor, and unsafe, and at times intolerable, working conditions. The pollution of water and air brought about by the growing industries gave evidence of health hazards that, until then, were unknown. The people who had crowded into the towns frequently suffered from inadequate diet, a lack of fuel, crowded living conditions, inadequate ventilation, an infiltration of pests and rodents, and inferior drainage systems. In the early nienteenth century, for example, Mathew Carey, a highly successful businessman in Philadelphia, brought such matters to the attention of his fellows by the issuance of a tract entitled "Appeal to the Wealthy of the Land," in which he attempted to make people aware that in that city "55 families, containing 253 individuals, are huddled together in 30 tenements without the convenience of a privy."

In 1832 Kay reported on the sanitary status of cotton spinners in England. In 1837 a paper was published in the United States on "The Influence of Trades, Professions, and Occupations in the United States in the Production of Disease," and in 1842 Chadwick made a report on the health of the English laboring classes. In 1850 the anthrax bacillus was discovered (for which Pasteur produced a vaccine in 1881), and in 1856 Perkins obtained aniline dyes. In 1859 Dr. John Snow investigated the circumstances of an epidemic of cholera in London and concluded that the water supply from the community pump had become polluted with human wastes, which, in turn, transmitted the disease. This epidemiological approach to disease led to confirmation that the disease was due to a parasitic microorganism. In this instance, however, the epidemic was brought under control by simply putting a lock on the pump handle. Two years later the records of an insurance company collecting agent in England revealed that in one industrial town, one-room dwellings were providing housing for anywhere from 10 to 18 persons each. The Seventh Report of the British Health Service in 1864 revealed the consequences: smallpox, typhus, scarlet fever.

About this time specific investigations, at an increasing rate, were also bringing to light the health hazards inherent in the new industrial processes. In 1845 the incidence of scrotal cancer among men working in copper smelters and exposed to inorganic arsenic was noted. In 1860 J. Addison Freeman, in the United States, published a study of mercurialism in the hatters' trade, and in that same year Dr. Greenhow issued a report on the potters in certain sections of England, which maintained that those workers suffered a shortened average length of life as a result of pulmonary diseases and that they increasingly became dwarfed and less robust. In that same year, as a result of the adulteration of food products, the British Parliament enacted a law "for preventing the adulterations of articles of food and drink." The adulterations noted at the time included perspiration, discharges from abcesses, cobwebs, soot, black beetles, and putrid yeast.

In 1863 Dr. J. T. Arledge did a further study of the English potters stating that these workers—men, women, and children—had become phlegmatic and bloodless, accompanied by attacks of dyspepsia, rheumatism, disorders of the liver and kidneys, pneumonia, phthisis, bronchitis, and asthma. In 1867 Karl Marx, in *Das Kapital,* described the working conditions in industrial England: child labor, 16-hour working days, overcrowded and poorly ventilated work rooms, and disease-breeding and life-shortening processes. He pointed out that whereas in some sections of England there were "only 9,000 deaths in the year" for every 100,000 children alive under the age of one year, in such industrial areas as Preston, Nottingham, Wisbeach, and Manchester the number of such deaths ranged from 24,000 to 26,000 in a year. Among the female and child lacemakers in Nottingham the number with tuberculosis rose from one in 45 in 1852 to one in 8 by 1860.

Other findings of the harm being done by the new industrial processes continued to be made. About 1875 Eulenburg published a handbook on industrial hygiene, and at that same time it was observed that skin cancer was caused by certain chemical substances. In 1880 it was observed that the miners in the Erz Mountains of Central Europe were dying of a lung malignancy. By 1910 it was found that the cause of this condition was radioactivity in those mines. Meanwhile, in 1886 Lehman investigated the effects of industrial poisons, and in 1893 it was discovered that aromatic amines were causing cancer of the bladder among German dye workers. Some years later certain of these amines were eliminated in Britain, Italy, Switzerland, and Japan. About 1900 asbestos was found to be a cause of fatal fibrosis of the lungs. Not until 1974 was it learned that vinyl chloride, a petrochemical gas used for the previous 35 years in the manufacture of many plastics, produced cancer of the liver and perhaps other organs.

Toward the end of the nineteenth century perhaps the most significant development occurred: the enactment in 1884 in Bismarck's Germany of the first workmen's compensation law. With direct financial responsibility placed on the employer for accidents and diseases in the workplace, and with the recovery of financial awards removed from the processes of legal liability, the employer, probably for the first time, became aware of the financial costs to business operations that resulted from accidents, diseases, and premature death. From that point in time employers would become increasingly interested in such financial costs and would of their own volition take steps to reduce them, if only out of self-interest. That was a great step forward. Industrial health, in a positive sense, had been born.

Ten years later Norway followed the example of Germany and by 1913 essentially similar laws were enacted in 41 nations. In the United States the movement of such laws through the states commenced in 1911, after earlier attempts dating from 1906 were declared to be unconstitutional. The enactment of workmen's compensation laws in the United States gave considerable impetus to the role of places of employment and insurers in the safety and health of workers. The onset of World War I, cutting off the steady flow of immigrants to America and taking large numbers of able-bodied men into the armed forces, required further interest on the part of employers in the health of their employees. The development of health insurance programs, starting in the late 1930s, with their costs to employers, created additional concerns. Interest increased rapidly in World War II, with millions of men in the armed services, and impetus was provided by the nature of the wartime wage-freeze legislation and income tax incentives.

THE GROWING ROLE OF GOVERNMENT

Increasingly through this period government at all levels in the United States came to play a role of significance in the control of diseases, improved sanitation, water purification, elimination of harmful influences in the environment, housing, food and drugs, and in working conditions and the health of workers. In 1780 the first board of health was established, at Petersburg, Virginia (by 1873, 134 cities in the United States had established boards of health). In 1816 Massachusetts established a Department of Factory Inspection and in 1836 enacted the first child labor law requiring that every employed child under the age of 15 attend school at least three months each year. The first action to control the hours of work took the form of a presidential order by President Van Buren limiting work in the navy yards to ten hours a day. In 1847 New Hampshire established a ten-hour working day as a general standard for all workers unless there was "an express contract requiring greater time." (In England at the time, factory working hours were reduced from 74 to 60 hours a week for adults and from 72 to 40 hours for children.) Pennsylvania, in 1848, prohibited the employment of children under age 12 in textile plants. The first federal law regarding working hours was passed by Congress in 1868, establishing for the first time the concept of an eight-hour working day. In 1877 Massachusetts enacted the first law in the United States requiring safeguards in factories (the Factory and Workshop Acts were passed in England in 1879), and in 1878 passed a law requiring the inspection of plumbing.

In 1881 New York State enacted the first Food and Drug Law in the United States, following the Adulteration of Food Act in the England in 1860, compulsory meat inspection in Germany in 1875, the Sale of Food and Drug Acts in England in 1875, and a German pure food law in 1879. In 1885 the U.S. Army constructed the first incinerator in the United States. In that same year Alabama enacted the first employers' liability law. The earliest bacteriological laboratory in the United States was set up in 1887 at the Marine Hospital Service on Staten Island, which eventually became the National Institutes of Health. Then, in 1890, Massachusetts passed the first law regulating the employment of women, and in 1893 California made mandatory one day of rest each week for female workers.

From 1900 onward a major period of formulation of social policy and legislation affecting the health of people gained momentum. In 1903 Illinois established an eight-hour working day for children under age 16. In 1906 the first Federal Food and Drugs Act became law. In 1908 Chicago made compulsory the pasteurization of milk, and disinfected the water supply in the stockyards with chlorine gas.

At this same time the first requirement for the reporting of occupational diseases was drafted by the American Association for Labor Legislation. In 1910 the U.S. Bureau of Mines was created and the Illinois State Commission on Industrial Diseases was appointed. The following year Illinois required monthly physical examinations for workers using lead, zinc, arsenic, brass, mercury, or phosphorus, and in 1912 the Division of Industrial Hygiene was established in the U.S. Public Health Service. In 1913 New York State appointed a Commission on Ventilation, and in 1914 the Public Health Service established the Office of Industrial Hygiene and Sanitation.

The onset of World War I brought about several developments of its own. For the first time there was compulsory delousing of the troops on a broad scale. For the first time the United States armed forces were required to be inoculated for typhoid fever. And for the first time there was a concentrated attack on venereal disease, another traditional crippler of fighting forces.

The period following World War I was marked with other government developments of significance in an attempt to cope with the problems of industrial expansion and growing urbanization, including the regulation of the employment of children and women; compulsory general education; the establishment of clinics, hospitals, and public health nurses; the prevention of occupational diseases and accidents; traffic regulation; food and drug surveillance; the provision of school lunches; housing regulations; compulsory immunization; rodent and pest control; and provisions for economic security.

THE ROLE OF VOLUNTARY AGENCIES

Toward the end of the nineteenth century voluntary agencies of various types, universities, and foundations in the United States also came to play a significant role in the prevention of diseases and accidents and in the improvement of industrial health. The American Medical Association was founded in 1847, holding its first symposium on industrial hygiene in 1915 and establishing a Council on Industrial Health in 1937. In 1872 the American Public Health Association was formed and in 1909 established a section on preventive medicine, industrial hygiene, and public health. In 1886 the New York Cancer Hospital was founded. In 1888 the American Association of Railway Surgeons came into being, followed by the American Society of Heating and Ventilating Engineers in 1894.

Following the turn of the century, the Rockefeller Institute for Medical Research was founded in New York in 1901. Three years later the National Tuberculosis Association, the first of this type of national voluntary organization, came into being. In 1905 the first university instruction in industrial hygiene was offered at the Massachusetts Institute of Technology. In 1906 more complete instruction in industrial hygiene was instituted at the University of Pennsylvania Medical School (followed by Johns Hopkins in 1916 and Harvard in 1918). In 1906, also, the American Association for Labor Legislation was organized and in 1910 convened the first national conference on industrial diseases. In 1910 the first occupational disease clinic was opened by the Cornell Medical School in New York City, and that same year cancer research was commenced at Memorial Hospital in New York City. In 1910 also the Carnegie Foundation sponsored the Flexner Report, which served to improve the level of medical schools in the United States tremendously.

In 1911 the National Council for Industrial Safety was formed, perhaps in reaction to the enactment of the first workmen's compensation laws in that year, and in 1915 the National Safety Council came into being. In that same year, S. S. Goldwater established an occupational disease clinic in New York City, and the following year the Industrial Medical Association of Preventive Medicine was established.

By 1919 the first issue of the *Journal of Industrial Hygiene* appeared. In 1921 the Industrial Health Service Bureau was formed in Chicago. By 1923 the National Health Council had launched a campaign to encourage periodic health examinations. By 1938 the American Conference of Government Industrial Hygienists was formed, and in 1955 the American Board of Preventive Medicine was founded. By this time several private foundations had developed with an interest in health, including the Mellon Institute, the Industrial Hygiene Foundation, the Melbank Memorial Foundation, and the Johnson Foundation.

The foregoing is by no means a complete recounting of the developments in the early part of the twentieth century in the United States that had the common purpose of controlling or eradicating certain diseases, accidents, or living or working conditions harmful to the health of people. It does, however, serve to indicate the breadth of the approaches taken, as well as the velocity in the growth of interest in the subject.

THE ROLE OF EMPLOYERS

During this period employers gradually developed a concern for the health of their employees. This concern was out of personal and humane considerations for their employees and at the same time reflected a growing realization that such matters had a direct relationship to productivity and the costs of operation. However, this interest was not universal. Prejudices and reactions to the newer immigrant population, upon which much of industrial development depended, ran high. Labor union pressures, mounting as the labor movement grew in importance, frequently resulted in open friction between labor and management. Yet the beginnings of a concept of industrial health and its importance to the process of production came into being. The following are just a few examples.

In 1869 the Southern Pacific Railroad opened a hospital and arranged for occupational health services for its employees. Shortly thereafter other railroads made provisions for health services, as did some of the mining companies. In 1886 Western Electric engaged a part-time physician to treat on-the-job injuries.

In 1890 the first employee dental care program was inaugurated at the Barber Match Company, closely followed by a similar program at the Armstrong Cork Company. In 1895 the Vermont Marble Company at Proctor, Vermont, was the first known instance of an employer engaging an industrial nurse to serve its employees; this approach being followed shortly afterward by the lumber industry at McCloud, California. In the early 1900s this development gained support when the Visiting Nurse Association of Chicago provided a nurse for the McCormick Printing Company, probably the first instance of a visiting nurse agency making available industrial health services to places of employment, and then nurses were employed by the Emporium in San Francisco, the Plymouth Cordage Company in Massachusetts, the Anaconda Mining Company in Montana, the Chase Metal Works in Connecticut, the Broadway Stores in Los Angeles, the Lowney Chocolate Factory in Boston, and the Pfister and Vogel Plants in Milwaukee.

In 1906 a large saw works at Providence, Rhode Island, is thought to be the first instance of an employer providing physical examinations for its workers. Psychological testing for employees was first introduced by the Boston Elevated Railway in 1911. In 1911 also the Norton Company at Worcester, Massachusetts, developed an industrial health department, said to be the first of its kind, to provide first aid, complete physical examinations, control of contagious diseases, and the means for the proper placement of workers in relation to any physical impairments they might have or develop.[4] In 1914 the Conference Board of Physicians in Industry was established by the National Industrial Conference Board (now The Conference Board).

By 1915 interest in, and specialized knowledge of, industrial health on the part of employers and unions had grown to the point where the American Association of Industrial Physicians and Surgeons was organized. Engineers, chemists, and architects also came to play roles of significance. An example of such interest was the formation in 1918 of the American Standards Association (now the American National Standards Institute—ANSI) to serve as the clearinghouse for industrial, technical, and governmental groups and to coordinate their standardization programs. Early efforts in safety and health standards on behalf of employees and the consuming public are exemplified by the promulgation of a safety code for lighting factories, mills, and workplaces in conjunction with the Illuminating Engineering Society in 1921; the development of a safety code for the protection of industrial workers in foundries in conjunction with the National Foundrymen's Association in 1921; and the promulgation of a safety code for the protection of the heads and eyes of workers in 1921.

This effort continued. For example, in 1935, a safety code was established for industrial sanitation in manufacturing establishments, and in 1938 a code was developed for the protection of respiratory organs of workers. To date ANSI, as the coordinating agency for voluntary consensus standardization in the United States, has approved some 6,500 standards in such areas as acoustics, air conditioning, boilers, construction, drawings, electrical equipment and electronic devices, energy conservation, environmental protection, fuel gas, graphics, heating, information systems, instrumentation, lamps, library work, materials and testing, measurement, mechanical devices, medical devices, nuclear energy, office equipment, photographic devices, physical distribution, piping, process equipment, quality assurance, refrigeration, safety and health, solar energy, testing methods, textiles, and welding.

By 1943, 113 of the larger places of employment, recognizing the problems of industrial health, had established in-plant medical services, 71 of these employing full-time physicians.

THE ROLE OF UNIONS

In recounting this history of developments in public and industrial health it is to be noted that one of the factors in the progress gradually made lay in the demands and activities of organized labor. Organized labor, among others, commencing in 1827 in Philadelphia, argued for free public education, and later for compulsory education as a means of combating child labor. It was one of the forthright advocates of child labor legislation and of laws to protect women workers. It has fought, bargained, and struck for higher wages, and for shorter work hours and workweeks. It initiated and supported workmen's compensation and unemployment compensation legislation. It has urged improved working conditions, safety protection against work accidents and diseases, inspection of workplaces, and many other measures that not only benefited the worker, but ultimately society and the economy. That such efforts have been fraught with conflict and strong differences of opinion is readily evident. In historical perspective, however, it would be difficult indeed to deny that the overall result has been to the good of people and the nation.

THE CONCERN OF INSURANCE COMPANIES

Insurance companies also played an active role in the prevention of disease, accidents, and premature death, as well as in the development of occupational health programs for their own employees. As the companies grew in size and importance they came to recognize that the health and safety of people were important to the soundness of their operations.

The interest of property insurance companies in the United States in the prevention of loss dates to 1866, when the National Board of Fire

Underwriters was founded in response to the Chicago conflagration. The results of its efforts have been striking, as demonstrated by the fact that a modern factory often pays not over 5 percent of what it would have paid a hundred years ago for fire insurance.[5]

Life insurance companies also were alert to the possibilities inherent in loss prevention approaches. As far back as the 1850s, insurance companies such as New York Life and Manhattan Life were contributing funds for sanitary improvements, including the construction of drains and sewers. Beginning in 1871, Metropolitan Life began a series of health and safety messages for employees and policyholders. In the 1890s, New York Life achieved a significant medical breakthrough: the development of the portable blood pressure gauge (the sphygomanometer), created by the company's chief medical examiner, which enabled standardized readings, thereby contributing to the study and knowledge of circulatory diseases.[6] In 1909, Metropolitan Life established its Health and Welfare Division, largely through the efforts of Dr. Lee K. Frankel, with the concept: "Insurance, not merely as a business proposition but as a social programme, will be the future policy of this company."[7]

At the same time Metropolitan instituted a program of preemployment physical examinations and an emergency medical service for its employees; in 1914 it commenced annual physical examinations for its employees through the Life Extension Institute, formed in 1913 following an offer of free periodic health examinations by the Provident Savings Life Association in 1909 for its policyholders. These developments created similar interest among other life insurers and, as will be shown, led to cooperative efforts on the part of many life insurance companies. In 1917 Dr. Louis I. Dublin of Metropolitian Life made a study of the causes of death by occupation.

In the area of accident prevention, the first organized effort on the part of insurance companies was the formation of the Workmen's Compensation Service Information Bureau in 1910. The ensuing prevention of work injuries developed largely as a result of research in engineering: placing guards on machines, designing safer machines, inspecting operations, and creating safer workplaces. Refusal to insure those who would not follow prescribed patterns of safety created an incentive for safety. The result was a 50 percent reduction in the fatality rate from work injuries between 1938 and 1958.[8]

Interest then developed in highway safety. Vehicle inspection was successfully promoted by insurance companies in the 1920s, and driver education was given an impetus as far back as 1928. Joint activities with professional groups and other safety organizations is a continuing effort to improve highway design, soundly conceived traffic regulation, and all other factors entering into safer driving. More recently, some 500 automobile insurers organized the Insurance Institute for Highway Safety with an initial annual fund of $1 million. The Insurance Information Institute was also formed about the same time. Through such organizations, research is carried on and technical publications and various visual aid materials are produced for public information.

Directly and indirectly, through making available funds, personnel, and facilities and through cooperation with other agencies, insurance companies have contributed to research efforts devoted to preventing or reducing the incidence of illnesses, injury, and premature death. In some areas their efforts have made direct and original contributions. The compiling of data on blood pressure by 20 life insurance companies some years ago, for instance, led to important findings in that field.[9] Then the Society of Actuaries drew on company records of millions of policyholders to establish a desirable range of body weight in relation to build, which completely revised the standard height–weight tables. A related actuarial study of hypertension called attention, perhaps for the first time, to the dangerous potential to human health of slight elevation of blood pressure combined with slight overweight.

In 1909, a joint effort was begun: the Medico-Actuarial Mortality Investigation. It was addressed to the question "Can insurance experience be applied to lengthen life?" Forty-three companies in the United States and Canada pooled their claims experience. The records of 2 million policyholders for a 25-year span were correlated and tabulated. The published report was issued in five volumes. Significant classes of experience included occupational hazards, family history of tuberculosis, rheumatism, syphilis, alcoholism, overweight and underweight, circulatory impairments, and family longevity.

Subsequently, Drs. Armstrong, Dublin, and Wheatley, of Metropolitan Life, first called attention to the apparent relationship between excessive weight and excessive mortality, especially from cardiovascular-renal disease. The ensuing campaign to control overweight, joined in by the American Medical Association, the U.S. Public Health Service, many local and state health departments, nutritionists, women's clubs, the Y.M.C.A., and science writers, led to the concept that overeating by adults is the major cause of malnutrition in this country. Then during World War II, Dr. William J. McConnell, of Metropolitan Life, was largely responsible for reducing deaths from TNT poisoning almost to the vanishing point, a far cry from the record in the First World War.[10]

Other examples are found in the Ungerleider-Gubner (of Equitable

Life) tables for heart size; the use of insurance company statistics as yardsticks by which the medical profession and public health and social workers could measure the effectiveness of their activities; and pioneering in electrocardiogram tests by Dr. Henry Blackburn, Jr., of Mutual Service Life. An outstanding example of accomplishment was the Framingham experiment conducted by Metropolitan Life at a time (1916) when tuberculosis was the largest single cause of death in the United States. Believing that most of these deaths were unnecessary, the insurance company undertook a demonstration at the industrial town of Framingham, Massachusetts, aimed at putting to use what was already known about the diagnosis, prevention, and treatment of tuberculosis, including the first use of radiography as a diagnostic tool. In seven years the death rate from this disease in Framingham was reduced 68 percent.[11]

Other demonstrations followed. There was the three-year Thetford Mines Demonstration in Canada, which proved that the infant mortality rate of a community could be reduced two-thirds simply by educating mothers in modern methods of maternity and child care. There was the twenty-five-year record of the Metropolitan Influenza-Pneumonia Commission, known for its studies of colds, influenza, antipneumococcic serum, virus problems, and chemotherapy. There was the five-year study of silicosis and pulmonary duct diseases among miners, carried on jointly in Picher, Oklahoma, by the United States Bureau of Mines, the Tri-State Zinc Mine Operators, and Metropolitan Life. From this study, valuable data on occupational hazards were collected and analyzed.

In the field of health education, some insurance companies came to play an active and continuing role. As early as 1909 the Provident Saving Life Assurance Society issued booklets on preventive health care to its policyholders. At the same time, Metropolitan Life commenced a program in health education in the form of a continuing flow of informative pamphlets.

Later several other insurance companies became involved in health education information programs. Information with respect to nutrition, overweight, occupational health, school health, childhood diseases, coronaries, alcoholism, drug addiction, syphilis, teeth care, first aid, home nursing, cancer, appendicitis, immunization, mental health, and planning for parenthood were included in such health educational activities. The media employed include pamphlets, folders, posters, cartoons, exhibits, displays, films, and other visual aids. Radio, and later television, also was used. Distribution was made to policyholders, to school children, to agencies interested in health and education, to professional groups, to the general public, and to places of employment.

THE BEGINNINGS OF HEALTH INSURANCE

It is interesting to take a brief look at the early beginnings of the means for financing medical care and disability. They, too, have a long history, extending to the ancient Chinese culture, where people paid the physician to keep them well. Mutual aid and a form of insurance were employed by the artisans of ancient Rome. In England, after Henry VIII had abolished the monasteries, which furnished health care, the medieval guilds provided financial relief for the care of sickness, for loss of sight or limb, and in cases of leprosy.

Then, with the dawning of industrialization, the concept of prepayment, or insurance, began to take hold. About 1750 a system of Friendly Societies developed in England. By 1800 there were 7,200 such societies with a membership of 648,000. Similarly purposed organizations sponsored by workers, employers, townspeople, religious groups, or physicians began to develop in the Scandinavian nations, the Low Countries, and Germany, the benefits being concerned with the payment for certain types of medical care and the replacement of income lost as a result of disability.

In Germany, in 1883, the first compulsory health insurance program for workers was enacted into law at the insistence of Bismarck. Essentially similar legislation was enacted in the following few years in Austria, Sweden, Denmark, Luxembourg, and Norway. In all cases the preexisting funds or societies were used in various ways for the execution of the government program. In England the National Insurance Act became law in 1911, applicable to employed people only, to be superseded in 1946 by the National Health Service to provide health care for everyone.

In North America the first known prepayment scheme was in 1655 in Montreal, where a physician made a contract with 37 families to provide them all necessary care. In 1880, single hospitals in northern Minnesota developed prepayment plans for the lumberjacks in the area. In 1882 the first major employer-sponsored mutual benefit association was established: the Northern Pacific Railway Beneficial Association. This was followed by the Macy Mutual Aid Association in 1885, the Benefit Association of the John Wanamaker Company in 1897, and a similar development at the Fredrick Loeser department store in Brooklyn. Meanwhile, in 1887, St. Mary's Hospital in East Saginaw, Michigan, contracted with people to provide hospital, medical, and surgical care at a cost of $5 a year, the only exceptions to the care to be provided being for contagious diseases and drunkenness. In 1911 Montgomery Ward bought the first group life insurance policy from Equitable Life. In 1921 a hospital in Grinnell, Iowa, offered a prepayment plan to the general public.

Then in 1929 the Baylor University Hospital at Dallas, Texas, developed a prepayment plan for the teachers at the university that was subsequently extended to include other employed groups in the area. This plan is generally considered to have been the forerunner of the Blue Cross plans of today. (The Blue Cross symbol was first used by a plan for the prepayment of hospitalization that was formed in Minnesota in 1933.)

In 1939 the California Physicians Service was formed, closely followed by similar developments in Michigan and Western New York State. These prepayment plans were the beginning of the Blue Shield plans, that symbol being first used in 1939 by a plan in Buffalo, New York.

Insurances against the loss of income resulting from disability had their inception in the United States in 1847 by the Massachusetts Health Insurance Company of Boston. This was followed in 1850 by the Franklin Health Assurance Company of Massachusetts, and in 1863 by the Travelers Insurance Company and the National Union Life and Limb Insurance Company (later to become the Metropolitan Life Insurance Company). In 1942 the first compulsory cash sickness benefits law for temporary disability was enacted, in Rhode Island.

Insurance companies first experimented with coverages for the cost of medical care in the early part of the twentieth century. The first period of growth developed in the 1930s in relation to disability-income insurance. The introduction of hospital insurance on a broader scale came later in that decade. In the late 1940s many of the larger life insurance companies entered the field, the coverages at first largely imitating the Blue Cross–Blue Shield benefits in that they indemnified against the costs of hospital care, surgery, and in-hospital medical care. In 1950, however, these coverages were appreciably broadened to include both in and out of hospital care, in the form of major medical expense insurance. Later, coverages were to be made available for the treatment of mental illness and alcoholism, for dental and vision care, for the cost of prescribed drugs, for care provided in nursing homes, and for home care.

The provision of health care on a prepayment basis (today known as prepaid group practice, or HMOs) had its beginning with the formation of the Ross-Loos Medical Group in California and the Farmers Union Hospital Association at Elk City, Oklahoma, in 1929. In 1931 Frank M. Close made such an arrangement available in San Francisco. In 1937 the Group Health Association was formed in Washington, D.C., by a group of federal government employees. The Kaiser Plan was developed in 1938 for its employees in the state of Washington and was extended to its shipyard workers on the West Coast during World War II. In 1945 the Kaiser Foundation Health Plan was opened to the community at large. In 1946 the Health Insurance Plan of Greater New York was formed, principally to serve the needs of municipal employees. In 1947 the Group Health Cooperative of Puget Sound was organized with the support of cooperatives, unions, and farmers.

On the labor front, in 1877 the Granite Cutter's Union established the first national union sick benefit program, followed in 1895 by the barber's union, the iron molders, the printers, the tobacco workers, and by 1903 the plumbers. Then in 1913 the International Ladies' Garment Workers' Union (ILGWU) established a health care program for its members. This was followed by the establishment of other labor-sponsored programs, such as the AFL Medical Services Plan in Philadelphia, the St. Louis Labor Health Institute, the Union Health Service, Inc. in Chicago, the International Longshoremen Workers' Union-Pacific Maritime Union on the West Coast, and the Amalgamated Clothing Workers Sidney Hillman Center in New York.

In 1926 the first collective bargaining agreement with a health and welfare clause came into being: the Amalgamated Association of Street and Electric Railway Employees with the Public Service Corporation of Newburgh, New York. The first medical benefits in a collective bargaining agreement were brought about by the Philadelphia Waist and Dress Joint Board of the ILGWU.

Several factors spurred the development of health insurance in the United States. One, unquestionably, was the increased public confidence in the technological development of modern medicine and in the concept of the modern hospital. This, resulting in a greatly increased demand for care, made the public aware of its cost. Another factor was often the absence of needed facilities for health care in certain locations where people worked. Yet another was the effect of the Depression on hospital financing, to be resolved by prepayment programs. Another was the growing interest on the part of both employers and labor unions in such insurance protection.

A final development, and one of very considerable importance, was the joint effect of federal government actions in the area of labor relations and federal income taxes. In World War II the wage stabilization policies of the War Labor Board (exempting fringe benefits from the wage-price freeze), the subsequent decisions of the National Labor Relations Board, and the rulings of the Wage Stabilization Board all served to stimulate the establishment of group health insurance programs for employed groups of people as well as collective bargaining for health insurance benefits. At the same time, amendments were made to the Internal Revenue Code allowing employer contributions to employee welfare funds to be tax-deductible as a business expense. In combination, these determinations by the federal government were in large part accountable for the tremendous growth of health insurance coverages for employed people and their dependents that commenced in the mid-1940s.

BIBLIOGRAPHY

J. F. Follmann, Jr., *Medical Care and Health Insurance* (Homewood, Ill.: Richard D. Irwin, Inc., 1963).

The National Health Council, *The Health of People Who Work*, New York, 1960.

The Office of Health Economics (London) — various publications were used for source material.

George Rosen, *Preventive Medicine in the United States, 1900-1975* (Boston: Science History Publications, 1975).

The U.S. Public Health Service, *Management and Union Health and Medical Programs* (Washington, D.C.: HEW, June 1953).

REFERENCES

1. J. Bronowski, *The Ascent of Man* (Boston: Little, Brown and Company, 1973).
2. *United Nations Bulletin*, December 1951, New York.
3. J. F. Follmann, Jr., *Alcoholics and Business* (New York: AMACOM, 1976).
4. Karl T. Benedict, M.D., "Industrial Health Protection," *National Safety News*, June 1975.
5. S. Bruce Black, "The Insurance Industry and Disability Control," *Geriatrics*, July 1960.
6. J. Edwin Matz, Statement before the Senate Subcommittee on Antitrust and Monopoly, April 1973.
7. Carl Cramer, "A Tower of Strength" (New York: Metropolitan Life Insurance Company, 1959).
8. Black, op. cit.
9. Edward D. Freis, "The Role of Hypertension," *American Journal of Public Health*, March 1960.
10. William P. Shepard, M.D., "Insurance as a Social Program" (New York: Metropolitan Life Insurance Company, 1959).
11. Thomas M. Ribers, "Medical Research 1909 to 2009" (New York: Metropolitan Life Insurance Company, 1959).

3

Health Status in the United States

Since places of employment do not function in a vacuum but find that the health of their employees is directly affected by the health of the community in which they operate, it is necessary to examine the health status in the United States today.

In reviewing the data, a word of caution is in order. Lerner and Anderson have commented that "the compilation of morbidity statistics is a difficult process; in contrast to mortality, morbidity often has a subjective component and is difficult to define precisely . . . the general state of morbidity statistics has been very unsatisfactory [being] collected at irregular intervals and usually for limited populations."[1] This position is well taken. It might be noted further that both mortality and morbidity statistics can be confused by changes in the classification of diseases from time to time and by differences in diagnostic practices and customs in both time and place. Recording practices can also differ in time and place. Regardless of these qualifying notes, however, the data shown will give at least a sense of proportion, an order of magnitude, which in itself is significant.

MORTALITY

Life expectancy today is 72.4 years in the United States. For females the average is 76.4 years; for males. 68.5 years. The average span of life is shorter among nonwhites. One result of this lengthened span of life is that today one out of every ten Americans is 65 years of age or older.

Some 1.9 million persons die each year in the United States. The causes of death in 1968 and their proportion of the total of deaths in that year were:

Heart diseases	38.4%	Cirrhosis of the liver	1.7%
Cancer	17.8	Emphysema	1.3
Stroke	10.9	Suicide	1.1
Accidents	5.9	Congenital anomalies	0.9
Influenze and pneumonia	3.2	Homicide	0.8
Diseases of infancy	2.3	Kidney infections	0.5
Diabetes	1.9	Peptic ulcer	0.5
Arteriosclerosis	1.7	Other	11.1

From this it can be seen that 66 percent of all deaths today are accountable to diseases of the heart, cancer, and stroke—diseases frequently attributable to advancing age, but by no means always. Accidents, usually attributable to younger ages, account for 5.9 percent of all deaths. Of all deaths today, 60 percent occur after age 65.

That the predominant causes of death vary among different age brackets of the population and by sexes is readily evident from mortality data. For males, accidents are the leading cause of death under age 35, accounting for half of the deaths between ages 15 and 24. For females between ages 15 and 19, accidents are the cause of two-fifths of all deaths. The cardiovascular-renal diseases are the major cause of death among males over age 35 and account for over half of all male deaths after age 55. For females these diseases do not become the leading cause of death until after age 65. Cancer causes a fifth of male deaths between ages 45 and 74; for females it is the principal cause of death between ages 25 and 64 and accounts for one-fifth of female deaths between ages 25 and 29 and two-fifths of deaths between ages 40 and 64. Suicide is the second-ranking cause of death among males aged 15 to 29 and ranks third for females age 20 to 24 and fourth for those aged 25 to 29.

There are also broad differences in the causes of death among the races. Nonwhite males between ages 15 and 24 have 100 times the incidence of death from homicide as do whites, two and a half times the deaths from heart diseases, twice the incidence of fatal attacks of influenza and pneumonia, a 40 percent higher incidence of fatal accidents, and for all causes of death they have one and a half times the incidence of death. They do, however, have 20 percent fewer deaths from automobile accidents, 20 percent fewer deaths from neoplasms, and 25 percent fewer suicides. Between ages 35 to 64 nonwhite males have one and a half times the incidence of death from neoplasms as do white males (twice the rate for lung cancer) and twice the rate of accidental deaths. At ages 35 to 44 they have twice the incidence of fatal heart disease as whites, but from ages 55 to 64 the difference is slight. At ages 35 to 44 they have three times as many deaths from cirrhosis of the liver as do whites, but by ages 55 to 64 the difference is nil. Deaths from homicide are over ten times as great from age 35 on, but the suicide rate runs roughly half that among white males.[2] Infant mortality has diminished rapidly in relatively recent years. As late as 1915 one in every 10 live births failed to survive the first year; by 1960 the death rate was one in 40. Since then the decline has continued steadily at the rate of about 4 percent annually.

DISEASES

The effects of various diseases on mortality have been seen. As will be discussed subsequently, they also result in disability. In this section the major diseases will be discussed briefly to indicate something of their nature and prevalence. Their economic impact, their effects on places of employment, and the potential for the prevention of these diseases will be discussed later.

Heart Diseases

Heart diseases take more than 600,000 lives each year and result in considerable disability. They account for 50 percent of all male deaths under age 40. although risk of the incidence of these diseases increases with age. It is estimated that 1.8 million people under age 65 in the United States have coronary heart disease and that another 1.6 million might with reason be suspected of having the disease. In this century the prevalence and incidence of these diseases have risen very sharply, in part perhaps because people are living considerably longer.

The past 15 years, however, have witnessed some diminution in fatalities from ischemic heart disease, particularly under 65 years of age, probably as a result of some improvement in living habits such as good diet, abstinence from smoking, and proper rest and exercise, coupled with improvements in the early detection of the disease and medical care techniques, and in preventive measures. Ischemic heart disease today is reported to account for over 90 percent of the deaths resulting from heart diseases.

Hypertension

It is estimated that 20 to 25 million persons in the United States (10 percent of the population) have high blood pressure, that half of these are undetected, that a fourth are known but untreated, and that only one in four are under any form of therapy. The five-year survival rate for serious cases is less than 20 percent. Each year 60,000 deaths result from hypertension, although the death rate has been reduced 50 percent in the past decade. Prevalence of the disease is more than double among blacks than among whites.[3] Factors contributing to the development of hypertension are tension, malfunction of the kidney, family history, diet, obesity, and lifestyle.

Today certain drugs are quite effective in controlling hypertension. Uncontrolled, hypertension is said to play a role in more than 1.5 million heart attacks and strokes.

Stroke

Stroke results in over 200,000 deaths in the United States each year; 80 percent of these victims are over age 65. It is estimated that two million living persons have suffered strokes. Some 80 percent of these will survive, but many will be disabled. The death rate from stroke is six times as great in cases where hypertension is also present. Stroke claims more female lives than males, and more nonwhites than whites.

Cancer

Cancer is the second leading cause of death in the United States today, taking 370,000 lives in 1976. It is the primary cause of death among women between ages 25 and 64 and the second leading cause of death among males over age 35. Except for accidents, it is the primary cause of death of children. For the entire population it causes one of every six deaths, more than half of which occur among people age 65 and older. Each year 665,000 new cases of cancer are diagnosed and at any given time over one million Americans are under treatment for cancer. The disease, it is estimated, will strike 53 million Americans living today, or one out of every four. The survival rate for all types of cancer today is one out of three. Today 1.5 million persons who have been stricken with cancer are considered cured.

The cancer death rate for women has been declining steadily, but males have shown a 40 percent increase over the past 40 years, with the growth particularly apparent among blacks.

Today there is increasing interest in the relationship of the incidence of cancer to such matters as nutritional factors (including pesticide residues in food, certain food additives, trace elements from the soil, aflatoxin produced by mold, incorrect dietary habits, and certain preservatives), the use of certain drugs (such as estrogen used for menopausal symptoms), environmental factors, occupational hazards, and personal habits (such as smoking and drinking). The evidence of the effects of such possible influencing factors is not yet clear, however, and strong differences of opinion on such theories prevail. Future research should resolve the issue eventually.

Diabetes

Diabetes mellitus is most frequently encountered in urban and industrial nations and generally more affluent societies. It is more prevalent among older populations and among females. In the United States it remains a major health problem, although since 1940 the trend in mortality from the disease has been downward. It is estimated that today some 4 million Americans are afflicted with the disease, and that another 5.6 million might be considered potential diabetics. Each year 38,000 people die as a result of diabetes, the rate being higher among older people and among nonwhites. Deaths ascribed to diabetes only represent a portion of the deaths among persons with the disease, since diabetes contributes to the fatality of other diseases.

Respiratory Diseases

Respiratory infections are the most frequent cause of disability and contribute heavily to mortality. They cover a wide spectrum of ailments ranging from the common cold to influenza, bronchitis, bronchiectasis, pneumonia, pulmonary tuberculosis, and fungus infections. Respiratory allergies include hay fever and asthma. Other respiratory diseases include pulmonary emphysema, tumors of the lung, pneumoconiosis, pulmonary manifestations of general diseases, and diseases of the pleura.

It has been reported that chronic respiratory conditions occur with an incidence of some 2 million cases each year and that 197 million cases of acute respiratory conditions occur annually. They result, respectively, in 9.6 and 1.5 bed days of disability per case, on the average. Generally the incidence of respiratory diseases has been increasing, owing in large part to urban living, air pollution, and smoking.

Influenza and pneumonia are the fifth-ranking cause of death in the United States, causing some 80,000 deaths each year. Influenza has its greatest incidence among children and adults between ages 25 and 35. Fatalities are heaviest among the very young and the aged, with death usually occurring in about one week. Mortality among males is two or three times the rate among females. It is an infectious disease carried by viruses and frequently occurs in epidemic proportions, moving from one part of the world to another. The 1918 epidemic killed over 20 million persons, a catastrophe second only to the Black Plague in the fourteenth century. In 1957 an epidemic of the Hong Kong, or Asiatic, flu was severe, with a reoccurrence in 1960. In January 1975 influenza also reached epidemic proportions in the United States, causing 745 deaths. Epidemics usually run from four to six weeks.

Pneumonia at one time was a leading cause of mortality, but today the incidence of the disease in its severe form is relatively rare as a result of the use of penicillin and other drugs. Most usually the disease takes its heaviest toll between ages 15 and 45.

Chronic bronchitis and emphysema are the ninth-ranking cause of death in the United States today, accounting for some 25,000 deaths each year. They account for 7 percent of all deaths of men between the ages of 45 and 64, and 3 percent of deaths of women in the same age brackets. There are some two million cases of these diseases each year, 3,000 of which are hospitalized. On the average, 5.3 bed days of disability occur from these diseases. The toll is two and a half times as great among men as among women. For either sex it is severest among heavy smokers. The incidence of these diseases has been increasing.

Tuberculosis has dropped sharply in incidence in this century. Today the rate of mortality is 2.8 per 100,000, compared with 194 per 100,000 in 1900. A third of a million cases develop each year, however, which are medically attended, and each year ten thousand persons die of the disease. It is an infectious disease, the incidence of which is considerably higher among blacks and Spanish-Americans.

Another classification of respiratory diseases is *respiratory allergies*, principally hay fever and asthma. Ten percent of the population is said to be affected by these diseases at some time in life. Hay fever is seldom fatal or disabling; it is annoying. Bronchial asthma, which is decreasing in incidence, does cause some 5,000 deaths each year, and is the cause of about 2,000 cases of hospitalization each year.

Pulmonary emphysema has been increasing in incidence in recent years, particularly among white males between the ages of 50 and 70. *Pneumoconiosis* most usually results from occupational hazards—silicosis among miners and stone workers, anthracosilicosis among coal miners, benign pneumoconiosis among iron workers and farmers (from hay), and asbestosis and berylliosis among workers who are in contact with such materials.

Urinary Tract Infections

Urinary tract infections are the commonest of bacterial infections. Their incidence is eight times as great among females as among males, occurring at all ages but rising with age. Pregnancy appears to play a role in this incidence. Treatment, basically, takes the form of an antibacterial agent. In 80 percent of the cases, a cure results, but the frequency of recurrence of the infection is high. The bladder and the kidneys are affected.

Infections of the kidney are the fourteenth-ranking cause of death in the United States, causing some 6,000 deaths yearly. Today there is some dramatic improvement in the survival rate from renal failures. Transplants, first demonstrated by Dr. John P. Merrill in Boston in 1954, are today being done at 250 health centers. Hemodialysis, developed by Dr. Belding Scribner in Seattle, today is being maintained on over 10,000 patients. The treatment is costly, but is often supported by federal or state governments. In a high proportion of cases the patient may live a useful life for several years. The mortality rate, however, is high and suicide is not infrequent.[4]

Viral Diseases

Among the viral diseases are hepatitis, encephalitis, and meningitis. The incidence of *hepatitis,* which had been increasing, has declined somewhat since 1971. Recorded cases in 1974 totaled 59,200. This figure, however, is a 75 percent increase over similar data for 1965. In 1970, the latest year for which mortality information is available, there were 1,014 deaths from the disease, an increase of 40 percent over 1965. The highest incidence is in ages 15 to 19 and drops sharply after age 29.

Encephalitis is associated with such childhood disease as measles, chickenpox, and mumps. There were 1,302 cases reported in 1972. A form of the disease, arbovirus, is transmitted by mosquito bites. Another form is herpes, which is frequently fatal or crippling. In 1972, 45 cases were reported, 9 of which were fatal.

Meningitis is an infection of the central nervous system. In 1973, 4,846 cases were reported, considerably higher than the reported cases in 1966. Recovery is usually complete after two weeks of treatment. Untreated, it is inevitably fatal. Incidence of the disease is greatest in areas with poor sanitation and a high density of preschool children. Sulfa drugs provide considerable success in coping with the disease.

Sclerosis

Multiple sclerosis and amyotrophic lateral sclerosis are diseases whose etiology remains obscure. There are an estimated 250,000 Americans with the former disease and some 10,000 with the latter. Both have a low mortality rate. *Multiple sclerosis* is the most common of the chronic diseases of the nervous system, and cases of relapse are common. Both incidence and mortality increase with colder climates, Wyoming, Montana, and South Dakota having an incidence considerably above the national average. Ages 20 to 40 are the most vulnerable. Incidence of the disease is roughly one-half among nonwhites.

Amyotrophic lateral sclerosis is a degenerative disease more common among males between the ages of 40 and 70. It most frequently occurs among farmers and others who do heavy work. Average life expectancy after onset is about two to four years. There were 1,441 deaths from the disease in 1969, a decided increase over 1960.

Cerebral Palsy

While no exact figures are available, it is estimated that some three-quarters of a million Americans are affected by cerebral palsy and that each year some 25,000 babies are born with the disease. The resulting disabilities can be varied and permanent, although today medical approaches can lessen the extent of the damage. There is no definite cause in almost half the cases. Premature birth, prenatal injury, or measles or radiation exposure during pregnancy contributes to the onset of the disease.

Mental Retardation

There are 5 to 7 million mentally retarded persons in the United States. Half are over age 20; a quarter million are institutionalized and 96 percent live at home or live independently. Almost a quarter are physically dependent. Some 40 percent have had no schooling, the remainder usually having eight grades of education. About 90,000 of the retarded people are classified as "profound," 210,000 as "severe," 360,000 as "moderate," and 5.3 million as "mild." The majority of the latter group are over age 20 and 87 percent of them are gainfully employed, usually having earnings equal to 85 percent of the average employed person.

Arthritic and Rheumatic Diseases

These diseases are seldom fatal, but are disabling in about 20 percent of the cases. While estimates of the extent of these diseases vary widely, it is thought that at present there are some 13 million victims in the United States. The cause of the diseases is unknown, although heredity is a factor. There is no successful cure. The diseases can strike at any age, and 15 percent of the cases are said to recover for no known reason. Relief is obtained from certain drugs; removal to a hot, dry climate; proper exercise; and rest.

Communicable and Infectious Diseases

Communicable diseases have reduced very sharply in incidence in this century through the use of inoculation and other preventive measures, although they continue to occur with greater incidence than should be the case. Whereas in 1900 they caused 262 deaths per 100,000 population under age 15, today that ratio is one per 100,000. Nonetheless, they continue to account for 140,000 deaths a year.

In recent years the incidence of diptheria, tetanus, mumps, measles, rubella, scarlet fever, and pertussis has declined greatly, although recurrences are always possible. For example, the incidence of measles had dropped from 500,000 cases yearly to 22,000 cases in 1968, but rose to 75,000 cases in 1971. Smallpox has just about disappeared in the United States, with no deaths reported since 1949. Cholera, too, has just about disappeared in the United States, only two deaths being reported in 1966 and one in 1973. Typhoid fever has also just about been eliminated, although in 1973 there was an outbreak at the South Dade Labor Camp in Florida. Plague has also almost disappeared.

Polio increased rapidly in the United States in the twentieth century, with 37.2 cases per 100,000 population in 1952. Of these, one-fourth were unable to walk as a result, one-third could walk with aids, and the remainder could walk without aids. In the 1916 epidemic, 7,000 persons died of polio. About a third of a million Americans living in 1956 had been stricken with polio. Then in 1955, the Salk vaccine became available and by 1962 the incidence of the disease had dropped to 0.5 cases per 100,000 population. In 1963 only 450 cases were reported, with 45 deaths.

Venereal Diseases

The venereal diseases, primarily syphilis and gonorrhea, have been greatly reduced in incidence in the United States since the development of penicillin and other antibiotics in 1943, although recently there has been some upsurge in the number of cases reported. Those infected are predominantly young people, most usually in their twenties. The incidence is very considerably (over ten times) greater among nonwhites, and is highest in the southern states and the District of Columbia. *Syphilis* can result in death and disability. In 1917 the reported death rate from syphilis was 19 per 100,000 population. Today the death rate from syphilis is about 1.7 per 100,000, although there are some 250,000 cases of syphilis each year in the United States. With the sharp reduction in the disease, infant deaths have been reduced 50 percent. *Gonorrhea* today has an estimated incidence of from 600,000 to 2.5 million cases yearly.

Leprosy

Leprosy, or Hansen's Disease, has never had a great degree of incidence in the United States. Its epidemiology is unknown. It results in disfigurement and an incapability to work, or an inability to obtain work as a result of the stigma that attaches to the disease, the fear of contagion, the reaction of other employees and customers, and the concern over increased insurance costs. Those most susceptible to the disease are characterized as having low incomes and low levels of education.

MENTAL ILLNESS

Mental illness is estimated to afflict 10 percent of the American population to the point where some type of care is warranted. Only 20 percent of these, however, are presently receiving care—some 2.5 to 4 million persons. Over a half million people are in long-stay mental hospitals, the median stay in such hospitals being six and a half years, although hospital stays for the treatment of mental illness vary considerably. The median age for hospitalized patients is 54 years. The trend in the incidence of mental illness among Americans is not known.

The incidence of mental illness increases with age, and its incidence is higher among women, unmarried persons, nonwhites, and ethnic minorities. The death rate is higher among the mentally ill.

The forms of care for mental illness are various, and frequently are in short supply and erratically distributed. The 414 state and county hospitals provide the care for 24 percent of mental patients (88 percent of hospitalized patients); the three federal mental hospitals provide the care for one percent of the patients (10 percent of the hospitalized patients); the 41 VA hospitals for mentally ill veterans care for 119,000 patients (4 percent of the mentally ill); and the 1,316 general hospitals and 180 private psychiatric hospitals provide the care for 7 percent and 2 percent, respectively, of the mentally ill. In all, then, some 560,000 of the mentally ill are patients in mental hospitals. The remainder of the mentally ill receive care on an outpatient basis, the 17,000 psychiatrists in private practice providing the care for 41 percent, and the 351 publicly funded community mental health centers providing the care for the remaining 21 percent.

Care is also provided by clinical psychologists, psychological social workers, and psychiatric nurses as well as by day or night hospitals, skilled nursing homes, halfway houses, and homes for the aged. Recovery rates differ by types of illnesses, but it has been estimated that 80 to 85 percent of the treated cases eventually return to the community or to work, although some 25 to 30 percent are said to have relapses.

ALCOHOLISM

There are estimated to be 9 million alcoholic persons in the United States. More than half of these are in the workforce. Alcoholics have a 12-year shorter life span than other Americans, at least 86,000 deaths each year being due to the abuse of alcohol. Alcohol abuse is seen as causing half the highway fatalities, half of all suicides, a third of all homicides, and the majority of drownings. Alcoholism ranks with heart disease, cancer, and mental illness as a cause of medical disability. It is the major cause of cirrhosis of the liver. Alcoholics are more susceptible to illness and are more accident-prone than other people.

The majority of alcoholics and problem drinkers are men in their most productive years, between ages 35 and 55, although the incidence among women is increasing. Identification can be difficult, and diagnosis can, for several reasons, be evasive. Alcoholism today is generally, although not universally, recognized to be a disease. The treatment of alcoholism occurs within the entire spectrum of health services, including specially purposed modalities such as detoxification centers, alcoholism treatment centers, and the highly successful Alcoholics Anonymous. There is no cure for alcoholism. The aim of treatment is to bring the disease under control. In recent years the alcohol control programs established in an increasing number of places of employment are proving quite successful, with recovery rates of from 60 to 80 percent or higher.

DRUG ABUSE

Drug abuse or drug addiction became an increasing cause of ill health in the United States in the 1960s. It results in premature death, increased morbidity, and suicide. The primary increase in abuse has been among the young, usually between ages 15 and 24. The number of drug abusers is not known, particularly because many do so as an experiment or under peer pressure. One estimate is that there are 400,000 drug addicts in the United States; other estimates ranging from 150,000 to 600,000. Half are said to live in New York City, where it is reported that 4,271 deaths occurred as a result of addiction in the 1960s. Almost 85 percent of addicts are male, and 60 to 70 percent are black or Spanish-American. Most are unemployed and most are unmarried. While heroin is the principal drug used by addicts, multiple drug use is frequent.

In 1968 alone in New York City, 1,692 deaths from drugs were reported, 70 percent of which resulted from overdoses. Of these, 544 were from the use of barbituates, 175 from aspirin and related drugs, 229 from opiates, 646 from tranquilizers and related drugs, and 376 from other drugs. Of these deaths, 227 were between ages 15 and 19.

Treatment modalities for drug addiction take many approaches. One is detoxification, although one such approach has found that only 24 percent of those patients who do not have other services remained drug-free for any length of time, and another found that 56 percent of the patients dropped out of the program, 38 percent made only one visit, 5.5 percent were clean for one month or more, and the remainder decreased the habit. Another approach is a variety of institutions called by the generic term "therapeutic community," which provide inpatient care. Yet another approach is outpatient abstinence programs. Today a popular approach is methadone maintenance. The approach employs detoxification, chemotherapy, group work, psychotherapy, medical-surgical treatment, and education and counseling. The cost appears to range from $500 to $2,500 per patient. Such programs have not been successful in attracting younger addicts, and the methadone approach remains a controversial subject. One program reports, however, that for the patients who remained in the program, the degree of gainful employment rose sharply, even though only 26 percent of the patients were employed at the time of their entry into the program.[5, 6]

PHYSICAL IMPAIRMENTS

Hearing

Hearing impairments are reported to afflict about 14.5 to 17 million people in the United States. The incidence is much higher among institutionalized persons, males, white persons, people of lower economic and educational levels, and those not living in metropolitan areas. It increases with age. Of those afflicted, the degree of impairment differs, but 2.5 million Americans cannot hear or understand any speech at all, most of these being over age 65. Causes of the impairment are not known. For many of those affected, hearing aids are beneficial.

Vision

From 5 to 9.5 million Americans have sight impairments of some type. Subnormal binocular visual acuity is said to be experienced by 46 percent of the population, rising to 95 percent after age 65. The incidence of subnormal vision is higher among lower-income families, people with a lesser degree of education, females, and nonwhites. For the adult population of the United States, 60 percent wear glasses, 32 percent wearing them full time.

HEALTH HAZARDS

Excessive Smoking

Today smoking, particularly excessive cigarette smoking, is recognized quite generally to present a health hazard. There is, however, a wide difference of opinion, and even controversy, as to both the nature and the extent of the hazard as well as the relationship of smoking to other factors that can be deleterious to health.

In 1964 a panel of advisers to the Surgeon General of the United States reported that cigarette smoking was a cause of lung cancer, of cancer of the larynx, and of chronic bronchitis. They also suspected it as a cause of heart disease. The Advisory Committee noted, however, that "statistical methods cannot establish proof of a causal relationship in an association." A decade later it was stated that cigarette smokers have higher mortality rates from lung cancer, cancer of the lip and throat, from coronary heart disease, as well as from cardiovascular disease, malignancies of all sites, cirrhosis, ulcers, accidents, murder, and suicide than do nonsmokers.

Surveys conducted by the National Center for Health Statistics in 1965 indicated, on the basis of household interviews, that cigarette smokers of both sexes reported a higher rate of chronic conditions than did people who had never smoked, although the differences were not great. For heavy smokers (41 or more cigarettes a day), however, the difference was much greater. The difference between smokers and nonsmokers was more pronounced for chronic bronchitis and emphysema, peptic ulcer, heart condition, and sinusitus. These surveys also found that the presence of acute conditions among smokers was 14 percent higher for men and 21 percent higher for women than among nonsmokers.[7, 8]

Concerning the relationship of smoking to heart disease, it is maintained that cigarette smoking is a major risk factor in the development of coronary heart disease (CHD), and that for men under age 50 it is a greater risk factor than any other (for pipe and cigar smokers it is only a slightly greater risk principally because they do not usually inhale).[9]

The evidence from various studies is not always consistent with what has been shown, however. Several studies have failed to find an increased prevalence of heart disease among smokers, or the difference found was not statistically significant.

Cigarette smoking has been identified as a major cause in the growing incidence of lung cancer (the relationship of pipe and cigar smoking is in doubt). Cigarette smokers have been found to have over five times the mortality from lung cancer as nonsmokers, although the evidence is not always consistent. The risk of contracting lung cancer appears to increase for persons exposed to air pollution.

In addition to the foregoing, it is reported that the death rate from all causes increases with the amount of smoking and is very much higher among heavy smokers.

There is no unanimity of opinion, however, with what has been shown. For example, one examination of the subject considers that much of the research methodology establishes an oversimplified relationship between smoking and the incidence of disease and mortality, failing to consider other related factors such as the age, height, and weight of the person involved; the use of other tobacco products; and such factors as the number of years the person has smoked, the extent of inhalation, the number of cigarettes smoked before breakfast, the use of cigarette holders and filters, the volume of smoke in a puff, and the size of the butt.[10] One medical statistician has observed: "The idea that cigarette smoking causes all these deaths from all these many causes does indeed seem seriously questionable. There is not any scientifically known pharmacologic or physical explanation for so widespread and multifarious an effect."[11]

Malnutrition and Obesity

Malnutrition and obesity present another health hazard of consequence. In general terms *malnutrition* might be considered as taking two forms: (1) under- or subnutrition and (2) overnutrition. Either can do harm. Undernutrition (less than two-thirds of the U.S. Agricultural Department recommended dietary allowance for protein, vitamins, and minerals) is reported to have increased in from 15 to 21 percent of U.S. households between 1955 and 1965. Calcium, vitamin C, vitamin A, and iron were the nutrients most often found to be below the recommended allowances. Poor diets are most usually found in families with low incomes and seldom occur alone, often being associated with poor housing, family disorganization, a low level of education, and a climate of apathy and despair.[12]

Dietary inadequacies can be related to the incidence of heart disease, respiratory and infectious diseases, infant mortality, hypertension, pernicious anemia, diabetes, skin diseases, pellagra, scurvy, and general health status and vulnerability to disease. They can affect physical size, appearance, weight, and energy. They can result in apathy, reduced intelligence, a greater difficulty in learning, and general behavioral disturbances.[13] Albeit, it is recognized that the nutritional status of people in the United States today is generally much better than it was at any time in the past. Furthermore, the ready availability of pure vitamins has made it possible to halt the progress of many diseases that are caused by inadequate diet.

Obesity (or overnutrition or overweight) can present a problem of definiton, and can be a highly subjective matter. There are, however, readily available tables of desirable weight ranges in relation to height, size of frame, sex, and age. The causes of obesity can be several, including not only the amount and types of food eaten but the habit of nibbling, inadequate exercise, psychological factors, ethnic mores, and genetic factors.

Statistical evidences indicate that overweight occurs more frequently among women than men, more frequently among blacks than whites, more frequently among the poor than the nonpoor, and that its incidence increases with age. One in 5 men age 20 and over is reported to be 10 percent overweight; one in 20 is reported to be 20 percent overweight.

The effect of overweight on mortality has been clear for quite some time from life insurance experience collected by the Society of Actuaries. These data show that men 20 percent or more overweight have a 31 percent greater mortality rate and that those 10 percent overweight have a 20 percent greater mortality rate. A study of over 73,000 overweight women found that those with severe obesity between ages 30 and 49 years increased the risk of diabetes 4.5 times, high blood pressure 3.3 times, gall bladder disease 2.7 times, and gout 2.5 times (although it is not implied that obesity is the cause of those conditions).[14]

ACCIDENTS

It has been shown that accidents are the fourth-ranking cause of death in the United States today, taking over 105,000 lives each year. They are the leading cause of death for people between the ages of 1 and 36. Accident mortality is greater among males, double the rate for females. The rate for females has been rising, however, principally as a result of motor vehicle accidents. The rate of accident mortalities among nonwhites exceeds that for whites by 40 percent. Accidental death rates decrease with age.

As will be shown in the following section, over 10 million disabling accidents occur in the United States each year, some 400,000 of which cause some degree of permanent impairment. Restricted activity requiring medical attention results from 48 million accidents each year. The costs to the economy and the consequent work loss are discussed later. Of all injuries, half result in sprains and strains of the back and lacerations and abrasions (about equally divided); the remainder are made up principally of contusions, fractures of the skull and other parts of the body, burns, poisoning, and adverse effects of medical and surgical procedures, in that order.

By the site and type of accident, some 30 percent of all accidents occur in or around the *home*, causing about 26,000 deaths each year and over 20 million injuries; 4 million of these being to some degree disabling. The incidence of such accidents is relatively higher in rural-farm areas. Over half are due to falls, principally among older people and children. A quarter result from burns and fire; 5,400 deaths each year are caused by fires. Other home accidents result from poisoning (1,400 deaths a year), firearms (1,200 deaths a year), poisonous gas (800 fatalities yearly), drownings in tubs or pools (500 yearly deaths), and from playthings, doors, power lawn mowers, and snowmobiles. Fortunately, the incidence of home accidents has been decreasing rather steadily, with a 21 percent drop since 1949.

In 1973 *motor vehicles* were the cause of 55,000 deaths, although in 1974 this number dropped to 46,500, presumably because of the gasoline shortage and reduced speed limit. In 1975 the number of such deaths dropped further, to 42,599, the lowest in 14 years. One-half of all motor vehicle deaths are said to be alcohol-related. The death rate today in terms of 100 million vehicle miles, however, is one-third the rate in 1934. Motor vehicles also cause over 2 million disabling injuries yearly.

Males have four times as many motor vehicle accidents as females, and half of such accidents involve persons under 35 years of age. Married persons have a lower rate than other people. The majority of motor vehicle accidents and more of the serious accidents happen at night. Pedestrians are involved in one-quarter of these accidents, but this rate is showing the greatest decrease. Speeding is involved in one-third of all motor vehicle accidents; one-fifth occur in bad weather. Other causes include faulty maintenance of the vehicle and sleepiness on the part of the driver.

Accidents at *places of work* account for about 8.9 million accidents each year, resulting in some 13,000 deaths and 2 million disabilities. The rate of fatal industrial accidents today, however, is half the rate in 1937. The principal causes of work accidents are handling objects (22.6 percent), falls (20.4 percent), falling objects (13.6 percent), and machinery (10.2 percent).

The rate of work accidents varies considerably with the type of work. It is highest in mining, followed by lumber, construction, transportation and transit, quarrying, agriculture and food production, wholesale and retail trade, the production of clay, wood, and mineral products, foundries, leather production, meat packing, and federal jobs employing civilians, in that order.

The data in Table 3-1 indicate a relationship between numbers of workers and the consequences of accidents for different types of employment. Smaller places of employment (less than 50 or 100 employees) have a higher accident rate than larger places of employment. It is of interest that for employed people the rate of accidents off the job is more than twice the rate of on-the-job accidents.

Accidents in *public places* (other than on highways or at work) cause some 24,000 deaths and over 2 million injuries each year. The fourth-ranking cause of accidental deaths is *drowning* while boating or swimming. Each year some 8,100 persons in the United States drown, the majority being males. The rate of such fatalities among nonwhites is twice that of whites. The rate of drownings is increasing.

Table 3-1. Relationship between number of workers and the consequences of accidents for different types of employment.

Industry	No. of Workers (in Millions)	Deaths	Disabling Injuries
Trade	17.5	1,200	390,000
Manufacturing	19.7	1,900	470,000
Service	16.5	1,900	350,000
Transportation and utilities	4.6	1,700	200,000
Agriculture	3.8	2,500	210,000
Construction	4.0	2,800	240,000
Mining and quarrying	.6	600	40,000
Government	12.3	1,600	290,000

Published by the National Safety Council, 1970.

Firearms cause 2,400 deaths each year, about half of which are in or about the home, and the rate of such mortality is increasing. Of these fatalities, 90 percent are males, and the rate is higher among nonwhites than whites. Each year some 3,800 deaths occur from *poisoning*, and the rate is increasing. The incidence is higher among males, and one-quarter of the victims are children. The overuse of barbituates accounts for 25 percent of such deaths; other drugs for 27 percent. Lead poisoning is also a recurrent problem. *Falls* cause over 16,000 deaths a year.

INCIDENCE OF DISABILITY

Data concerning the incidence and duration of disability are gathered for various purposes. Some existing data are not relevant to the incidence or duration of disability among the workforce since they include disability among children, students, housewives, retirees, the unemployed, the unemployable, and other non-income-producing persons. In some instances the available data are noncomparable and difficult to relate. In others, there are significant gaps. In still others, the data are based on household interviews, which can be subject to respondent error or subjective individual evaluation. The definition of disability, or of the degrees of severity of disability, can also vary.

The following are some reported findings of the incidence of disability among the general population of the United States. The incidence of disability in the workforce is shown in a subsequent chapter.

Data gathered by the Social Security Administration indicate that four out of every five people in the United States are disabled for one day or longer each year. The National Health Survey has estimated that 18 million Americans of working age are disabled to some degree each year. Of these over 6 million are unable to work or to work regularly (3.8 million are women), over 5 million are unable to work full time or have had to change the nature of their job (over half are women), and over 6 million experienced work limitations in the kind or amount of work they could do, although they could work full time.

This incidence of disability is estimated to result in 3 billion days of restricted activity and 1.2 billion days of bed disability—an average of 15.6 days of restricted activity per person per year and 6.3 days of bed disability per person per year. For this latter category, the average of bed-disability days per person per year ranged from 4.4 days for persons between ages 6 and 16 to 18.7 days for those classified as retired, age 45 and over.

Of those categorized as totally disabled, 20.9 percent have been disabled for less than 2 years, 21.5 percent have been disabled from 2 to 4 years, 20.5 percent from 5 to 9 years, and 37.1 percent for 10 years or more. Of those categorized as "severely disabled," 22.4 percent have been disabled less than 2 years, 21.7 percent from 2 to 4 years, 20.8 percent from 5 to 9 years, and 35.1 percent for 10 years or more.

The incidence and duration of disability were higher among the unemployed, those with low incomes, and those earning more than $10,000 a year in 1966. They were higher among children under 5 years of age, persons over 45, and particularly retirees and the aged. They were higher among females from ages 15 to 44 and age 75 and over. They were higher among the lesser educated and nonwhites. They were higher among alcoholics and problem drinkers and among heavy cigarette smokers.

For the totally disabled, according to the Social Security Administration, 24.8 percent are disabled as a result of circulatory disorders, 30.9 percent as a result of infirmities to the bones and organs of movement, 10.8 percent due to allergic disorders, 6.3 percent to mental disorders, 7.2 percent to digestive system disorders, 5.2 percent to disorders of the nervous system, 3.7 percent to respiratory disorders, and the remainder to sense organ disorders, infective and parasitic diseases, neoplasms, genitourinary disorders, and others. Presumably mental retardation resulting in total disability (affecting approximately 700,000 persons) is included under mental disorders, and presumably vision and hearing impairments are included under sense organ disorders. For persons having limited mobility, according to findings from the National Health Survey, the two principal causes are arthritis and rheumatism (24 percent of all such cases) and heart conditions (12.6 percent).

The incidence of *acute illnesses and injuries* in 1974 was 175.7 per 100 persons, respiratory conditions accounting for 94.4 per 100 persons and accidents 30.4 per 100. These illnesses and injuries were reported by the National Health Survey to result in 4 days of bed disability per person per year among the U.S. civilian population not in institutions and 9.2 days of restricted activity per person per year. Of these, 54.5 percent were medically attended. The incidence of such conditions was higher among persons under age 44, females, white persons earning over $10,000 a year, and the better educated.

BIBLIOGRAPHY

J. F. Follmann, Jr., *Medical Care and Health Insurance* (Homewood, Ill.: Richard D. Irwin, Inc., 1963).
―――, *Insurance Coverages for Mental Illness*, AMACOM, 1970.
―――, *The Mentally Retarded and Insurance Protection*, The President's Committee on Mental Retardation (New York: The Free Press, 1976).
Public Health Reports (Washington, D. C.: HEW).

Material from the following organizations was also used in the preparation of this chapter:
    American Cancer Society
    American Diabetes Association
    American Heart Association
    Arthritis Foundation
    Bureau of the Census
    Metropolitan Life Insurance Company
    National Cancer Institute
    National Center for Health Statistics
    National Safety Council
    National Tuberculosis and Respiratory Disease Association
    Office of Health Economics (London)
    Social Security Administration
    United Cerebral Palsy Association, Inc.
    U.S. Department of Health, Education, and Welfare.

REFERENCES

1. Monroe Lerner and Odin W. Anderson, *Health Progress in the United States: 1900-1960* (Chicago: University of Chicago Press, 1963).
2. Victor Fuchs, *Who Shall Live?* (New York: Basic Books, 1975).

3. Jeremiah Stamler, M.D., "Hypertension," an undated publication of the Northwestern University Medical School.
4. Richard B. Singer, M.D., "Chronic Renal Failure—New Hope for the Patient," *Best's Review*.
5. Ford Foundation Report, "Dealing with Drug Abuse" (New York: Praeger Publishers, 1972).
6. Mel Segal, National Institute of Drug Abuse, paper presented at meeting of National Association of Drug Programs, June 4, 1975.
7. HEW, "The Health Consequences of Smoking," 1974.
8. "Report of the 1972 APHA Smoking Survey," Public Health Nursing Section, *American Journal of Public Health*, October 1973.
9. American Heart Association, "Smoking and Heart Disease," New York, 1974.
10. P. M. Moody, J. Averitt, R. B. Griffith, and J. F. Benner, "Quantitative Characterization of Human Smoking Behavior," Tobacco and Health Research Institute, University of Kentucky, Lexington, circa 1964.
11. J. Berkson, "Smoking and Lung Cancer," *American Statistician*, October 1963.
12. J. B. Cordoro, "Nutrition: A Neglected Goal," *AFL-CIO Federationist*, July 1975.
13. Herbert G. Birch, M.D., "Malnutrition, Learning, and Intelligence," *American Journal of Public Health*, June 1972.
14. Rimm, Werner, Yserloo, and Bernstein, "Relationship of Obesity and Disease in 73,532 Weight-Conscious Women," *Public Health Reports*, January-February 1975.

4

The Economic Costs

The economic costs of death, illnesses, and accidents are important to individuals and to family units. They are important to employers, labor unions, health and welfare funds, and to insurers. They are important to government at all levels in many ways: the effect on the economy; the effect on tax receipts; the effect on various types of government-sponsored programs, including general welfare, aid to dependent children, aid to the totally and permanently disabled, aid to the blind, Social Security old age and disability benefits, benefits for the temporarily disabled, unemployment compensation, workmen's compensation, and the Medicare and Medicaid programs; the effect of health care provided by the government on a direct basis; and the effect on government as a large employer.

Unfortunately, the measurement of such economic costs is extremely difficult. The underlying concepts that enter into the measurements are complex, and there is not always unanimity as to their details. Methodology can differ, and the available data can be sparse or incomplete, frequently lacking in refinement. Yet such information is necessary for the formulation of national policy on any elements of such economic costs. Thus Burton Weisbrod commented over 15 years ago that "estimates of losses from disease involve questionable, misleading, or simply incorrect procedures . . . costs, real costs, that is, and monetary expenditures are not synonymous terms. There may be expenditures without real, or social costs (for example, production loss). And there may be social costs without expenditures."

A few years later Dorothy P. Rice wrote that "a variety of studies and figures relating to the costs of illnesses have been generated in the health field. The methods used range from broad statements of costs, with no apparent basis, to detailed, sophisticated economic analyses . . . costs are often estimated in gross terms: average earnings are applied to overall morbidity (or mortality) data, often with no adjustments for labor force participation and unemployment rates; age and sex differences in earnings may or may not be taken into account; public payments in the form of pensions or public assistance and taxes on earnings are sometimes incorrectly included in the total; and direct costs are a conglomeration of educated guesses and available statistics on institutional care, with noninstitutional costs neglected."

About that same time, Mrs. Rice noted: "The value of human life expressed in terms of lifetime earnings is a basic tool of the economist, program planner, government administrator, and others who are interested in measuring the social benefits associated with investments in particular programs. For public programs . . . the valuation of human lives is a basic requirement for the proper calculation of the benefits to be derived." She then attempted "improved, refined, comprehensive, and up-to-date estimates of the present value of lifetime earnings," with consideration of age, sex, color, and educational level. In so doing she examined the approaches taken by others and noted that all were devised for a specific use and are not readily adaptable for other purposes. She then reviewed the basic assumptions and economic concepts employed as respects life expectancy, labor force participation, earnings, housewives' services, productivity increases, allowances for consumption, the pattern of lifetime earnings, male and female differentials, and educational differences. She also examined the practice of discounting for the value of money changes with time, noting the general agreement among economists on the use of discounting but their lack of agreement on what the rate of discount should be.

The economic costs of death, illnesses, and accidents are usually divided into direct and indirect expenses. The direct costs are generally limited to health care expenses. The indirect costs might include such matters as lost production, earnings, and employment; losses to the gross national product (GNP); the loss of housewives' services; transfer payments by government (welfare, disability payments, medical expense benefits, unemployment compensation, and workmen's compensation); and the loss of taxes to government. The directness of such costs indeed calls into question the use of the phrase "indirect costs." Entering into the projection of "indirect" economic costs are such matters as life expectancy, the degree of labor force participation, potential increases in productivity, and allowances for consumption.

Using such an approach, it was estimated in 1963 that the economic toll of illnesses, accidents, and death at that time was $105 billion—$34.3 billion of that figure was for health care, $21 billion for morbidity loss to the economy, and $49.9 billion for the present value of lost output for those who died. These dollar amounts would quite obviously be considerably larger today as a result of the growth in the population since 1963, the growth in size and productivity of the labor force, and inflation. (For example, it will be noted that by 1975 health spending alone considerably exceeded the above total economic cost.)

The indirect costs in lost production in 1963 amounted to $21 billion resulting from morbidity and $2.7 billion resulting from mortality. Persons between ages 45 and 64 accounted for almost half of the total: $10.7 billion. Those between ages 25 and 44 accounted for another $7.1 billion. The younger ages accounted for $1.1 billion, and the older ages for $4.7 billion. The currently employed accounted for $16 billion, those unable to work $5.3 billion, and housewives an estimated $862 million. Chronic illnesses were the cause of 60 percent of these economic costs. Mortality and morbidity in 1963 are estimated to have cost the nation 4.6 million lost production years.

COSTS OF HEALTH CARE

Expenditures for health care are influenced, over time, by a great many factors. Included are the population growth, the increased longevity of the population, increased urbanization, the growing number of working wives, the improved standard of living, increases in real income levels, more equitable distribution of wealth, increases in savings and pensions, and most certainly inflation. They are also influenced by the rising levels of education, an ever increasing health awareness, and a growing confidence in the processes of health care, with a resultant increased demand for health services.

The presence of health insurance and such government-sponsored financing programs as Medicare, Medicaid, and workmen's compensation, which ease the financial burden on the individual, also produces a marked effect on both the costs and the expenditures for health care. The costs of care, meanwhile, are affected not alone by inflation but also by such developments as improved technology (cobalt machines, transplants, hemodialysis, pacemakers, more complex surgical procedures, laboratory tests, anesthesia, transfusions), increased specialization among the health professionals, improvements in both diagnostic and treatment techniques, the increasing number of wonder drugs, the ever improving quality of the care provided, and even by malpractice litigation.

These are all dynamic factors, subject to continually shifting forces. The degree to which each affects the growing expenditures for care has not been adequately evaluated scientifically. In toto, however, their effect is unquestionably strong.

In fiscal year 1976 spending for health care in the United States reached an unprecedented $139.3 billion, or $638 per person. This amount represented 8.3 percent of the GNP. That these expenditures have been growing rapidly is clear from the fact that in 1921 they were $3.4 billion, 3.5 percent of the GNP. By 1950 they had increased to $12 billion, 4.6 percent of the GNP. By 1960 they were $26.9 billion, 5.2 percent of the GNP. By 1965 expenditures had grown to $38.9 billion, 5.9 percent of the GNP. By 1970 the amount spent was $69.2 billion, 7.2 percent of the GNP. By 1972 it was $86.7 billion, 7.9 percent of the GNP. The Department of Health, Education, and Welfare has estimated that by 1980 health care expenditures will be $156 billion (based on a lower range of assumptions) to $189 billion (based on higher assumptions).

In reviewing the foregoing data it is important to recognize that they include estimates for expenditures for nonprescribed drugs, vitamins, and weight-reducing remedies. They include luxury hospital accommodations, telephones, television and radio sets in hospital rooms, and guest meals and gift shop purchases in hospitals. They include such items as sunglasses, toothpaste and brushes, toilet tissue, scales, and baby food when purchased in a drugstore. They include unnecessary visits to a physician. They include care provided for cosmetic rather than medical reasons. They include the net cost of health insurance, even for people who receive no medical care. They include the administrative costs of government financing programs and government public health activities (the latter totaled $3.5 billion for fiscal 1975). They include research costs of $2.7 billion, the $3.2 billion spent for the construction of privately owned facilities, and the $3.2 billion spent for the construction of publicly owned facilities.

The utilization of certain health services in 1973 was as follows: 10.7 percent of the population had one or more short-stay hospital episodes. The average length of stay was 8.1 days. In all, there were 32 million admissions to hospitals during the year. The number of short-stay hospital episodes was 13.9 for every 100 persons. Physician visits averaged 5 per person. Of all Americans, 48.9 percent had at least one visit to a dentist.

In fiscal 1975, hospital care accounted for 39 percent of the total expenditures for health care: $46.6 billion. Expenditures for physicians' services accounted for 19 percent of total expenditures: $22.1 billion. Dental care cost $7.5 billion, drugs and sundries cost $10.6 billion, eyeglasses and appliances $2.3 billion, nursing home care $9 billion, and other professional services $2.1 billion.

Since hospital care accounts for such a large proportion of the total expenditures for health care in the United States, it is interesting to note the changes that have taken place in the per diem cost of hospital care in relatively recent years. The average per diem cost of care in short-term general hospitals in 1974 was $127. In 1929 this cost was $4.53 a day. By 1945 it was $8.60; by 1955 it was $24.51; by 1960 it was $34.55; and by 1968 it was $66.83. The principal item in the cost of hospital care is labor; the number of employees required per patient has increased from 1.5 employees per patient in 1946 to 2.5 per patient today, reflecting the increasing degree of technology in the modern hospital. Other operating costs that, of necessity, reflect the national inflation include food, fuel, equipment, supplies, rent, interest, and depreciation.

COSTS OF DISABILITY

The incidence of disability among the United States population was shown earlier. There is a severe sparcity of information on the total economic costs of disability. One estimate, for the year 1963, was that the total cost of disability was at that time $94 billion. Of this amount, $50 billion resulted from mortality, $21 billion from morbidity, and $23 billion from the cost of care. In 1960 the estimated economic loss from nonoccupational short-term illness was estimated to be $8.6 billion (an illness that lasts not more than six months is defined as short-term; the first six months of a long-term disability is also considered short-term). Obviously these amounts would be considerably greater today.

The incidence and cost of disability among the workforce are shown later.

COSTS OF CERTAIN ILLNESSES

The following estimates of the economic costs of mortality, morbidity, and care resulting from certain specific illnesses are of interest.

In 1962 *cardiovascular diseases* had a total economic cost of $32 billion. Included here are heart disease, arteriosclerotic disease, hypertension, and stroke. Of the total, $31 billion resulted from heart disease and $1 billion from stroke. The cost of care was $3.1 billion, $2.6 billion of which was for the treatment of heart disease, and $1.2 billion of that going for hospital care. Included here are the costs of research, personnel training, construction, and the net cost of insurance. Indirect costs (lost earnings and production) were estimated at $28.8 billion, of which almost $26 billion resulted from mortality and $3 billion from morbidity. From heart disease alone, 542,000 man-years were lost in 1962, and 132 million workdays were lost, valued at $2.5 billion. The cost of lost output resulting from 250,000 premature deaths from heart disease was valued at $1 billion. The loss of output from disability and premature death resulting from stroke was estimated at $700 million at that time. Obviously these figures would be considerably greater today.

The economic costs of *neoplasms,* or cancer, in 1962 amounted to an estimated $11.2 billion. Greatest contributors to these costs were cancers of the intestines and rectum, lungs, digestive organs, and breast, in that order. The cost of care for cancer in 1962 was $1.2 billion, of which hospital care accounted for $920 million. In 1962 cancer accounted for 14 million days of hospital care, or 8 percent of the days of hospital care from all causes. These health care costs included the costs of research, training of professional personnel, construction, and the net cost of insurance. Indirect costs to the economy (lost earnings and production) were estimated at over $9 billion. The 290,208 deaths in that year accounted for 4.8 million lost man-years, or a loss in lifetime earnings of over $8 billion. Lost earnings from morbidity were estimated at $1 billion, with 54 million workdays being lost in 1962.

In 1963 *acute diseases of the respiratory system* were estimated to produce an economic cost to the nation of $7.4 billion. Direct costs of care were placed at $1.6 billion. The indirect costs of morbidity from this cause were estimated at $3.2 billion. The indirect costs of 105,235 deaths in that year, resulting in 2.2 million man-years lost, were estimated at $2.7 billion.

At that time the economic costs of *tuberculosis* were estimated to be $968 million, the cost of care being $241 million, the indirect costs of morbidity being $385 million, and the cost of 9,306 deaths, resulting in 168,000 man-years lost, costing $341 million in lost lifetime earnings.

In 1963 *acute diseases of the digestive system* were estimated to have an economic cost of $7.8 billion. Of this, $4.2 billion resulted from the costs of care, indirect costs of morbidity producing a cost of $1.2 billion, and the cost of 71,700 deaths, with 1.4 million man-years lost, was $2.5 billion.

Allergic, endocrine, metabolic, and nutritional diseases in 1963 were estimated to produce an economic loss of $2.6 billion, of which $902 million were for costs of care, $539 million were the indirect costs from morbidity, and the cost of 43,414 deaths, resulting in 690,000 man-years lost, was $1.2 billion from lost earnings.

In 1964 an epidemic of *rubella* (measles) resulted in 30,000 babies being born dead and another 20,000 being born with serious congenital defects—an estimated economic cost of $2 billion.

At that time the costs of *arthritis and rheumatic diseases* were estimated to be $2 billion (including $310 million for worthless drugs and treatment). The costs of *cerebral palsy* were estimated at $1.3 billion in lost productivity alone. In 1958 loss of income resulting from *syphilis* was estimated to be $86 million, and the economic costs of the syphilitic blind were placed at $15.7 million. The cost of maintenance for those with syphilitic psychoses was estimated to $50.2 million. The World Health Organization has estimated that *noise* costs the United States $4 billion yearly in accidents, absenteeism, and workmen's compensation claims.

Diseases of the nervous system and sense organs in 1963 were estimated to produce an economic cost of $6.8 billion. Of this, $1.4 billion was for the cost of care; $1.5 billion was the indirect cost of morbidity; and the 215,648 deaths resulting from this cause, with 2.6 million man-years lost, cost an estimated $3.9 billion in lost earnings.

The costs of *mental illness* in 1963 were estimated to be $7.3 billion. Health care and institutionalization cost $2.5 billion; indirect costs resulting from morbidity, including 609,000 institutionalized persons, were placed at $4.6 billion; and the 4,651 deaths resulting from this cause, with a loss of 121,000 man-years, cost $251 million in lost earnings. Other estimates in 1970 placed the cost for care at from $3 billion to $6 billion, with $1.7 billion being paid from public funds, $1.5 billion resulting from loss of earnings, and $14 billion going for other economic costs.[1] By 1975, the total economic cost for mental illness was estimated to be $26.2 billion.[2] The components of these costs were as shown in Table 4-1 in 1975.

Mental retardation in 1963 was estimated to result in a loss of 107,000 man-years, at a cost of $494 million. Of all the retarded, many of whom are employed, only a small number (some 174,000) are institutionalized.[3]

The economic costs of *alcoholism* today have been estimated by the National Institute on Alcohol Abuse and Alcoholism to exceed $25 billion annually. Of this amount over $9 billion represents lost production (not including the reduced production of female workers and housewives). Over $8 billion is the cost of health and medical costs for the treatment of alcoholism (hospital care alone costing $5.3 billion in 1971). Alcohol-involved motor vehicle accidents accounted for more than $6 billion ($3.56 billion for fatal accidents, $2.38 billion for injuries, and $500 million for property damage). Costs to the criminal justice system resulting from the abuse of alcohol was placed at $500 million, the costs

Table 4-1. Components of costs for mental illness in 1975.

Expenditure		Cost
Direct care		$10.1 billion
Public hospitals	$ 3.1 billion	
Private hospitals	281 million	
General hospitals	981 million	
Community centers	285 million	
Outpatient clinics	481 million	
Drugs	538 million	
Medical services	358 million	
Nursing homes	2.7 billion	
Rehabilitation	195 million	
Children's programs	304 million	
Private psychiatrists	833 million	
Private psychologists	79 million	
Training of personnel		187 million
Research		399 million
Construction		604 million
Management		788 million
Indirect costs		14.1 billion
From death	3.5 billion	
From disability	7.4 billion	
From patient care activities	3.2 billion	

of social services related to alcoholism were estimated to be $135 million, and approximately $640 million was spent on alcoholism programs for treatment and rehabilitation of the alcoholic and for public education and research. Fire losses resulting from alcohol abuse were not included. Thus the total economic cost to the nation resulting from alcohol abuse and alcoholism could very well exceed $30 billion yearly. Alcoholism is estimated to account for 12 percent of all health expenditures and 20 percent of all hospital expenditures (13 to 22 percent of all people in public mental hospitals are there because of alcoholism).[4]

COSTS OF ACCIDENTS

The economic costs of mortality, morbidity, and property damage resulting from accidents in 1963 were estimated to be $11.8 billion. The costs of care at that time amounted to $1.7 billion. Indirect costs of morbidity were $1.8 billion; from mortality, with 130,665 deaths from injuries resulting in 4.2 million lost man-years, $8.3 billion.

By 1974 the National Safety Council had estimated the economic costs of accidents in the United States to be $43.3 billion. Wage loss resulting from accidents was $13.2 billion; the cost of medical care, $5.7 billion; motor vehicle property damage, $6.5 billion; property damage from fire, $3.7 billion; insurance administration and claims settlement costs, $5.7 billion; and the money value of time lost by workers other than those with disabling injuries, $6.8 billion. In 1974 motor vehicle accidents cost $19.3 billion; work accidents, $15.3 billion (not including property damage); accidents in the home, $5.1 billion; and accidents in public places (other than motor vehicle accidents), $4.4 billion.

Thus accidents alone account for about one-eighth of the total economic losses resulting from diseases and accidents.

CONTROLLING COSTS AND EXPENDITURES

One question that always arises in any discussion of the economic costs of illnesses, accidents, and premature death is the extent to which the cost (prices) and expenditures for health care can be controlled or contained, and the extent to which the total economic costs can be controlled.

It has been seen that many of the elements in these economic costs are influenced by such matters as the population size and composition, the result of a longer life span in the nation, the state of the national economy, inflation, real income levels, and the level of general education. It has also been seen that many other elements enter into the costs of and expenditures for health care, some of which are quite socially and personally desirable: improvements in the quality of the care provided, technological advancement, greater availability of health services, growing public confidence in those services, and more general means for financing care.

Within the realm of both costs of and expenditures for health care, however, several approaches to the containment of costs and expenditures are evident.

At the outset, it must be recognized that there is no easy or simple way in which such containment can be accomplished. It is a subject over which health economists ponder at length, but with no ready answer, and certainly none that would receive anything approaching unanimity of accord. Any accomplishment requires a complex, multifaceted approach, with the various elements unproved to date. In the interim, private health insurers and government have been experimenting with several approaches to the subject but, as the foregoing figures make readily evident, with little success.

On one point, however, there is public unanimity. No one wants to reduce costs at the sacrifice of the quality of the care provided, and no one wants to reduce expenditures through a reduction in the opportunity to obtain needed care.

The following are some of the areas in which a degree of solution might lie. Each could contribute its part. In the aggregate an impact could conceivably be felt that would provide some relief to the continuous sharp rises in both the costs of and expenditures for care—relief in the sense of slowing down future increases.

National inflation. Since the medical care complex must live in the world around it, and react to it, it goes without saying that a first step in a solution to rising health care costs would be to bring under control the degree of national economic inflation. Until this is accomplished the costs of care must rise, regardless of any other influencing factors.

Effect of the Economic Stabilization Program. On August 15, 1971, President Nixon announced an immediate 90-day freeze on prices, wages, salaries, and rents, to be administered by the Cost of Living Council. On November 14, 1971, Phase II of the Economic Stabilization Program commenced. The purpose was to reduce the rate of inflation. A Committee on Health Services was appointed in an advisory capacity. On December 30, 1971, this committee promulgated regulations for (1) institutional providers (which permitted price increases to produce a 2.5 percent increase in revenues without prior approval, to produce up to 6 percent increased revenues subject to justification, and above 6 percent if permitted) and (2) noninstitutional providers (which permitted price increases of no more than 2.5 percent).

The period of mandatory controls ended January 11, 1973, except for certain elements of the economy, including health care. The effect of this development on increases in medical care prices is indicated by the fact that whereas during the two years preceding the freeze, medical care prices rose 6.7 percent (the Consumer Price Index generally rose 5.6 percent), from August to November 1971 medical care prices fell .8 percent (while the CPI rose 1.6 percent), and from November 1971 to January 1973 medical care prices rose 3.4 percent (while the CPI rose 3.6 percent). Quite obviously, then, inflationary controls would serve to contain medical prices and, therefore, the expenditures for medical care. The degree to which such containment would be effective over a long period, however, is a matter of speculation. This is said in the sense that over a short period certain expenditures, such as capital investment, repairs, equipment replacement and improvements, can be postponed, but over a longer period their effect would eventually become evident.

Unnecessary care. Unquestionably the elimination of unnecessary care would reduce medical expenditures. Hypochondriacs waste doctors' time and their own money over imaginary complaints. They buy quantities of well-advertised nostrums over the counter. They engage in no end of nonsensical practices over imaginary or minor complaints that some applied common sense would probably dispose of. Physicians, on their part, might be more forthright and blunt with such patients.

Superfluous or duplicated laboratory tests are another form of excess that could well be eliminated. The threat of malpractice suits, however, can very well thwart any effort in this direction.

Elimination of excessive hospital stays, at the insistence of the physician or of the patient, would serve to cut costs. This can be accomplished by careful surveillance of each case. Considerable savings have resulted where this is done through utilization review committees. Again, however, the threat of malpractice suits can thwart such potential savings.

Most certainly the elimination of unnecessary surgery would produce significant savings. The record is clear that from time to time, in various geographic areas, what amounts to "epidemics" in certain types of surgery occur. These operations have included such procedures as hysterectomies, gall bladder removal, hernia repair, hemorrhoidectomies, prostatectomies, appendectomies, and tonsilectomies.[5] The correction of this practice could lie in the effective, active functioning of tissue committees in the hospitals, in surveillance on the part of medical societies, and in the collection of statistics on the number of each type of surgical procedure performed in each geographic area per 100,000 population by some central agency, such as the National Center for Health Statistics, or insurers acting cooperatively. Most certainly the decision to perform surgery can be influenced by honest differences of professional opinion. A combination of the foregoing corrective measures should, however, stabilize such differences somewhat. Professional policing should be diligent on this score.

One way of coping with elective surgery is thought to be the obtaining of a second opinion when surgery is recommended. One labor union in New York City has been requiring a second opinion under its health insurance plan, and the practice has resulted in a reduction of such surgery by 17.5 percent among its 11,000 members. Applied nationally, this approach is estimated to have the potential of very considerable savings, perhaps $5 billion a year.[6] By 1977 several Blue Cross plans had adopted this approach. Here it is of interest that the Health Insurance Benefits Advisory Council, which offers counsel with respect to the Medicare program, in its 1966-1967 report takes cognizance that there were cases of "evidence of fraud," "repeated overcharging," and a "pattern of rendered services substantially in excess of those justified by sound medical practice." Also of interest is the publication in 1974 by one State Insurance

Commissioner (Pennsylvania), "14 Rules on How to Avoid Unnecessary Surgery."

Hospital management and pricing. Hospital care is an extremely costly form of health care. Today, however, considerable thought is being given to the means by which these costs can be contained. One approach considered is that of combining hospitals, particularly in a given geographic area, under a common management. This could bring about the joint purchasing of such things as linens, foods, and basic equipment. It could also bring about the sharing of expensive equipment.[7]

Another approach taken to reduce hospital costs is that of utilization review, mentioned earlier, whose purpose is to reduce the unnecessary utilization of hospital care, principally by reducing the lengths of hospital stays. Some savings do result from this approach, although a report of the staff of the U.S. Senate Committee on Finance in 1970 stated: "Utilization review requirements have, generally speaking, been of a token nature and ineffective as a curb to unnecessary use."

The pricing practices of hospitals also call for reevaluation. In the past, hospitals have made cost reimbursement arrangements with certain sources of patients (Blue Cross subscribers and Medicare and Medicaid recipients) at prices not adequate for their total financial requirements. To compensate for this, the established charges had to be increased for other patients, a discriminatory and inequitable system.

More recently many have come to recognize that the existing approach to hospital reimbursement does not provide any incentive to contain costs. The solution proposed is referred to as a system of prospective rating, which in turn would subject hospital financing to public accountability. Under such a system, hospitals would have to justify in advance before a state rate review commission any increases in their rates and to furnish a certification of need for any rate increase. The objective would be to bring about increased efficiency, to limit plans for expansion or modification in plant or services consistent with community needs, to discourage the duplication of expensive equipment and services, and to bring about price equality among all patients. The result should be a reduction in the rate of escalation of hospital costs, one estimate being a saving of more than $500 million a year.[8]

Health insurers are quite involved in this development. To date, 24 states have enacted certificate of need legislation, although the North Carolina law was declared unconstitutional. All are so recent that the effectiveness of this approach cannot yet be proved. In addition, in some states the applicability of the law is limited to only certain patients such as those being paid for by Medicaid. The experience in Connecticut is of interest. In fiscal year 1974-1975, total hospital costs in the nation increased 13.6 percent; in Connecticut the increase was only 8.3 percent. There are some, however, such as Clark C. Havighurst, professor of law at Duke University, who feel that this approach is unlikely to appreciably affect cost control in the long run.

Supportive of this development is the process of comprehensive community health planning, encouraged by the Partnership for Health Act of 1967. In 1974 the federal government lent further encouragement to this approach in the National Health Planning and Resources Development Act, which provides assistance to state and local health planning agencies. The purpose of community health planning is to reduce health care expenditures by restricting the unjustified expansion and duplication of health facilities and services. Again, there is a serious involvement on the part of health insurers in this development. To date, it is probably fair to say that, except for a few isolated instances, the effort has not been too successful.

PSROs: For many years the concept of peer review by and of physicians has been advocated and urged, but with little in the way of concrete success. Recently, encouraged by a federal law in 1972, the concept has been expanded to bring about the establishment of professional service review organizations (PSROs). Unfortunately the federal concept is presently limited to the Medicare and Medicaid programs, although the principles involved could well have general applicability. The purpose of this approach is the development of a mechanism for determining the medical necessity of institutional care and for the review and evaluation of health services that are provided on an ambulatory basis.

Failure or success of the PSRO development will, in the last analysis, be determined by public cooperation, understanding, acceptance, and support of the concept. Health insurers are deeply involved in the development. By early 1976 there were 12 PSROs operational and 90 in the planning stage. It is hoped that this effort will reduce hospital admissions and reduce the length of hospital stays. The full cooperation of physicians is a sine qua non.

Medical foundations. A forerunner of the PSRO concept, and closely related to it in certain ways, was the development of medical care "foundations." The first such foundation was formed in San Juaquin, California, over 15 years ago. The purpose was to develop screening guidelines, to monitor health services, to evaluate the care rendered, and to engage in claims surveillance. Wasteful practices were identified and discouraged. Experience has shown that foundations have the ability to generate savings, and this has been demonstrated by the Foundation for Health Care Evaluation in the Twin Cities. Today such foundations have the potential of becoming an integral part of the PSRO development.

Alternative forms of care. One important aspect in any consideration of reducing expenditures for hospital care is the general availability of alternative and less costly forms of care. Particularly in cases requiring a long period of convalescence or chronic illness, or in the treatment of such diseases as alcoholism, costs could be cut if there were readily available acceptable forms of hospital outpatient services, nursing home care, home health care services, homemaker services, meals on wheels, or accredited and acceptable clinics and treatment centers.

Such services are not yet available in a great many communities. If they were, it should be possible to reduce a certain degree of hospital care, and to that extent reduce health care expenditures. Much, however, would depend on the acceptance of such alternatives by the public.

Insurance benefit patterns. It has long been maintained that the pattern of health insurance coverages, in those cases where the benefits are restricted to in-hospital care, in effect compels the use of costly hospitals, whether such use is necessary or not. Recognition of this situation led the insurance companies to develop, about 1950, major medical expense insurance coverage for both in- and out-of-hospital care. That form of coverage now predominates the coverages provided by insurance companies.

Clear evidence is lacking that medical care expenditures have thereby been reduced. A Rand study, presently in process, might throw light on this proposition. It should also shed some light on whether deductibles and coinsurance, which give the patient a stake in both costs and expenditures, produce the hypothesized result.

Insurance claim surveillance. The day-to-day claim surveillance and analysis process of insurers also serves to contain medical costs somewhat, although more cannot be expected of this process, standing alone, than is warranted. Other simultaneous and complementary efforts, such as those previously mentioned, are requisite. The development of a national data bank on physician charges, by procedure and geographic area, is one such complementary approach. Thereby, unusual charges could be more readily identified. Such a data bank would also relate closely to the PSRO development.

HMOs. A final development in which there is considerable interest from the cost-saving standpoint is the provision of care on a prepayment basis through Health Maintenance Organizations (HMOs), or prepaid group practice plans as they were once known. This development was encouraged by congressional enactment of the Health Maintenance Organization Act in 1973. In early 1976 there were 173 HMOs in the nation, with an enrollment of 5.7 million persons. The Kaiser Foundation Health Plan is the largest, with 2.6 million members. Other larger, and long-established, plans are the Health Insurance Plan of Greater New York, the Group Health Cooperative of Puget Sound, and the Group Health Association of Washington, D.C. Most of the others are relatively more recent, and most are quite small in terms of the size of membership. Practically all HMOs give evidence of reducing costly hospital utilization, usually by 20 to 40 percent. On the other hand, the costs for their services seem to balance out these savings. The result is that the case is not yet clearly proved, although the evidences continue to be hopeful.

Prevention. A basic approach to the containment of the total economic costs resulting from illnesses, accidents, and premature deaths is to prevent their occurrence to the maximum extent possible. Another is to bring about an optimum of early detection and treatment of diseases so that a minimum of economic losses will result.

BIBLIOGRAPHY

Theodore Allison, "The Costs of Health Care," *Best's Insurance News,* April 1964.

Rashi Fine, *The Economics of Mental Illness* (New York: Basic Books, 1958).

J. F. Follmann, Jr., *Medical Care and Health Insurance* (Homewood, Ill.: Richard D. Irwin, Inc., 1963); "Rising Health Costs: What Can Be Done About It." *Employee Benefits Journal,* Fall 1975.

Herbert E. Klarman, *Economics of Health* (New York: Columbia University Press, 1965).

Klarman, Rice, Cooper, and Stettler. "Accounting for the Rise in Selected Medical Care Expenditures," *American Journal of Public Health,* June 1970.

Lerner and Anderson, *Health Progress in the United States, 1900-1960* (Chicago: University of Chicago Press, 1963).

Charlotte Muller, "Economic Analysis of Medical Care in the United States," *American Journal of Public Health,* January 1961.

Dorothy P. Rice, "Estimating the Cost of Illness," *American Journal of Public Health,* March 1967; "The Economic Value of Human Life," *American Journal of Public Health,* November 1967; "The Direct and Indirect Cost of Illness," material sent to the author by Mrs. Rice.

Material from the following organizations was also used in the preparation of this chapter:
American Cancer Society
American Diabetes Association
American Heart Association
Arthritis Foundation
National Cancer Institute
National Safety Council

National Tuberculosis and Respiratory Disease Association
Office of Health Economics (London)
U.S. Department of Health, Education, and Welfare

REFERENCES

1. J. F. Follmann, Jr., *Insurance Coverages for Mental Illness* (New York: AMACOM, 1970).
2. E. James Lieberman, M.D., "Mental Health: The Public Health Challenge," *American Journal of Public Health,* 1975.
3. J. F. Follmann, Jr., "The Mentally Retarded and Insurance Protection" (Washington, D.C.: Health Insurance Association of America, 1973).
4. _____, *Alcoholics and Business.* AMACOM, 1976.
5. *The New York Times,* January 27, 1976.
6. *The New York Times,* August 7, 1975.
7. *The Washington Post,* August 26, 1973.
8. *The New York Times,* August 8, 1975.

Part II

The Health of Employed People

Part I showed the health status of people in the United States and noted the economic costs of diseases, accidents, and premature death. The question is now posed: How do these matters affect employed people and places of employment? This question is of concern to employers, to labor unions, to insurers, and to government. Each has become more and more active in coming to grips with the problems presented. Not that their concerns always coincide, not that the approaches taken are always a harmonious endeavor. Yet the end result is increasingly to the advantage of employed people and, therefore, to the community and the nation.

In discussing the health of employed people and its costs, it is necessary to bear in mind that their health is not an entity unto itself. There is a direct correlation between the health of workers and the health of a community as a whole. For example, if an influenza epidemic strikes a community, it also takes its toll among employed people. This discussion, therefore, is by no means so simple that it can be limited to work accidents or diseases brought about by working conditions, although these, obviously, must be included.

5

Incidence of Diseases and Accidents

Today more than 90 million Americans are employed in civilian occupations; 2.4 million are federal government employees. This total comprises two-thirds of the adult population. Some 4 million of these hold two or more jobs. The majority of the workforce are between ages 17 and 44 and the majority are males, the proportion of females being greater in the age 17-to-44 category than in the later years. Married women in the workforce have increased considerably since World War II, so that today over 20 million of the currently employed are married females. In 1961 this number was 13 million. Of the employed married females, 40 percent work full time during the entire year. The labor force participation of married women rises with the level of education and is highest proportionately among those who completed college (52.7 percent). Of married women age 45 to 54, 47.8 percent are employed, for ages 25 to 34 the proportion is 38.3 percent. Today in over 44 percent of the American families both the husband and wife are wage earners; in 1960 this proportion was 34 percent. Jobs classified as "white collar" increased from 43 percent of the total to 48 percent since 1960.

According to the Bureau of Labor Statistics, some 16 million civilian workers are unionized. The majority of union workers are males, the exceptions being in wholesale and retail trade, in the service industries, and in professional and related services. By industries the following were the totals of unionized workers in 1970 (in thousands):

Agriculture	46
Construction	1,948
Manufacturing	7,600
Transportation and public utilities	2,527
Wholesale and retail trade	1,709
Services and finance	2,103

These employment data are shown not only to indicate the size and composition of the labor force but also to indicate certain trends that have been taking place recently. These trends can produce significant effects on the incidence of diseases and accidents among employed people in both private and government places of employment. Unfortunately, certain data on the incidence of illness, accidents, and premature death among the workforce are not current; other data are incomplete or spotty. Regardless, they can most certainly be considered indicative.

In 1976 the Institute for Occupational Safety and Health was reported in *The New York Times* as estimating that each year 100,000 people die as a result of work injuries and diseases and that 5.9 million workers experience work injuries and diseases. Of these latter, 30 percent result in loss of time, a loss of some 25 million workdays each year. Studies by the Bureau of Labor Statistics show that the incidence of occupational injuries and illnesses per 100 employees is greatest in the contract construction industry, followed by manufacturing, agriculture and forestry, transportation and public utilities, wholesale and retail trade, service industries, and finance, in that order. The incidence was lowest in places employing less than 20 employees or over 2,500 employees.

DISEASES

In 1974 acute disease conditions among employed people that required medical attention and/or that resulted in a restriction of the person's usual activity occurred as follows. Respiratory conditions affected 78.6 of every 100 workers (influenza, 37.7 of every 100). Infective and parasitic diseases affected 13.8 of every 100 workers. Digestive conditions affected 7.2, and other conditions affected 15.9. The incidence of all acute conditions was 144.4 for every 100 employees.

In 1963, 3.2 percent of the employed people were unable to work because of chronic conditions, 9.5 percent had some limitation of their activities, and 47.3 percent had no limitations as a result of chronic conditions. The two leading causes of limitation of major activity were (1) arthritis and rheumatism and (2) heart conditions.

Unless adequate safeguards are taken, industrial processes can induce diseases that range in seriousness from discomfiture, to brief or extended periods of disability, to permanent disability, to premature death.

The U.S. Department of Labor, in unpublished data, has estimated that in 1973 there were 200,000 cases of occupational illness. Of these, 44.5 percent were skin diseases, 13.7 percent were disorders due to physical agents, 11.8 percent were disorders due to repeated trauma, 5.7 percent were respiratory conditions due to toxic agents, 3.4 percent resulted from poisoning, 0.7 percent were due to diseases of the lung caused by dust, and 20.2 percent were due to other occupational illnesses. These illnesses might be contracted by skin contact, by inhalation, through the mouth and digestive tract, and from stress. They can emanate from the use of chemicals or toxic substances, air pollution, welding processes, excessive noise, vibrations, temperature extremes, faulty lighting, or radiation. The respiratory system, heart and blood vessels, blood, liver, kidneys, digestive tract, eyes and ears, or the nervous system all can be affected. Some diseases become evident in a short period, or even suddenly; others require many years before their effects become evident.

Many industrial hazards do not in themselves cause disease but can be contributory factors in relation to a person's physical condition, drinking and smoking habits, and lifestyle, and to harmful factors in the home, on the highway, or in the community. For example, certain occupational exposures that result in chronic bronchitis and emphysema have more pronounced effects among individuals who are heavy cigarette smokers.

Some of the disease hazards of industry are, today, becoming clearly evident. In a great many cases the presence or the degree of the hazard is suspect but is as yet lacking in scientific verification or sufficient statistical evidence. One reason is that modern industrial processes change so frequently that there is not sufficient time for scientific investigation or for the effects of the processes on health to become clearly evident. To take just one example, each year some 3,000 new chemicals are said to be introduced into American industry. Time and research investigation are required to adequately determine their effects on workers.

Many safety standards, existing or in the process of development, protect workers against occupational disease hazards. Many of these standards are arrived at by industry itself through research and voluntary consensus; others are government-imposed, and violation is subject to penalty.

Of pertinence here is a summary of what is known and what is speculated about industrial diseases and their effects on the health of employed people. Unfortunately, hard facts concerning industrial diseases are relatively rare, most of the data available being confined to accident reports. Unfortunately, also, there is a very considerable difference of opinion about what some people suspect are, or might be, occupational disease hazards. It is difficult, then, to find unanimity as respects the possible inherent dangers in many industrial processes.

It has been estimated that 100,000 coal miners are affected by black lung (pneumoconiosis), that 200,000 former miners are disabled from the disease, that each year 4,000 die of the disease, that over the next 20 years 600 to 1,100 uranium miners are expected to develop cancer, and that of the 800,000 textile and cotton processing workers, 12 to 30 percent have brown lung (byssinosis), a disabling disease.

Of the 500,000 persons who work with asbestos, it is estimated that 100,000 will die of lung cancer (other projections estimate 300,000), 35,000 of abdominal or chest cancer, and 55,000 of asbestosis. It is further estimated that half of these workers already suffer from asbestosis. One study of 689 asbestos and textile workers over the period 1959 to 1971 revealed 72 deaths from cancer, compared to a rate of 27.8 for the general population. Many of the effects of exposure to asbestos are said to date back to the 1930s, revealing the slow-acting nature of this hazard. Exposure to asbestos dust can occur in such varied occupations as the manufacture of asbestos textiles, valve packings, asbestos sheets, brake linings, boiler blankets, building materials, and electrical appliance cords.

Among steelworkers who use coke ovens, three times the incidence of cancer has been found as compared with other steelworkers. Workers who are employed in nickel refining are reported to have five times the incidence of cancer found in the general population. Migratory farm workers have a life expectancy ten years less than the general population, and their mortality rate from respiratory diseases is 250 percent of the national average.[1] Work-noise levels at which hearing loss occurs are said to exist for 14 million workers. More than 1.6 million people are employed in many industrial manufacturing processes and service industries (including printing type, plumbing, bearings, solder, paints, and cables) where they are exposed to lead poisoning, which can result in gastrointestinal disorders (one study found that 100 to 145 lead workers have lead contamination in the bloodstream).

Workers exposed to carbon disulfide are reported to have twice the risk of coronary heart disease as other workers. Those working with chromium are said to have 29 times the rate of lung cancer as other workers. It is reported that 15 per 1,000 of manganese workers suffered from pneumonia over a seven-year period compared to a rate of 0.73 for the general population. Workers in chromate-producing industries have been found to have 25 times the incidence of cancer as the general population. Today there are some 50 nuclear-power reactors built or under construction. Ionizing radiation (as, for example, from X rays) can produce cancer, genetic damage, and damage to the bone marrow. It is estimated that 1.1 million workers are exposed to silicosis resulting from the inhalation of silica dust in iron and mineral processing, mining, abrasive-soap production, glass manufacturing, potteries, foundries, stone-cutting and -finishing industries, and the use and manufacture of abrasives.

Today there is growing concern over the exposure of a considerable proportion of the workforce to significant levels of known toxic substances. Some of the reasons for this concern are the incidence of cancer among arsenic workers, the role of vinyl chloride in cases of angiosarcoma of the liver, and the possibly hazardous effects of lasers, microwaves, and polychlorinated biphenyl. In 1970 the American Conference of Governmental Hygienists estimated that from 6,000 to 12,000 toxic industrial chemicals were in common use in the United States. At present very little is known about the effects of many chemicals, gases, dusts, and metals; however, some substances can damage the lungs, liver, kidney, bladder, eyes, and skin and cause pulmonary fibrosis, chronic bronchitis, emphysema, diarrhea, lead poisoning, and pneumonia. In addition, some chemicals are flammable or explosive.

One final occupational hazard warrants mention—vibration and shock. The effects can be a disturbance of sensory and neuromuscular bodily functions, including speech communication and vision; lung damage, injury to the inner wall of the intestine, or cardiac damage; or Raynaud's disease, bone change, or ear damage. They may stimulate nervous system or hormonal activity, or produce fatigue, inattentiveness, or changes in work capacity. Some of the sources of potentially harmful vibration are pneumatically or electrically operated tools and devices, vibration machinery, heavy vehicles, and farm tractors. Among the industries affected by vibration are mining, foundries, transportation, textile, construction, lumbering, machine tool, and agriculture.

Many of these occupational disease hazards are presently receiving attention from a variety of sources. It is hoped that in the future much more in the way of definitive knowledge of potential hazards will exist. Meanwhile, safety standards bearing on industrial disease hazards are being developed or refined on a continuing basis.

ACCIDENTS

Accidents are one of the principal causes of disability and premature death. Some 112,000 persons are killed yearly by accidents of various types, and in 1974 24.2 million persons were reported to have suffered some degree of injury. Eleven million accidents resulted in disabling injuries, many of them affecting people in their productive years.

With the exception of workmen's compensation claims, off-the-job accidents have essentially the same effect on places of employment as do occupational accidents. They result in absenteeism, lost production, labor turnover, disability-insurance claims, health insurance claims, and the payout of salary-continuance benefits.

In 1969 there were 3.1 million injuries among employed people resulting from off-the-job accidents. Of these 1.2 million occurred in or about the home and 1 million in public places other than the highway; slightly less than 1 million resulted from motor vehicle accidents. In 1960 the reported total of such injuries was 2.3 million.

Of all accidental deaths to gainfully employed people, 70 percent happen off the job. For disabling injuries the proportion is about half. The ratio of off-the-job injuries to work injuries ranges from 2 to 1 to 45 to 1, depending on the type of employment. One employer in New Jersey reported that over a six-month period industrial accidents resulted in 16 days' lost time, while over the same period off-the-job accidents cost the company 3,200 days of work. Another employer reported that over a 19-year period, seven times as many employees were killed away from work as were killed in the course of employment. The similar experience of another employer showed the off-the-job death rate from accidents to be eight times the job rate. A further study has shown that in 1958 of the 4.7 million persons injured in motor vehicle accidents, 50 to 60 percent were employed people.

Accidents in the workplace, despite very considerable gains in the adoption of safety measures in relatively recent years, continue to take a serious toll. They have many causes: faulty equipment, inadequate safeguards and failure to use safety devices, explosions, and fires. Contributing factors include inadequate or faulty lighting, noise, vibration, excessive heat or cold, and job monotony. Workers themselves contribute to the cause by having faulty vision, hearing defects, ill health, fatigue, stress, and poor concentration and by being careless, all of which may be factors in the occurrence of accidents. One study found that 45 percent of work injuries were due to the carelessness of the worker or misuse of tools (particularly among inexperienced younger workers), 30 percent to momentary hazards, such as wet floors, and 20 percent to permanent physical factors.

Most frequently work injuries are caused by handling objects, representing 22.6 percent of all injuries. Falls account for 20.4 percent of all work injuries and falling objects 13.6 percent. The most costly injuries result from falls and motor vehicle accidents. James H. Mack of the National Machine Tool Builders Association was reported in *The New York Times* on December 5, 1976, as maintaining that 75 percent of all insurance claims involve a machine that is ten years old or more. Only 13 percent of the cases, he said, involved mechanical failure or malfunctioning of the machine. He has found that poor maintenance of machines by employers, failure on the part of the employer to provide adequate guards for machines, failure to require employees to use safety equipment, and lack of adequate training of the employee in the safe use of machinery account for the majority of machine-involved accidents.

Today, according to the National Safety Council, there are some 13,400 deaths yearly as a result of work injuries, some 4,000 of these involving motor vehicles. That very considerable progress is being made to reduce this toll is evident from the fact that in 1937 there were 43 work-injury deaths per 100,000 workers; today the corresponding figure is 15 per 100,000. In 1975, according to the Bureau of Labor Statistics, there were 4.8 million occupational injuries, a decrease of almost one million from the 5.7 million such injuries in 1974. Of these work injuries 2.3 million were disabling injuries.

This reduction in the accident rate can be attributed to a combination of a great many factors. The more significant of these factors include improved working conditions, shorter workweeks, improved living standards, higher levels of education, improved workplace and machinery design, the increasing installation of safety and protective devices, improved ventilation and temperature control, the eradication of unsafe conditions in the workplace, more effective lighting and noise control, increased awareness on the part of more employers, heightened concern among labor unions, safety campaigns in the workplace, the enactment of safety legislation, and the effects of workmen's compensation laws.

There are those, however, who contend that data such as have been shown are misrepresentative since, they argue, the majority of work injuries remain unreported. In 1973, according to the Health Interview Survey of the National Center for Health Statistics, 9 million persons reported having been injured while at work.

Back injuries are a common cause of disability in workplaces. About one-half of all back injuries are said to occur in the course of work. They can result from incorrect work posture, lifting, carrying, slipping, or falling. In 1970, according to estimates of the National Safety Council, injuries to the body trunk accounted for 26 percent of all workmen's compensation cases and 36 percent of the dollars paid in claims. That year there were 570,000 injuries to the body trunk. Table 5-1 shows the injury rate, by industry, per one million man-hours worked in 1962.

Table 5-1. *Injury rate, by industry, per one million man-hours worked in 1962.*

Industry	Injury Rate	Industry	Injury Rate
Communications	1.01	Printing and publishing	7.18
Electrical equipment	1.56	Railroad equipment	7.22
Aircraft manufacturing	1.75	Iron and steel products	8.52
Automobile	2.20	Federal civilian employees	8.52
Rubber	2.68	Mining, surface	8.52
Storage and warehousing	2.77	Meatpacking	8.71
Cement	3.11	Leather	9.13
Steel	3.14	Foundry	9.27
Chemical	3.71	Wood products	9.79
Textile	4.10	Clay and mineral products	10.28
Sheet metal	4.19	Wholesale and retail trade	10.38
Shipbuilding	4.33	Food	11.06
Miscellaneous manufacturing	4.39	Quarry	12.23
Machinery	4.51	Transit	13.75
Electric utilities	5.82	Air transport	16.16
Tobacco	6.06	Construction	18.11
Nonferrous metals and production	6.38	Marine transportation	20.61
		Lumber	23.33
Petroleum	6.89	Mining, underground, except coal	23.56
Pulp and paper	6.93		
Gas utilities	7.01	Mining, underground, coal	25.20
Glass	7.12		

Some studies have shown that the injury-frequency rate from on-the-job accidents decreases with the size (number of employees) of the place of employment. Thus in manufacturing the injury rate was 12.82 for places with fewer than 50 employees compared to 3.57 for those employing over 500 employees. In chemical production the relative rates were 7.05 and 2.36; in pulp and paper production they were 17.12 and 5.25; and in the production of electrical equipment they were 3.36 and 1.36.

The office accident, incidentally, is an often-overlooked form of occupational accident. While seldom fatal, some 80,000 such injuries occur yearly. The principal causes of office accidents are slipping or tripping, the mishandling of doors, and the misuse of equipment. A study made at the Equitable Life Assurance Society of the United States revealed that from 1962 to 1966 there was a total of 195 disabling office work accidents that resulted in 1,158 days lost from work. Injuries sustained from falls accounted for more than half the days lost.

INCIDENCE AND DURATION OF DISABILITY

Disability results in lost production, wage loss, disability-insurance claims, and workmen's compensation claims. A study in 1974 showed that 299 million days were lost from work as a result of illness or accident, an average of 3.5 days per person per year. The rate was higher among females than males, higher among those under age 44, higher among whites who earned over $10,000 a year, and higher among the better educated. The peak period for the incidence of disability was January to March. Respiratory conditions were the greatest single cause of disability, accounting for 125 million days lost from work (influenza alone caused 66 million days of work loss).

Although other studies show different findings, the magnitude of the problem is always great. A study in 1973 reported 5.4 days of work lost per worker per year due to illness and accident. In this study men age 17 to 44 averaged 4.6 days lost; women of the same ages 5.8 days; lower-income persons averaged 6.8 days lost; upper-income 5 days; nonwhites averaged 6.7 days lost; whites 5.3 days.

According to the Bureau of the Census, 12 percent of all employed males and 10 percent of females had some work disability in 1970. The proportion of males with complete work disability was 3.8 percent; for females it was 4.8 percent. The proportions with partial disability were 8 percent and 5.1 percent respectively. The incidence of work loss was higher among blue-collar and service workers than among white-collar workers. Differences in work-loss days by type of industry were not significant, although the rate ranged from an average of 4.8 days in wholesale and retail trade and agriculture to 6.3 days in manufacturing. By type of occupation a range is shown from an average of 8.3 days a year among laborers to 4.4 days for professional, technical, and kindred occupations and 4.9 days for clerical and kindred occupations.

It has been estimated that one out of every six men presently age 35 will be disabled six or more months before he reaches age 65 and that 10 percent of these men will be permanently disabled. A 1970 study showed that work-disabling conditions are more prevalent and of longer duration among persons with little formal education.

A household survey in 1974 of adult persons age 18 to 64 who had recently become disabled found that 90 percent had received medical care at the onset of disability. For men with occupational disabilities the proportion was almost 100 percent. More than half had been hospitalized, the mean number of hospital days being 25. For those persons who had been employed at the time disability occurred, two-thirds had either continued working or returned to work, in most cases with the same em-

Table 5-2. Number of new cases per 1,000 personnel and cause of disability.

Cause	Males	Females
Diseases of respiratory system	13.6	50.2
Infective and parasitic diseases	2.6	5.5
Diseases of digestive system	18.0	16.0
Diseases of circulatory system	15.9	11.0
Diseases of heart	10.4	2.7
Diseases of genitourinary system	5.4	23.7
Diseases of bones and other organs of movement	8.6	11.6
Neoplasms (total)	4.8	14.3
Benign	3.6	12.3
Diseases of eye and ear	2.5	4.5
Diseases of skin and cellular tissue	1.2	3.2
Diseases of nervous system	2.4	2.9
Psychoneuroses and psychoses	4.3	5.4
Accidental injuries	17.7	21.8
All other causes	4.5	8.9

ployer. For the large majority, return to work was within 13 weeks after the onset of disability. For those who had been severely disabled, 37 percent had either continued working or returned to work. On being asked whether work requirements interfered with or prevented working at the same job, two-thirds of the disabled reported such problems as lifting or carrying weights (almost half), standing for long periods, stooping, crouching, or kneeling, keeping up with the work schedule, dealing with people, traveling to and from work, and sight and hearing difficulties. Persons severely disabled were largely those with low-skilled jobs or in agriculture; manufacturing had a relatively low proportion of severely disabled employees.

The Metropolitan Life Insurance Company made studies of the incidence of disability lasting eight days or longer among its own employees (both office and sales). In 1973, for each 1,000 employees there were 101.4 new cases of disability from all causes among male employees age 17 to 64 and 178.8 among female employees of the same ages. For employees age 45 to 64 the incidence of new cases was 141.5 per 1,000 males and 246.6 per 1,000 females. The number of new cases in that year is listed in Table 5-2 by cause of disability. The annual days of disability per 1,000 personnel and the average duration of disability in days for Metropolitan Life employees in 1972 are shown in Tables 5-3 and 5-4.

This study also found that high-salaried employees were disabled less frequently than those in lower-paying positions; when they were disabled,

Table 5-3. Annual days of disability per 1,000 personnel.

Cause	Males	Females
All causes	6,282	6,580
Diseases of respiratory system	399	666
Infective and parasitic diseases	104	152
Diseases of digestive system	819	710
Diseases of circulatory system	1,616	685
Diseases of heart	1,264	319
Diseases of genitourinary system	200	639
Diseases of bones	593	487
Neoplasms (total)	297	738
Benign	169	524
Diseases of eye and ear	169	158
Diseases of skin and cellular tissue	49	92
Diseases of nervous system	266	227
Psychoneuroses and psychoses	453	492
Accidental injuries	898	948
All other causes	419	585

Table 5-4. Average duration of disability in days.

Cause	Males	Females
All Causes	64.6	36.1
Diseases of respiratory system	31.9	12.5
Infective and parasitic diseases	36.1	23.5
Diseases of digestive system	48.7	40.6
Diseases of circulatory system	107.6	64.9
Diseases of heart	141.0	104.0
Diseases of genitourinary system	45.0	31.2
Diseases of bones	78.8	50.9
Neoplasms (total)	66.0	49.0
Benign	48.6	39.5
Diseases of eye and ear	77.2	36.3
Diseases of skin and cellular tissue	37.6	32.8
Diseases of nervous system	128.9	83.9
Psychoneuroses and psychoses	135.7	92.9
Accidental injuries	49.4	42.2
All other causes	63.8	49.8

they lost fewer days. It found that sales employees had longer periods of disability than office workers, and older employees had longer periods than younger people. One percent of all disabilities terminated in death.

Another study of interest, made by the Society of Actuaries in 1972, concerned workers with group insurance coverage for long-term disability. The study found that 5 out of every 1,000 of these workers are disabled so severely in the course of a year that they are unable to work at their jobs for at least three months. Of these, 10 percent result from accidents. The frequency of such disabilities was greatest for those from 45 to 65 years of age, three times the average of long-term disabilities among employees covered by this type of insurance. The rate for females under age 45 was 50 percent higher than for males in this age bracket.

Other studies have shown that of work absences due to disability, 89.5 percent lasted less than one week, 8 percent from one to four weeks, and 2.5 percent more than four weeks. For short-term disabilities (those lasting less than six months) the average loss of time has been found to be 3.3 days per worker per year.

Studies of the cause of disability among employed people have shown various findings, dependent upon the purpose of the study and the methodology employed in conducting it. The following are some of the pertinent findings.

In 1972 *acute conditions* resulted in 985,000 workers being absent each day. Of these, 444,000 were age 45 or over. The average work loss in that year resulting from acute conditions was 3.7 days per employed person. It is estimated that on any one day 1.5 percent of the currently employed are absent from work owing to acute conditions and that acute conditions account for 65.3 percent of the total of work-loss days. The common cold accounts for 40 million workdays lost each year. In 1974 the days lost from work per 100 currently employed persons per year, by types of acute conditions, were as follows: infective and parasitic conditions, 22.9; respiratory conditions, 148.5 (influenza, 85.9); digestive system conditions, 17.7; and miscellaneous injuries 100.6.

Chronic conditions are responsible for the loss of 224 million workdays each year, or an average loss of 2.9 workdays per employed person. On any one day 1.3 percent of the currently employed are absent because of chronic conditions. The rate of absence is higher among blue-collar workers than among white-collar workers, and is highest among farm workers and service workers.

The greatest toll is taken by orthopedic impairments, heart conditions, genitourinary conditions, emphysema, arthritis and rheumatism, peptic ulcers, diabetes, mental and nervous disorders, and asthma or hay fever.

One study, made in 1965, shows the incidence of work loss for heavy smokers in comparison to nonsmokers. This is not to say that cigarette smoking is the cause of disability, for there can be many related factors resulting in disability other than cigarette smoking per se. Nonetheless, the study showed that during that year currently employed males who had never smoked cigarettes averaged 4.6 work-loss days. For those who smoked under 11 cigarettes a day, the work-loss average was 4.8 days. For those who smoked 11 to 20, the average was 6 days; for those who smoked 21 to 40, the average was 6.7 days; and for those who smoked 41 or more, the work-loss average was 8.4 days. For female employees the relative findings for work loss was higher in each category.

In 1971 the work-loss days per 1,000 employed persons because of *specific impairments* were shown to be as follows:

Visual impairments	0.4
Paralysis	1.6
Back or spine impairment	1.4
Upper-extremity and shoulder impairment	1.1
Lower-extremity impairment	1.2
Other limb and trunk impairment	2.8

The effect of *air pollution* on employee absenteeism has been demonstrated in two studies. A study of work absences lasting seven days or more was conducted among women hourly employees of the Radio Corporation of America at five urban locations over the period 1955 to 1958. The findings showed a high correlation between the average concentration of suspended particulate sulphates in the air of those communities and the incidence of work absences. A comparative study over the years 1936 to 1958 of the employees of the Bell Telephone Companies in Pittsburgh and Philadelphia showed that prior to 1946, when Pittsburgh instituted a vigorous smoke-abatement program, the Pittsburgh employees had had a higher incidence of illness. After 1946, the incidence of illness was higher in Philadelphia.[2]

Accidents are one of the predominant causes of disability. In 1974 accidents resulted in 98 million work-loss days, one in every four employed people having experienced an accidental injury during the year. Of this total, 52 million man-days of production were lost because of off-the-job accidents and 46 million man-days were lost because of work injuries. In that year there were 9.3 million work accidents, resulting in 116.4 days lost from work per 100 employees. In 1970 accidents were estimated to be the cause of 20 million bed-days in hospitals.

MENTAL ILLNESS

Mental and emotional disturbances have a decided impact on employed people and places of employment in a great many ways. They are reported to be the cause of more absenteeism than any other illness except the common cold.

There is general agreement that mental and emotional disturbances are primary or contributory factors in a large proportion of industrial accidents. One source has maintained that 80 to 90 percent of all industrial accidents have a mental base. There is equally general agreement that mental and emotional disturbances are the primary or contributory factors in many of the illnesses among employed people.

Some insight on work loss from mental disorders, as diagnosed by attending physicians, is gained from the experience of employees of the Metropolitan Life Insurance Company. New cases of disability lasting eight or more days due to psychoneuroses and psychoses occurred at the rate of 4.4 per 1,000 male employees and 6.2 per 1,000 female employees in 1966. For both sexes the incidence of these disabilities increased regularly with age, women experiencing a higher rate in each age group. While psychoneuroses and psychoses were not a common cause of disability, they ranked high as a cause of prolonged absence. For men of all working ages, 549.2 days of disability due to psychoneurosis and psychoses (excluding the first seven days of each illness) per 1,000 personnel were reported annually. The corresponding figure for women was 870.1 days. However, the average length of disability for which a claim was submitted was higher for men than for women at all working ages combined. There was a wide differential at the older ages, where men recorded an average of 150.5 days per claim as compared with 116.9 days for women.

The average length of hospital stay due to mental disorders was appreciably longer than for other causes of disability—averaging 19.8 days for men and 39.6 days for women. The rate of hospitalization per 1,000 employees for psychoneuroses and psychoses for the years 1963 to 1967 was 2.6 for males and 2.0 for females. The incidence of such hospitalization rose with age, being highest between ages 45 and 64.

In 1968 another insurance company conducted a study among insured employees covered for long-term disability to identify the conditions causing such disability, with special attention directed to claims arising from mental and nervous disorders. Claims that resulted from mental, psychoneurotic, and personality disorders represented 5.7 percent of the survey. The claims for mental illness were subdivided into the following five categories, using the International Classification of Diseases as a guide:

Chronic brain disorders	6.2%
Psychotic disorders	12.3
Psychophysiological autonomic and visceral disorders	7.7
Psychoneurotic disorders	55.4
Personality disorders	18.5

In a 1963 survey, this same insurance company had found that mental, psychoneurotic, and personality disorders accounted for 10.1 percent of all long-term disability claims. For disabilities lasting more than 6 months, the proportion due to these disorders rose slightly, to 10.3 percent. For disabilities lasting longer than 12 months, however, the proportion rose to 12.9 percent, and for those lasting over 24 months, the proportion rose to 14.3 percent. By age categories, 25.7 percent of all long-term disabilities due to mental disorders occurred in the 18-to-44 age bracket; 25.7 percent for ages 45 to 49; 8.6 percent for ages 50 to 54; 17.1 percent for ages 55 to 59; and 22.9 percent for ages 60 to 64. Of all long-term disability claims, 15.7 percent of claims by those who were age 49 or under were due to mental disorders; this proportion dropped to 7.4 percent for persons age 50 or over.

It has been estimated that *alcoholism* results in 36 million man-days of disability each year. The alcoholic has two and a half times the absenteeism of other employees, averaging 22 days lost each year. The alcoholic experiences more illnesses, more on- and off-the-job accidents, and more premature deaths than other people.

One study has found that 12.8 percent of alcoholics are disabled 30 days or more each year (compared to 6.5 percent for other employees) and that 3.8 percent of alcoholics are disabled 90 days or more a year (compared to 0.6 percent for other employees). In addition, once under treatment, 21 percent of alcoholics are disabled for 30 days or more, and 9 percent for 60 days or more.

PREMATURE DEATH

Premature death affects business in many ways. First, of course, it results in labor turnover and the concomitant expense of replacement. Other costs include the cost of group life insurance, the expense of medical care, and the lost time and productivity caused by the disease or accident that ultimately resulted in the employee's death.

The causes of premature death are evident throughout this entire discussion. Most certainly accidental fatalities all result in premature death. The same is true of occupational diseases that have their onset in the course of employment. Alcoholic persons, as has been shown, have a life span that is 12 years shorter than that of other people.

In 1969, 14,200 employed people died as a result of on-the-job accidents. Table 5-5 shows the deaths resulting from work accidents in 1960 by types of employment. Notable here is the large number of deaths occurring in agriculture when it is considered that only a small proportion of the workforce is employed in agriculture. Larger numbers of accidental deaths occur in the construction and service industries, and both these in-

Table 5-5. Deaths from work accidents, by type of employment, in 1960.

Industry	Deaths	Percentage Change from 1950
Trade	1,200	−20
Manufacturing	1,700	−41
Service	2,800	+27
Transportation and utilities	1,600	—
Agriculture	3,300	−23
Construction	2,400	+ 4
Mining	800	−20

dustries showed an increase over a ten-year period, whereas the number of deaths in other industries had declined.

It should also be noted that in 1969, 42,600 employed people died as a result of off-the-job accidents. Of these 27,000 resulted from motor vehicle accidents, 8,100 occurred in other public places, and 7,500 happened in or about the home.

BIBLIOGRAPHY

American Labor Health Association, "Proceedings of the National Conference on Labor Health Services," 1958.

American Public Health Association, *Mental Health: The Public Health Challenge* (Washington, D.C.: 1975).

Susan M. Daum and Jeanne M. Stellman, *Work Is Dangerous to Your Health* (New York: Vintage Books, 1973).

The Drug Abuse Council, "Employment and Addiction: Overview of Issues," June 1973.

J. F. Follmann, Jr., *Insurance Coverages for Mental Illness*, AMACOM, 1970; *Health Insurance and Vision Care* (New York: Health Insurance Association of America, 1975).

Herbert E. Klarman, *The Economics of Health* (New York: Columbia University Press, 1965).

Metropolitan Life Insurance Company, Statistical Bulletins.

Metropolitan Life Insurance Company and American Social Health Association, *Drug Abuse in Industry*, 1970.

National Health Council, *The Health of People Who Work*, 1960.

George Rosen, *Preventive Medicine in the United States, 1900-1975* (Boston: Science History Publications, 1975).

Material from the following organizations was also used in the preparation of this chapter:
  American Public Health Association
  Metropolitan Life Insurance Company
  National Center for Health Statistics
  National Safety Council
  Social Security Administration
  Society of Actuaries
  U.S. Department of Health, Education, and Welfare
  U.S. Department of Labor

REFERENCES

1. J. F. Follmann, Jr., *Migratory Workers* (Washington, D.C.: Health Insurance Association of America, 1974).
2. American Industrial Hygiene Association, *Air Pollution Manual,* Akron, Ohio, 1972.

6

The Economic Costs

The economic costs of illnesses and accidents among employed people often reach very serious proportions. Unfortunately, there is a paucity of factual information about such costs, either to the American economy as a whole or to places of employment. Where such facts or reasonable estimates do exist, they frequently are out of date, so that allowance must be made for the intervening inflation in the value of the dollar. Unfortunately also, many of the costs of illnesses and accidents among employed people are as yet unmeasurable.

These costs constitute a sizable part of the operating costs of any business or government agency. In the case of private business they must, of necessity, be ultimately borne by the consumer as part of the price paid for goods and services. In the case of public employment they must ultimately be borne by the taxpayer. Where the consequences of illnesses, accidents, or premature death result in the additional use of welfare or other publicly established programs, the taxpayer again must bear the cost. Where the consequences are reduced production, a reduced gross national product, and unemployment, society at large feels the impact.

MEASURABLE COSTS

Available data do not as yet show clearly the measurable costs of employee illnesses and accidents to places of employment. The data do, however, quite clearly indicate the magnitude of the problem and the challenge presented to industrial health programs.

Herbert E. Klarman has noted that business expenditures for health care include (1) the costs of workmen's compensation insurance, (2) contributions to voluntary health insurance programs on behalf of employees and their families, and (3) the costs of in-plant health services. Here we will discuss the extent of the costs to places of employment of lost production, wage loss, medical care, health insurance, disability insurance, accidental death and dismemberment insurance, salary-continuance programs, and workmen's compensation insurance.

In 1963 the cost of mortality and morbidity among the American population was estimated to be over $46 billion. Of this amount, $22.5 billion was estimated to be the direct cost of medical care (by 1976 this cost alone was over five times the 1963 estimate). Production loss resulting from illnesses, accidents, and disability in 1963 was estimated at $21 billion. The remainder of the cost was principally the value of those housewives who had become disabled. In that year the total loss of wages resulting from disability was estimated to be $5.8 billion. That such estimates can vary rather sharply, usually because of the approach taken and the factors used, is evident from a study in 1960 showing the wage loss from disability in that year to be $13.6 billion (this was 5.3 percent of the civilian payroll at that time). This estimate, however, was made up of 43 percent for the wage loss among currently employed people, 48 percent for those out of the labor force as a result of illness, and 9 percent for those who were patients in long-term hospitals.

In 1965 the wage loss for workers in private business alone resulting from nonoccupational short-term illnesses and injuries (those lasting not more than six months) was estimated to be $8 billion. A study in 1960 of the estimated income loss from nonoccupational short-term disabilities placed the total loss at $8.6 billion. Of this, persons in private employment accounted for $6.3 billion; federal employees for $400 million; state and local government employees for $762 million; and the self-employed for $1.2 billion.

Data such as these reflect the effects of inflation as well as the changes in the size and composition of the workforce. Regardless of any qualifications one might care to attach to such studies and estimates, they clearly present an order of magnitude of the effects of disability among the workforce of the nation and the costs to the operation of any place of employment.

Since the specific causes of disability, medical expenditures, and premature death as well as their costs influence the direction taken by an industrial health program and the establishment of priorities, the following information should prove useful.

In 1963 acute conditions were estimated to cost, in wage losses alone, $3.8 billion, or 1.5 percent of the total civilian payroll. Of this amount more than half was attributable to respiratory conditions (including colds) and one-fourth to injuries. In that same year the cost of chronic conditions in lost wages was estimated to be $3 billion, or 1.2 percent of the civilian payroll.[1]

Table 6-1 shows, by type of illness, the costs for 1963 as respects the entire population.

Table 6-1. Costs for American population for 1963, by type of illness (billions of dollars). *

Illness	Direct Expenditures	Indirect Costs	Total
Neoplasms	1.3	1.3	2.6
Circulatory diseases	2.3	4.1	6.4
Respiratory diseases	1.6	3.3	4.9
Digestive diseases	4.1	1.4	5.5
Mental illness	2.4	4.6	7.0
Nervous diseases	1.4	1.8	3.2
Diseases of bones and organs of movement	1.4	1.2	2.6

* Costs for those in the labor force were estimated to constitute 87 percent of these totals.

Today heart diseases are estimated to cost industry 542,000 man-years of work and $2.5 billion annually. For cardiovascular diseases the estimated yearly cost is $22.7 billion; $9.3 billion of this figure being lost wages and $12.3 billion the cost of care. Cancer is estimated to cost $15 billion annually in lost productivity and earning power, as well as for care ($3 billion). In 1975 black lung among coal miners, under the provisions of the Federal Coal Mine Health and Safety Act of 1969, was reported to be costing $1 billion a year.

Mental illness is also costly to places of employment; however, precise measurements of such costs are not possible since mental illness can have a direct yet undetected effect on accidents, illnesses, the incidence and duration of work absences, inefficiency, poor employee morale, labor turnover, and harmful customer and public relations. In the late 1960s the Opinion Research Corporation estimated the annual cost of mental illness to American industry to range from $3 billion to $10 billion. Other sources have estimated the loss of earnings to the mentally ill to exceed $1.5 billion yearly and the cost in terms of reduced marketable output to exceed $14 billion annually.[2]

In 1974 alcoholism was estimated to cost American business $8 billion a year, or $32 million for each working day. The wage loss alone was estimated at $432 million yearly. The loss in goods and services, poor productivity, faulty decisions, and personnel turnover cost approximately $1 to $2 billion yearly.

The North American Rockwell Corporation, with 100,000 employees, places the cost of alcoholism to its operation at $50,220 per year for each of its alcoholic employees. Gulf Oil Canada, Ltd., with 11,000 employees, estimates an annual cost of $400,000 due to alcoholism. The United California Bank of Los Angeles, with 10,000 employees, estimates such cost to be $1 million a year. The Kennecott Copper Corporation, with 8,000 employes, 7 percent of whom suffer from alcoholism, loses $500,000 a year because of this problem. The Scovill Manufacturing Company estimates the average cost of an alcoholic employee to exceed $4,550 a year for absenteeism alone. In 1964 the Illinois Bell Telephone Company experienced wage replacement costs of $418,500 for 155 cases of sickness disability caused by emotional illness, mostly alcoholism. According to the U.S. Postal Service, each alcoholic employee costs the service over $3,000 a year, a total annual cost of $168 million. The U.S. General Accounting Office estimates the cost of alcoholism among federal civilian employees to run from $275 million to $550 million annually.[3]

The cost of work accidents in 1974 was estimated by the National Safety Council to be $15.3 billion. Of this, $3 billion was estimated to be wage loss, $1.7 billion medical expenses, $2.1 billion administrative expenses, $1.7 billion fire losses, and $6.8 billion indirect work loss, the latter being the money value of time lost by workers other than those disabled and the time required to investigate accidents. This cost averaged $175 for every worker, and constituted one percent of the GNP.

Such findings prompted the American Public Health Association to comment in its publication *Health and Work* in 1976: "The costs of occupationally related disability and death are enormous: half a billion work-loss days occur annually, resulting in human suffering as well as a high social cost in terms of productivity losses for the United States economy, an estimated $9 billion annually. Investments in preventive and curative measures for disability associated with employment are not commensurate with the scope of the problem."

The costs of insurance, employee benefits, and health and welfare funds also constitute a significant part of operating expenses. Employers' contributions to group health insurance premiums for the medical expenses of their employees and their dependents have increased steadily and appreciably since World War II. As a result of wartime wage-price freezes and subsequent rulings of the National Labor Relations Board, health insurance became an important element in the processes of collective bargaining. Employer contributions to premiums were tax-deductible as a business expense. For employees the benefits constituted, in effect, tax-free wages. The results rapidly became apparent and today, as a consequence of collective bargaining, it is not infrequent to find that the employer pays 100 percent of the premium cost.

In 1951 it was estimated that employers, on the average, paid 37.5 percent of the group health insurance premiums. In 1952 the proportion was estimated to be 45 percent, totaling $750 million. In 1957 it was estimated that in 10 percent of the cases of group health insurance the employer paid 100 percent of the premium and that in 49 percent of the cases the employer paid part of the premium.

In 1970 an unpublished survey by the Health Insurance Association of America showed that for over half (52.2 percent) of the covered employees and their dependents the employer paid the entire premium. For 24.7 percent of the insured employees the employer paid from 75 to 99 percent of the premium; for 16.3 percent, from 50 to 74 percent; and for 6.2 percent, from 25 to 49 percent. For 0.6 percent of insured employees the employer made no contribution to the cost. The overall contribution of employers for all insured employees in 1970 was 80 percent of the cost. This survey also found that on the average the employer contribution increased with the number of employees. Thus in groups of less than 25 employees, less than 25 percent of such groups had the entire premium paid by the employer, whereas in groups of over 500 employees, over half had the entire premium paid.

These data clearly indicate the trend over the past quarter century toward an increased proportionate contribution by employers for employee health insurance protection. Since in 1974 the premiums for group health insurance coverages totaled more than $27 billion ($13.5 billion for insurance company coverages and $14.1 billion for Blue Cross, Blue Shield, and other types of prepaid medical care), it is quite apparent that the cost of health insurance alone to places of employment is very considerable. In that year, incidentally, health insurance benefits paid to employed groups exceeded $25 billion for medical expenses. Today some 160 million employed people and their dependents benefit from this form of protection.

The magnitude of these costs to large employers in 1976 is readily evident from the fact that General Motors Corporation, Chrysler, and Ford spent $1.4 billion for health insurance in 1976 and that the cost of health insurance for each worker in that year averaged over $150 per worker per month.[4]

Disability-income insurance for the partial replacement of income lost as a result of nonoccupational accidents or illnesses is written on groups of employees by insurance companies, or it might be self-insured. Such insurance might be for short-term disabilities ranging from three months to two years or for long-term disabilities ranging from two years to age 65. It might also be for temporary nonoccupational disabilities (26 weeks) as required by the laws of five states and Puerto Rico (in which case the insurer might also be the state).

Group insurance premiums for disability insurance on employed groups of people in 1973 cost $2.8 billion. The proportion of these premiums paid by employers is not known. In that year the benefits paid as a result of disability under group policies exceeded $1.8 billion. In 1973, 32 million workers had this form of insurance protection. In the interest of completeness, in that year 13 million employed people had individual disability-income insurance policies with premiums totaling $1.5 billion. Since the majority of these latter are self-employed people, a large part of this figure constitutes a cost of operation for the self-employed.

Closely related to the foregoing are the salary-continuance programs established by many employers; most of these are on a formal basis but some function informally. The purpose is to provide for the entire or partial payment of wages during periods of disability. Most programs provide for salary continuance for relatively short periods, frequently until eligibility for disability-insurance payments is established. Approximately 21 million employed people (approximately one-fourth of the workforce) are entitled to salary-continuance benefits. Employers usually pay the complete cost of these plans, although the amount of the costs is not known. However, the U.S. Department of Health, Education, and Welfare reports that 31.8 million workers (half the workforce) are protected by formal salary-continuance programs or the statutory temporary-disability benefits required by five states and that the cost to the employers exceeds $3.6 billion a year, $2.8 billion of which is the cost to employers in private industry.[5]

Variations in the benefits under salary-continuance programs are made evident by Labor Department publications. The Retail Food Industry in Southern California, for example, provides pay for six days commencing with the second day. The Texas Division of the Aerospace Industry provides full pay (on a 40-hour-a-week basis) from the first day of absence. Safeway Stores, Inc. provides full pay from the first day of absence to the extent of one day for each month of employment, up to 60 days. The Susquehanna Corporation pays full pay from the first day of absence up to 18 days. Johnson and Johnson provides full pay from the first day of absence up to 10 days per year. For hourly employees Lever Brothers Company pays from the first day of absence due to occupational causes and from the eighth day for nonoccupational causes. For salaried employees Lever Brothers pays the benefit from the first day of absence for a period of 13 weeks at full pay and an additional 13 weeks at half pay for employees with less than 11 years of service. Benefits then increase with longevity of employment up to 26 weeks at full pay for those with 23 years or more of service.

The Minnesota Mining and Manufacturing Company program commences on the third day of absence for those with less than 5 years of service; from the first day for those with 5 years or more with the company. The benefit continues for 32 hours of pay (with a maximum accumulation of 96 hours) for those employed by the company for less than 5 years, and for 80 hours (with a maximum accumulation of 240 hours) for those with the company 5 years or more. There is no benefit for employees with less than one year of service. For its salaried employees U.S. Steel provides full pay as follows:

1 month	2 months to 1 year of service
2 months	1 to 5 years
3 months	5 to 10 years
4 months	10 to 15 years
5 months	15 to 20 years
6 months	20 years or more

Cluett, Peabody and Company, Inc. provides salary continuance on the following schedule:

2 weeks	6 months to 1 year of service
1 month	1 to 2 years
2 months	2 to 5 years
3 months	5 to 10 years
4 months	10 to 15 years
5 months	15 to 20 years
6 months	20 years or more

The American Airlines program provides as follows:

1 week	6 months to 1 year of service
2 weeks	1 to 2 years
4 weeks	2 to 3 years
6 weeks	3 to 4 years
8 weeks	4 to 5 years
10 weeks	5 to 6 years
12 weeks	6 to 8 years
14 weeks	8 to 10 years
16 weeks	10 or more years

It is obvious that salary-continuance programs vary considerably and that in many cases they can be quite costly to a place of employment.

Accident insurance, written by insurance companies to provide death and dismemberment benefits on groups of employed people for nonoccupational accidents, in 1973 paid benefits of $271 million. The costs to employers are not known, although in 1972 the costs to both employers and employees were $253 million. It is estimated that 39.7 million workers have this form of protection.

Over 53 million workers are protected by life insurance or death benefits through the place of work. The total cost to employers and employees in 1972 was $4.3 billion (in 1950 this amount was $480 million). Benefits paid in 1972 exceeded $2.9 billion.

Workmen's compensation insurance, required to varying degrees by all states, is another health-related cost to employers. The benefits are for losses resulting from occupational diseases and accidents for specified types of occupations; they cover both medical care costs and the partial supplementation of lost income. About seven-eighths of all medical care expenditures under workmen's compensation are incurred by private business and one-eighth by government. In 1973 the cost of workmen's compensation insurance to employers was $5 billion, or 1.13 percent of payroll.

In 1968 workmen's compensation benefits totaled $2.4 billion. Of this, $1.5 billion was paid by insurance companies, $538 million by state funds, and $326 million by self-insurance. An indication of how these payments have increased is evident from the fact that in 1959 the total of such benefits was $1.2 billion ($415 million of which was for medical costs and $815 million for wage compensation); in 1939 the total was $235 million. This form of insurance, then, is a sizable and rapidly increasing cost to places of employment.

In 1970 the National Safety Council stated that 26 percent of all workmen's compensation cases were due to injuries of the trunk and that these accounted for 36 percent of the dollars paid in workmen's compensation benefits. There were 570,000 such injuries in that year. Injuries to the legs accounted for 14 percent of the benefits paid, injuries to the arms accounted for 12 percent of the benefits, injuries to the fingers accounted for 12 percent, and injuries to the head and eyes 9 percent. The average cost per case was highest when the injury was incurred in a motor vehicle accident. Next highest were injuries resulting from the use of elevators, hoists, and conveyors. Third highest were injuries resulting from harmful substances and from falls.

In 1973 the Chamber of Commerce of the United States estimated that the cost of all employee benefits, both public and private, to employers was 32.7 percent of payroll. This figure included Social Security, unemployment compensation, workmen's compensation, health insurance, disability insurance, and salary-continuance programs, life insurance, pensions, holidays, and vacations. The impact of injuries, illnesses, disabilities, and premature death, whether occupationally or nonoccupationally induced, then, is quite clear. Its severity argues for a reduction in such losses to the lowest possible degree.

One final cost to employers should be noted: the cost of pollution control. According to McGraw-Hill surveys capital expenditures for pollution abatement increased from $4.5 billion in 1972 to $7.9 billion in 1975, a 70 percent increase.[6] The economic impact of such costs on just the price of goods alone was shown by the EPA in 1972. The average increase in prices to offset the cost of meeting federal pollution standards alone in that year was 5.5 percent for large plants, 6.4 percent for medium-size plants, and 9.6 percent for small plants.[7] The inflationary effects are obvious.

UNMEASURABLE COSTS

The unmeasurable costs of employee illnesses and accidents are very considerable. These costs are noted here as unmeasurable in the sense that they have not yet been measured in a meaningful way. For the most part they actually are measurable, but intensive in-depth research would be required, and that in itself would be costly.

The so-called unmeasurable costs can, however, be identified. They include lost production, inefficiency with consequent waste of time and materials, declining employee morale and conflict between employer and employee, friction among co-workers, and deteriorated customer and public relations. They also include faulty decision making by executives and managers. All these problems can be very costly to any business operation.

Other costs flowing from employee illnesses, accidents, and premature death could be measured with sufficient study. One such area is the cost of replacing employees who can no longer work. Included here are the costs in both dollars and time of going into the labor market as well as the time consumed in training new employees and other expenses (such as clothing) for the new employee. Another cost is the overtime pay when employees must make up the lost production of absent employees.

Yet another area of costs arises from cases when, as a result of illness or accident, employees can no longer perform their original jobs and must be retrained for new jobs. These costs are part of the rehabilitative process.

In the aggregate, all such costs can constitute a sizable proportion of operating expenses for places of employment.

REFERENCES

Materials in Chapter 6 derive from the National Health Council's *The Health of People Who Work;* Herbert E. Klarman's *The Economics of Health;* the *Proceedings of the National Conference on Labor Health Services;* the Health Information Foundation's *The Economic Costs of Absenteeism;* and publications and studies of the National Center for Health Statistics, the U.S. Department of Health, Education, and Welfare, the U.S. Department of Labor, the National Safety Council, and the Health Insurance Association of America.

Specific references are:
1. *The New York Times,* January 27, 1976.
2. J. F. Follmann, Jr., *Insurance Coverage for Mental Illness,* AMACOM, 1970.
3. _____, *Alcoholics and Business,* AMACOM, 1976.
4. *Business Week,* May 17, 1976.
5. *Social Security Bulletin,* May 1974.
6. *The New York Times,* December 20, 1976.
7. Environmental Protection Agency, "The Economic Impact of Pollution Control," Washington, D.C., 1972.

Part III

The Potential of Preventive Medicine

The importance of preventing morbidity and premature mortality to the greatest possible extent is readily apparent from what has been shown. The relationship of preventive medicine to industrial health is direct: Beyond the provision of emergency care, industrial health is preventive medicine. Furthermore, today, when there is mounting public concern over the rising costs of health care, preventive medicine assumes particular significance. This concern is at once personal and social. It extends from individuals and families, to the providers of health services, to employers and labor unions, to health and welfare funds, to insurers, and to government at all levels.

The result is that there is general agreement throughout the industrialized world that the potential of preventive medicine is great. In our own country alone, the virtues of prevention have been espoused in any number of official or quasi-official reports. These include the report of the Committee on the Costs of Medical Care in 1932; "The Nation's Health: A Report to the President," in 1948; the proceedings of the National Health Assembly in 1949; the report of the President's Commission on the Health Needs of the Nation in 1951; the report of the Commission on Chronic Illness in 1956; the report of the President's Commission on Heart Disease, Cancer, and Stroke in 1964; the report of the National Commission on Community Health Services in 1966; and the report of the President's Committee on Health Education in 1973.

Yet, oddly, a dichotomy immediately presents itself. Despite such avowals of faith, all too frequently what is done is very considerably less than what could—and should—be done. There are several reasons for this, but the essence would appear to be that, on the part of all interested parties, curative medicine remains infinitely more appreciated, more desired, and more valued. Until this situation is reversed, preventive medicine presumably will not obtain its full potential.

Prior to a discussion of what can be done and what is being done in places of employment to reduce the toll of illnesses and accidents among employees and their families, it is well to take a look at the field of preventive medicine generally: its concepts, its many aspects, those engaged in various types of preventive measures, certain deterrents, and the potential.

7

The Concept

The concept of preventive medicine has two distinct facets. The first, referred to as primary prevention, is concerned with averting the occurrence of diseases and accidents. The second, known as secondary prevention, has as its concern the early detection of disease, with the purpose of halting the progress of the disease or of preventing the development of complications to the degree possible.

Today there is a vast body of knowledge of the cause of diseases, their prevention, and their cure. Advancement in both the early diagnosis of disease and in therapy would appear to make secondary prevention in many instances both a humane and an economically practical matter. Without applying the techniques of preventive medicine, it is increasingly doubtful whether a greater availability of medical and therapeutic technology on the basis of the present pattern of health care will continue to bring with it increased benefits.

"Preventive medicine" is an ill-defined term encompassing many areas of endeavor, all with the common purpose of reducing the incidence of accidents and disease, of detecting the onset of disease in its early states before too much damage has been done, of reducing the number of premature deaths, and of eliminating unnecessary expenditures for medical care. Thus Dr. Edward J. Stieglitz has written: "Preventive medicine may mean many things. Some definitions are sharply restricted and limit the scope of preventive medicine to certain specific activities. Other definitions go to the opposite extreme in including almost all of the medical and sanitary services. . . . Vagueness of the meaning of preventive medicine is increased by uncertainty of what it is intended to prevent."[1]

This view is made evident by the fact that preventive medicine, as often conceived, runs the gamut from a wide spectrum of public health efforts (for example, sanitation, sewage and garbage disposal, water and air purification, and pest control) to efforts at combating age-old enemies of health (such as poverty, inferior housing, malnutrition, low levels of education), immunization, eugenics, prenatal and postnatal care and pediatric management, the prevention of mental illness, accident prevention, occupational health, the alteration of deleterious lifestyles, health education, the employment of physical examinations or multiphasic screening, and research.

This wide spectrum of endeavor has caused Anderson and Rosen to caution that several different concepts of preventive medicine exist side by side and that they entail the use of different methods and resources and, in fact, represent different social philosophies. The authors describe much of the discussion of preventive medicine as Utopian and state that "prevention has a yet undetermined future role in the achievement of attainable health objectives."[2]

In considering the subject of prevention, it is well to bear in mind that measures aimed at promoting good health are visibly effective in an inverse ratio to the degree of good health of the population involved. In a population subjected to contaminated water, disease-carrying pests, inferior sewage disposal, and the like, noticeable gains might be expected from health-promoting measures. On the other hand, in a population where the contrary set of circumstances exists, further progress is more difficult to achieve and its effects are less noticeable.

This latter situation, in turn, requires a change or amplification in the nature of the preventive measures to be taken. The latter population, as has been shown, will have a lower death rate, with the consequence that people will live longer. Thus the disease pattern changes and preventive medicine must respond to this change. More people survive the early years only to encounter such chronic diseases as heart disease, cancer, hypertension, and diabetes.

There also will be an increase in frustrations and anxieties, as well as such detrimental health habits as tobacco and alcohol abuse. The toll of motor vehicle accidents will be severe. Air and water pollution will increase. People will be given to overeating and ignoring the need for physical exercise. They will move at a jet-age pace and function under abnormal stress. They will disregard the need for rest. Mental illness concepts will become broader and more widespread.

This transfer from one set of conditions to another calls for a reevaluation of preventive medicine. It also demands an increasing amount of individual initiative, effort, willpower, and frequently an expenditure of time and funds. For example, water purification usually results from a social act financed with public funds and requires no individual initiative; on the other hand, physical examinations require a deliberate action on the part of the individual and often at the person's own expense.

Therefore, despite all that is written and said on the subject, the concept of preventive medicine can remain hazy and imprecise. Its advocates talk continually among themselves, sounding and resounding essentially the same theme. Too infrequently do they reach the world around them with a convincing and telling logic. Not that they are denied or contradicted; to the contrary, like motherhood and the flag, they are applauded. But to bring about positive action is often quite something else. This can only mean that the logic cannot stand on its own legs: it needs the support of something tangible, such as economic persuasiveness.

REFERENCES

1. Edward J. Stieglitz, M.D., *A Future for Preventive Medicine* (New York: The Commonwealth Fund, 1945).
2. Odin W. Anderson and George Rosen, M.D., "An Examination of the Concept of Preventive Medicine," Health Information Foundation, University of Chicago, 1960.

8

The Many Aspects of Preventive Medicine

There are several approaches to preventive medicine. Each approach can have its own purposes, and there often is overlap of effort in the execution of these purposes. Some approaches focus exclusively on specific diseases; others on designated population groups. This, in turn, presents the problem of which approaches should be given priority of resources, manpower, and funds. The value of each approach is to be weighed economically in terms of what might reasonably be accomplished in reducing premature deaths, disability, absenteeism, lost wages, lost production, and medical expenditures. Such evaluations are at best difficult and must be balanced against the cost of the effort itself. Called for is clear policy direction, supported by conclusive research.

The many aspects of preventive medicine can be categorized, furthermore, in different ways and each can serve its own purpose. In any instance, as will be shown, they bring into play a host of disciplines, pursuits, and agencies, and can be financed with private funds, public funds, or both. Regardless of the means of categorization employed, the mere identification of the various aspects of preventive medicine serves to give evidence to the breadth, complexity, and comprehensiveness of the subject.

PUBLIC HEALTH MEASURES

Public health measures devoted to the prevention of disease encompass a wide range of activity, including pure-water supply systems and the availability of potable water; sanitation; sewage and refuse disposal; quarantine; control of pests and rodents; street paving and cleaning; and disposal of the dead. By no means unrelated to the health status of people is the concept of compulsory education.

Later came food and drug surveillance, attempts to eliminate toxic foods, inspection of public eating places and markets (including open markets), the reporting of deaths and certain diseases, building laws, housing standards, and safety precautions for highways, public places, and public means of transportation. Other measures provided for flood control, the prevention of earth slides and rock falls, surveillance of workplaces, control of work hours and child labor, minimum wages, construction standards, school lunches, food stamps, unemployment compensation, and a variety of means for the provision or financing of health care, including maternal and child-care clinics and psychiatric clinics.

That these measures have had a tremendous positive effect on the health of people and on the span of life is readily evident from what has already been shown. In many instances they involve public policy decision executed by government at all levels, and are financed with tax funds as a socially desirable pursuit. The accomplishment of the desired purpose often involves some form of compulsion on the individual for the protection of society as a whole. In other instances, all members of society enjoy their benefits, with no positive act or expenditure of funds required on the part of the individual. Frequently, compulsory immunization, compulsory education, and other preventive measures have been resisted as being in violation of the civil liberties of the individual.

Contributing to these efforts by public agencies are a host of nongovernment agencies, including foundations, universities, voluntary agencies, private industry, labor unions, and insurers.

More recently, public health concerns have extended to a broader area of interest, generally referred to as ecology or environmental health. For many millenia human beings, as has been shown, have had a fundamental concern for such matters, but as the population has grown, particularly in the past century, the once seemingly unlimited supply of health-inducing resources has been called into question. As a consequence, there is now considerable investigation concerning such matters as air and water pollution; the disposal of agricultural and industrial wastes; the use of pesticides, insecticides, poisons, and other chemicals;

the adverse effects of noise; and the effects of modern industrial processes. Knowledge developed by research and statistical evidences in relatively recent years has refined, deepened, and broadened these concerns and made them more relevant to modern conditions.

Many of these more recent areas of interest remain controversial, in the sense that their effects, or potential effects, on health are still speculative. In some instances, difficult decisions present themselves in weighing the good that emanates from a particular hazard against the potential harm that can result and in weighing the cost of an effort against the value it might be expected to produce.

Furthermore, as the Environmental Protection Agency has noted, the total impact of pollution is difficult to measure since not only do the physical and chemical qualities of pollutants vary, but the effects differ in relation to the length, intensity, and method of exposure, the ability of an individual to tolerate the pollutant, the climate, the weather, and many other variables. Beyond this, any harmful effects of pollutants may require years before they become apparent. Compounding the problem are the great deficiencies in our knowledge of the effects of pollutants on health.

Other problems also present themselves. Because environmental problems are interrelated, interjurisdictional conflicts result. This happens between the various departments of government, as well as between government jurisdictions (bordering communities, counties, or states). The result frequently is inefficiency, overlap, and ineffectiveness.

Where the public stands on environmental pollution, meanwhile, is always a question. If efforts to reduce the degree of pollution result in increased taxes, a rise in the cost of consumer goods, work layoffs, or inconvenience, people very often oppose such efforts. Most certainly the public indicates little desire to participate in such efforts if it means the elimination of unnecessary use of their automobiles, restrictions on disposing of leaves and trash, refraining from smoking in public places, control of their pets, or any other of the polluting acts that people are accustomed to doing.

A few words on each of these environmental areas is in order. Immediately it will be noted that the processes of private and public places of employment are frequently involved. These are discussed in more detail later.

Air pollution. According to the American Medical Association, research on air pollution has proved that long-term, low-level exposure affects health, particularly the health of high-risk persons and children. Others have maintained, however, that there is no solid evidence for such a claim. The classic examples of the contrary are, of course, the episodes in Donora, Pennsylvania, in 1948 and London in 1952, with severe occurrences in the Meuse Valley in 1930 and New York City in 1953. In all these instances there was a rise in the death rate, increased hospital admissions, a higher number of health insurance claims, and increased work loss. The incidence of accidents also rose.

The effect of air pollution on the nation's health was the subject of the National Conference on Air Pollution in 1962. About that time President Kennedy recommended legislation that would bring about more intense research in and studies of the subject and that would provide financial stimulation to the state and local air pollution control agencies. In 1970 Congress enacted the Clean Air Act.

The Environmental Protection Agency (EPA) has estimated that air pollution costs the nation $16 billion annually and that the cost of mortality and morbidity alone resulting from air pollution is some $6 billion a year. In 1976 EPA estimated that 30 percent of all respiratory morbidity and mortality, and 15 to 20 percent of all cardiovascular disease, resulted from air pollution. In that same year the Council on Environmental Quality, which is advisory to the president, reported that only 91 of the nation's air quality control regions met the statutory deadline for compliance with federal standards to counteract air pollution and that the full effects of variances granted some industrial establishments to burn high-sulphur coal and oil "are still uncertain."

That attempts to cope with air pollution are fraught with tradeoffs is evident from a report by Grace Lichtenstein in the April 16, 1976, issue of *The New York Times*. A plan by western utilities to build a large coal-burning power plant at Kaiparowitz, in southern Utah, was defeated with the assistance of environmentalists. As a result the utilities decided to abandon the project, which would have been the largest in the country. What would have been a needed source for electric power on the West Coast did not come to fruition. Local residents where the plant was to have been located had been counting on the project to inject millions of dollars into their economy. Swapoffs, then, present difficult choices.

With the continuing growth in urban living, the ever increasing use of the automobile, and the accelerating process of industry, air pollution can only become a more serious problem unless adequately controlled.

Furthermore, noxious odors of various kinds frequently accompany air pollution. These might emanate from a variety of sources, including food processing, coke ovens, the manufacture of iron and steel, oil refining, rubber processing, tanneries, paper mills, gasworks, breweries, or the manufacture of soap, detergents, or pharmaceuticals. They can result from the presence of fish, pigsties, manure, garbage, and other decomposing matter. Their source can be the burning of leaves and rubbish by individual home owners. The ill effects of malevolent odors can include appetite loss, weight loss, nausea, restlessness, sleeplessness, and allergic reactions.

Water pollution. Water pollution has become an increasing problem in contemporary society. A 1970 survey, for example, revealed that 56 percent of the public water systems in the United States had major deficiencies, that 25 percent contained bacteria or chemicals that exceeded safe limits, and that an additional 16 percent were distributing water whose pollution levels exceeded the federal drinking water standards and was considered dangerous. Some 8 million people are said to receive questionable water from public supplies.

The sources of water pollution are innumerable; some are inadequate sewage disposal; human waste; fertilizers; agricultural wastes; soil erosion; sediments; industrial wastes; oil spills; and swamps. Furthermore, water that is corrosive to pipes will ingest metals. Once an underground water supply has become contaminated it is virtually impossible to purify. On balance, it has been recognized that by 1976 the more serious sources of water pollution have been effectively controlled and that many of the polluted waterways were being cleaned up. Some difficult problems remain, however, including the presence of nutrients and trace metals and the consequences of land runoffs.

Disposal of agricultural and industrial wastes. The Environmental Protection Agency has noted that there are only three repositories for the wastes of society: the earth, its waters, and its atmosphere. A consideration of waste disposal, then, crosses over into the previous discussions of air and water pollution. For example, the use of open dumps breeds rats, flies, and other disease vectors. Livestock and poultry production create a problem of waste management. Waste-water treatment plants produce solid sludges, which present a problem of disposal. Banning incinerators or open burning of yard debris creates a greater burden of solid-waste disposal. Improperly designed landfills contribute to water pollution. In any instance, the health hazards created are considerable.

Pesticides, insecticides, poisons, and chemicals. Today there is considerable concern about the broad use of pesticides and insecticides and their effects on human health, EPA reports that certain pesticides can interfere with the functioning of the central nervous sytem, that high levels of exposure can cause several types of illnesses, and that 75 types of pesticides have caused damage to human health. In 1976 EPA banned the further production of virtually all pesticides containing mercury as a result of indications in Japan, Iran, and the United States that nervous system diseases result from their use. There is no indisputable evidence, however, to link pesticides with human malignancy or with an influence on human fertility.

On the other hand, pesticides have made possible the tremendous increase in food production in recent years, which has enabled food supply in a nation such as the United States to more than keep pace with a growing population. Thereby, such conditions as malnutrition, pellagra, rickets, goiter, protein deficiency, scurvy, and stunted growth have been reduced to a minimum. DDT has also been reported to have halted epidemics of typhus, yellow fever, malaria, onchocerciasis, and schistosomiasis by destroying the carriers of such diseases. It might well be that the good that is produced considerably outweighs the damage that is done, although this is by no means an argument for relaxing our surveillance.

There is also general concern over poisons. When it is observed that 3.7 percent of all infant deaths, 28.5 percent of all deaths before age one, 39.9 percent of all deaths before age 2, 20.4 percent of all deaths before age 3, and 7.5 percent of all deaths before age 4 result from poisoning, the seriousness of the problem becomes apparent. Such poisoning can result from any number of causes: items in the medicine chest, kerosene, gasoline, lighter fluid, fuel oil, disinfectants, turpentine, pesticides, detergents, furniture polish, drain pipe cleaners, and lead poisoning from paint and plaster. By 1976 much had been done to avoid the occurrence of poisoning through preventive packaging. Human carelessness, however, remained a problem.

The harmful effects of various types of chemicals used under many very different situations have received considerable attention during the past decade. Currently receiving appreciable attention, and not without a strong divergence of opinion, is the matter of the use of chemicals in modern foodstuffs. Thus in 1976 the Food and Drug Administration banned the use of Red Dye No. 2 in foodstuffs, drugs, and cosmetics as a possible carcinogen, and PCBs' (polychlorinated biphenal) residue attracted attention, although the PCB levels in most foodstuffs had dropped. Some environmental problems from the use of PCB, however, remained to be controlled.

Noise. The effect of noise, which has been increasing in modern urban and industrialized life, can be harmful. In large part the effect depends upon the magnitude of the noise, its decibel level (sound energy), and its frequency. Excessive noise might occur in the home (dishwashers, band saws, power tools, air conditioners, hi-fi equipment), in the general environment (airplanes, automobile horns, trucks, motorcycles, riveting, construction, hammering), or at the place of work (drop forges, knitting machines, power plants).

The primary effect of excessive noise is the creation of varying degrees of hearing impairment. Other effects on health remain controversial, but there is some opinion that other adverse physical effects may include increasing the incidence of accidents; aggravation of a heart condition, hypertension, nervous disorders, or ulcers; weight loss; and fatigue and sleeplessness. Continuous exposure to excessive noise often leads to irritability, increased susceptibility to infection, indigestion, and vertigo. By 1976 much was being done in places of employment to counteract such harmful effects. Meanwhile, in 1972 Congress passed the Noise Control Act.

COMBATING AGE-OLD ENEMIES OF HEALTH

Another area of preventive medicine is that of combating the age-old enemies of health that affect some segments of the population and that prevail regardless of knowledge, preventive techniques, or the availability of health care: poverty, inferior education, malnutrition, unsanitary living conditions, and poor housing.

Fortunately, the proportion of the U.S. population so affected is less than in many nations, but the American track record is by no means the best; much remains to be done to eradicate these harmful conditions. That the deleterious effects on health are interrelated and interacting is evident from the fact that malnutrition can result from low levels of education, as well as from poverty. It can, in turn, have an adverse effect not only on general health and growth but also on intelligence, learning capacity, motivation, income-earning ability, and general welfare. It can result in such health conditions as anemia, scurvy, pellagra, skin disease, tooth decay, apathy, and a lesser degree of disease resistance.

The solutions to such problems are many and essentially lie beyond the health field itself. Government has many roles to play here, not alone in the form of regulatory controls or in the provision of economic relief programs, but also in the encouragement of economic development. Certainly reducing unemployment to the lowest level possible would provide a significant form of preventive medicine.

IMMUNIZATION AND CONTROL OF INFECTIOUS AND COMMUNICABLE DISEASES

That great strides have been made in this century in controlling or, in some instances, practically eliminating certain diseases through immunization and other measures is evident from what has been shown earlier. In many instances such measures are brought about by government actions, sometimes through compulsion but more often by persuasion. Frequently the costs of such measures are borne entirely by public funds or by voluntarily contributed funds.

Despite the progress made, the potential outbreak of infectious and communicable diseases is ever present. Several somewhat recent occurrences bear testimony to this: The influenza virus that came out of southern China in 1957 resulted in an epidemic in Hong Kong, then the United States, and spread throughout the world; a second influenza epidemic in 1960; the more-than-double incidence of infectious jaundice in 1961 in the United States over 1960; the sudden and mysterious appearance in eastern New Jersey in 1959 of Eastern Equine Encephalitis taking 20 lives; the increase in certain types of staphylococcus that are real killers.

Enormous strides have been made, but infectious diseases have not by any means been brought completely under control. Constant alertness is called for. The public readily relaxes its concerns when the incidence of any such diseases becomes sharply reduces. This is to be guarded against with every available means. That such relaxation and carelessness do happen is evidenced by a 1973 report of the General Accounting Office that fewer persons were immunized against polio, diptheria, measles, whooping cough, and tetanus than was the case a decade earlier. In 1974, some 5 million of the 13 million children age one to four were unprotected by immunization from these diseases. One example is the fact that the proportion of children given the polio vaccine dropped from 84.1 percent in 1963 to 62.9 percent in 1972 (57.5 percent in the central cities) and that the majority of cases of paralytic polio occur among unvaccinated persons. Another example is the incidence of measles. A vaccine for this disease was introduced in 1963. By 1968 the number of cases of measles had dropped from 400,000 a year to 22,231. Thereafter, however, a resurgence occurred. In 1971 there were 75,000 cases, this outbreak being attributable to a 42 percent decline in the use of the vaccine between 1968 and 1970. In 1972 there were measles outbreaks in New Jersey, Ohio, and New Hampshire.

Examinations of immunization practices have indicated that the failure to obtain immunization is greater among lower-income segments of the population and can be related to cutbacks in or discontinuation of public programs to provide immunization. It is also to be noted that with a disease such as epidemic influenza the nature of the disease frequently changes, lessening the effectiveness of available vaccines.

Venereal disease also serves as an example of an infectious disease in which precautionary measures are required on a continuing basis. The rising incidence of VD in recent years has been noted earlier. Despite very considerable gains in bringing the effects of the disease under control by the end of World War II, the recent upsurge makes it clear that efforts at control cannot be relaxed.

EUGENICS

While eugenics is a touchy subject, some people consider it to be a factor in certain aspects of preventive medicine. At present, however, scientific opinion does not appear to be in agreement on just how significant a factor it is, whether diseases can thereby be avoided, or even whether the tendency toward them can be lessened appreciably. The evidence does seem rather clear, though, that such conditions as diabetes, hypertension, heart disease, sickle cell, and congenital malformations, are influenced by family history. Unfortunately, much remains to be learned about this aspect of preventive medicine. In 1976 the National Institutes of Health gave consideration to research in this area, including the possibility of replacing disease-causing defective genes with normal ones. Because of inherent dangers in the project, guidelines were prepared as safety precautions. The need for such was recognized in a program inaugurated in Contra Costa County, California, in 1962 based on pedigree-taking.[1]

PRENATAL AND POSTNATAL CARE AND PEDIATRIC MANAGEMENT

In the past 50 years great gains have been made in reducing infant mortality and in bringing to adulthood an increasing proportion of healthy children. Important contributions have been made through the knowledge developed in the fields of prenatal and postnatal care, pediatric management, well-child care programs, and school health programs. Many have contributed to these gains: the family physician, the specialists in obstetrics and pediatrics, hospitals, well-child and child-guidance clinics, public health nurses, and a variety of voluntary and public agencies. In many instances, the care is provided by the community through the use of philanthropic or public funds. It is a hopeful sign that, as a generality, public acceptance of these forms of care has increased very considerably in relatively recent years.

PREVENTION OF MENTAL ILLNESS

The prevention of mental illness is a difficult matter. There is, in fact, some question whether prevention can be accomplished at all. Efforts are considered elusive, impractical, and difficult to evaluate. Another question is what is it that is being prevented. The lack of specificity in the term "prevention" is critical in a field that already has a large element of vagueness. The question also arises whether, in approaching the prevention of mental illness, social matters such as injustice, discrimination, economic insecurity, poverty, and slum living should be the focus of concern, since environmental manipulation can be looked upon as the most effective approach to prevention. This question poses a tremendous challenge.

It has been noted that prevention activities have traditionally been delineated into three areas, whose boundaries are fuzzy in both theory and practice. The three areas are:

Primary prevention to remove the causes of illness; this area is considered to be the least-understood and most-neglected of public health concepts, and one in which mental health workers are not generally knowledgeable. This can include the correction of nutritional deficiencies, the development of mental stimulus, prenatal care, premarital counseling, parent education, ego-strengthening measures, preparation for retirement, and guidance in such matters as the use of leisure time, career decisions, and coping with family deaths and pregnancy. It can also include such matters as the eradication of venereal disease and the vaccination of child-bearing women to reduce mental retardation.

Secondary prevention focuses on early detection and treatment; it presents such problems as the accessibility of needed services, prompt assessment of the need for therapy, continuity of care, and the necessity for minimizing institutional confinement.

Tertiary prevention is concerned with rehabilitation and training, job placement, and adequate housing; the special services for these areas are not by any means readily available. Included here might also be the overcoming of stigmas that can attach to mental illness, alcoholism, or disfigurement resulting from diseases or accident.

An opportunity for prevention that is usually unexploited is the place of work. Today a good many of the larger employers have established "troubled employee" programs, which are proving to be of value. In the main, these programs have grown out of alcoholism control programs. The possibilities in this area would appear to be extensive if properly handled and fully exploited. Places of employment are a valuable source for health education and the early identification of health problems. All indications are that they can be more than self-supporting in the savings in absenteeism, labor turnover, inefficiency, accidents, insurance claims, and poor customer and public relations.

Another opportunity for the prevention of mental illness is the community mental health center. Primary prevention is often inherent in the basic philosophy of such centers, although the exact goal and function remain ambiguous. Nonetheless, a commitment to prevention is widely accepted. One example of this is evident from the functioning of the Hall-Mercer Community Mental Health Center at the Pennsylvania Hospital in Philadelphia, where attempts, particularly at secondary prevention, are made through contact with the community.

ACCIDENT PREVENTION

Since accidents constitute one of the major causes of death and disability, and since they are by definition largely avoidable, the importance of their prevention is readily apparent. Unfortunately most accidents are caused by human factors, and too little is known about this, particularly the effects of mental and emotional disturbances, anger, stress, fatigue, judgmental errors, confusion, frustration, or grief. In other instances accidents result from physical infirmities such as defective hearing, faulty vision, lack of motor control, or the effects of advancing age.

Many accidents are caused by direct indiscretions on the part of the individual. One of these is the ingestion of alcohol or drugs while driving, swimming, boating, or walking. Another is excessive speeding while driving. Another is the disregard of safety devices while driving or at work or the failure to adhere to traffic precautions. Still another is taking undue risks at whatever task one is doing. Others are simple carelessness—in the home, behind the wheel, at work, while handling firearms, or during recreation—such as failure to correct accident-inducing conditions in the home, smoking in bed, or the inadequate supervision of young children.

Some accidents, of course, result from conditions over which the individual has no control. Included are mechanical failures, falling objects, the skidding of an automobile onto a sidewalk, hazardous weather conditions, traffic congestion, faulty highway construction, or fires and explosions.

It has been aptly stated that "safety must be practiced and promoted by individual citizens, industrial plants, local officials, and local organizations."[2] It has also been noted that "there is no mystery about the accident; it belongs within the realm of natural and manageable phenomena."[3] It goes without saying that different types of accidents call for different approaches to prevention.

OCCUPATIONAL HEALTH AND INDUSTRIAL MEDICINE

Another important area in the field of prevention that has received appreciable attention is occupational health and industrial medicine. Many forces have served to bring about constructive developments in this area. Labor unions have long since included in their collective bargaining concerns measures for occupational safety and the elimination of health hazards. Some employers have become aware of the relationship of good occupational health conditions to production, efficiency, and absenteeism, and have voluntarily taken progressive steps. Workmen's compensation insurers have established a relationship between the cost of the coverage and the degree of occupational health measures employed. Through the years government has influenced developments through regulatory measures aimed at the protection of the public health. The Occupational Safety and Health Act of 1970 has had significant impact. Voluntary agencies also play an important role.

Occupational health and industrial medicine comprise several different areas of interest. These include accident prevention, the creation of healthful working conditions, and control of specific occupational disease hazards such as dust, mercury vapors, radiation, lead, carbon monoxide, actinic rays, and harmful chemicals. Other areas include preemployment examinations, in-plant physician and nursing care, the availability of physical examinations with referral to the family physician, and health and job counseling. Developments in this area will be discussed in detail subsequently.

HEALTH EDUCATION

Health education provides a long-range approach to prevention. A great many agencies contribute to this effort, including voluntary agencies, professional health organizations, labor unions, employers, insurers, churches, community groups, mass communications media, schools, consumer groups, and government. Through such means an attempt is made to increase public consciousness of health.

Health education can be developed along specific lines. One approach is to make people aware of microbiological agents of disease (such as polluted water and pests), of the importance of immunization and personal hygiene, of the consequences of certain lifestyles (discussed subsequently), of the need for protection against physical environmental factors (such as polluted air, noise, and fire), and of the need for periodic consultation with a physician. Another approach is to create an awareness of the hazards of accidents and certain diseases and to suggest what to do about them. Yet another approach is to identify target populations (mothers, school children, college students, employees, union members, people in rural areas, the middle aged, the aged, low-income populations, foreign-language-speaking people, racial minorities, or patients in hospitals). The many aspects of health education can be categorized as follows.

Information: the presentation of health facts on a mass basis. While reaching many people, this approach probably has relatively little effect in changing health attitudes or behaviors over any period of time.

Education: the detailed exploration of an organized body of information, generally with a homogeneous group of people. While reaching smaller numbers of people, this approach can be more effective in producing change.

Community organization: bringing together affected individuals or groups for the exploration of a health problem area. While difficult, this approach can be effective.

Counseling: a person-to-person discussion of facts and their relationship to the individual. While reaching much smaller numbers of people, and being time-consuming and therefore more costly, this is the most effective approach to health education.

Quite obviously, these various aspects of health education can serve to supplement one another. In any instance, the effort must proceed on a continuing basis to be productive.

The means employed in health education include pamphlets, posters, films, lectures, seminars, employer journals, union publications, radio, television, newspaper advertisements, and face-to-face contacts by a physician or qualified counselor.

Organized programs for health education of the public on any scale are relatively recent, emerging in the United States just before World War I. More recently, an active interest in the subject has been taken by the American Medical Association, the American Hospital Association, the National Commission on Community Health Services, the American Public Health Association, the Health Insurance Association of America, the President's Committee on Health Education, and the National Health Council, among others. The periodicity of these activities, as well as the diversity of the agencies, gives a rather clear indication of a growing interest in the subject of health education since 1964.

Health education, however, is a complex subject and much still needs to be learned to make such efforts more effective. Furthermore, the effort, as was recognized in the report of the President's Committee on Health Education released in 1973, is hampered by fragmentation. In 1972 the National Health Forum took cognizance of such additional problems in health education efforts as the lack of a focal entity, a need for priorities and goals, an absence of organized efforts, the inaccessibility and unacceptability of services, as well as public ignorance, indifference, inertia, and irresponsibility. Most certainly the success of health education efforts cannot be measured, as so frequently happens, in a recounting of the numbers of pamphlets distributed, or the number of posters displayed. These are simply bookkeeping entries. Success can only be measured in terms of improved health.

Another deterrent to efforts to develop an awareness of and desire for sensible health practices is what might be considered an element of quackery. The overt creation of health fads, the building up of false illusions and hopes, the offering of shortcuts to good health, the encouragement of self-medication, the installation of a belief in the occult, and the sale of nutritional supplements, medications, or devices of dubious merit — all create a negative attitude toward serious educational programs, waste people's money, and in some instances cause physical or mental damage.[4]

Once launched, a program should be subject to continuing reevaluation and changed in accordance with the findings.

That health education programs can increase public awareness of health is evident from several sources. In 1971, for example, a Gallup Survey commissioned by the American Cancer Society revealed that:

> About 62 percent of the respondents claimed to have had physical checkups without symptoms; about half of the population stated that these exams were obtained in a three-year period prior to interview.
>
> About 47 percent of women claimed to have had breast examinations; one-third stated that these examinations occurred in the year prior to interview.
>
> About 53 percent of women claimed to have had cervical cytological tests; better than one-third of the population stated that they obtained this test in the year prior to interview.
>
> Skin and proctoscopic examinations were claimed to have been obtained by less than one-fifth of respondents.
>
> More younger people, more people with higher incomes, and more people who had achieved high school educations rather than having dropped from school earlier claimed to have had each of these examinations.

What is sorely needed is an evaluation of the various types and approaches to health education and of their concrete accomplishments in the prevention of accidents and diseases and in the early detection of disease. In this way a great deal could be learned and the knowledge applied to make all such efforts more effective. These needs were recognized in the report of the President's Committee on Health Education. The report also recommended tax incentives to encourage business, industry, and labor organizations to undertake and evaluate comprehensive health education programs for employees and their families.

CHANGING LIFESTYLES

Another aspect of preventive medicine that has been receiving attention quite recently is the relationship of lifestyles to health and how they might be changed where they present health hazards. This area is particularly difficult to penetrate because people in general behave as if they were not only impervious to harm and disease but immortal. Thus the need is to find the means for motivating people to live more healthfully.

Victor Fuchs has said, "The greatest potential for improving the health of the American people is not to be found in increasing the number of physicians . . . or even in increasing hospital productivity, but is to be found in what people do, and don't do, to and for themselves."[5]

The fact is that industrialized, urbanized, automated, mechanized, affluent, and busy America has developed lifestyle patterns that function as a deterrent to good health and general well-being. Essentially, corrections rest with the individual alone, although many forces and agencies can provide assistance, guidance, and even motivation.

Imbalanced diets often have long-range harmful effects. Obesity may lead to diabetes, heart disease, hypertension, digestive and gall bladder ailments, and gout. Functioning under undue stress is bad for the heart, blood pressure, and digestion; it may produce insomnia. Insufficient exercise, rest, relaxation, and recreation also have long-range adverse effects. The abuse of alcohol, drugs, or cigarettes can have devastating effects. Carelessness, whether at home, on the highway, at work, or while engaging in recreation, frequently has crippling and deadly consequences.

Changing harmful lifestyles requires communicating knowledge that will create health awareness in the individual. This effort must employ a long-range approach encompassing all the means of health education already discussed. It demands a continuous effort. To the extent that people are motivated to change their lifestyles, to that extent will the effort be successful. Only by experimenting with approaches to the problem, and by adequate, meaningful evaluation of the efforts made, can progress be accomplished.

PHYSICAL EXAMINATIONS AND SCREENING

An important aspect of preventive medicine is that of physical examinations and screening. Here the emphasis is on secondary prevention: the early detection and treatment of disease. This matter is very important in the prevention of many premature deaths, in the reduction of disability, and, perhaps, in the reduction of medical expenditures.

Physical examinations and multiphasic screening have received considerable attention over the past two decades by such prestigious groups as the American Medical Association, the Commission on Chronic Illness, the Health Information Foundation, the President's Commission on Heart Disease, Cancer, and Stroke, the U.S. Department of Health, Education, and Welfare, the U.S. Congress, the World Health Organization, the Commonwealth Fund, and the American Public Health Association. Throughout this period employers and labor unions increasingly made provision for periodic examinations of their employees or members.

Physical examinations are brought about in different ways. There are preemployment examinations, premarital examinations, examinations for military service, and life insurance examinations. There are periodic school examinations. There are in-plant examinations available for employed persons on a periodic basis. There are mass examinations for specific diseases such as tuberculosis, diabetes, glaucoma, hypertension, venereal disease, sickle cells, various types of cancer, or heart disease. There are multiple-screening examination programs devoted to the simultaneous detection of several diseases. And there are periodic examinations brought about by the individual on his or her personal initiative.

Each type of examination has its own purpose. This, in turn, can dictate the nature and extent of the examination. It also can influence the attitude of the patient toward the examination and his reactions to its findings. Professional opinion differs with respect to the efficacy of each, the degree to which each is periodically warranted, and the degree to which each can do harm as a result of the misunderstandings that can surround its purposes. Misunderstanding also can occur with respect to what might be accomplished.

There appears to be general agreement that periodic examination by a physician is the best approach to the early detection of disease. The views of private practicing physicians on the subject differ, however. Anderson and Rosen in 1960 found that 80 percent of private physicians felt that people should have such examinations even though they are apparently well, but only 45 percent made a point of recommending them to most of their patients. A third of the physicians complained that patients do not follow their instructions. Pediatric care of presumably well children is, of course, an exception. This now appears to be a generally accepted practice.

Physical examinations, moreover, are time-consuming and, therefore, somewhat costly. This has led to the exploration of other means by which similar purposes might be accomplished. This approach, known as multiple screening, multiphasic screening, or multi-test, has developed largely since 1950. Multiple screening is not intended to be diagnostic, but provides a rapid and economical means of disease or defect detection for larger groups of apparently well people. Upon the detection of such presumptive evidence, referral is then made to the physician for definitive diagnosis and treatment if needed. Multiple screening has considerable potential if it can be perfected and if properly used, but it is not without its limitations.

It is of interest that in 1971 there were reported to be 172 multiphasic screening programs in the United States, compared to 15 in all of Europe, 5 in Japan, 2 in Canada, and 2 in Australia. From this it can be observed that the concept of multiphasic screening does not appear to have too much credibility outside the United States.

Of particular interest here are certain known findings resulting from the screening of employed groups of people:

- In September 1968 the International Business Machines Corporation initiated a voluntary health screening examination program for employees over age 40. By 1969, 2,500 examinations had been given each month, involving 80 percent of the eligible employees. The results of 21,000 examinations revealed that 40 percent of those examined had a diagnosed ailment, and that of these a third of the ailments were unknown to the individual.[6]

- In 1951, 583 insurance company employees over age 49 were screened at the Pratt Diagnostic Clinic of Boston. The result found cancer in 146 cases, significant tumors in 1 of 18 cases, significant heart disease in 1 of 25 persons, moderate hypertension in 1 of 20, diabetes in 1 of 48, obesity in 1 of 5, and psychiatric conditions in 1 of 15 examined employees.[7]

- A mobile multiphasic screening program of the Health and Welfare Department of the ILGWU designed to serve small outlying plants in 1952 found that of 13,000 screenings, 5,338 disclosed significant abnormalities.[8]

- A screening of ferrous foundry workers in 1951 revealed pulmonary tuberculosis in 0.7 percent of white workers and 1.7 percent of nonwhite workers, nodular pulmonary fibrosis or silicosis in 1.5 percent of the workers, and diffuse pulmonary fibrosis in 7.7 percent of the workers.[9]

The value of such screening examinations has been questioned, however, in the following sense:

- Is screening able to detect disease which is likely to have an important impact on health? One examination of the subject found that 43 percent of diagnosed cancer cases, 58 percent of diagnosed coronary heart disease, and 51 percent of all potentially fatal diseases diagnosed did, in fact, result in death.

- Will the treatment of risk factors have a major impact upon the subsequent development of the disease? It is felt that there is no guarantee of this.

- What are the prospects that patients will comply with therapeutic regimens initiated as a result of the screening program? It is concluded that ambulatory patients are unlikely to take more than 50 percent of the prescribed medications, and that early detection that requires high degrees of compliance is not to be expected to be forthcoming.

- Do existing screening programs really alter disease outcomes? It is maintained that the Kaiser Health Plan experience did not produce any favorable results of testing among women and most men (except men ages 45 to 54); and that of 400 deaths only 60, or 15 percent, were judged to have been "potentially postponable through optimal preventive measures." (These findings will be discussed further in later chapters.)

- Are we misled by the traditional methods used in evaluating the clinical effect of screening programs? An affirmative answer is doubted, and it is maintained that evidence which appears to be to the contrary is misleading.

- Have we considered the entire range of the possible effects of screening, of the labeling of persons as diseased, and of long-term therapy? It is argued that the result can be to lower the social, emotional, and occupational ability of the individual to function and thereby to prolong absenteeism.[10]

Anderson and Rosen, in their examination of the subject, have concluded: "Prevention has a yet undetermined future role in the achievement of attainable health objectives. Even among physicians and medical educators there is no consensus as to the theoretical and substantive nature of preventive medicine."

RESEARCH

Valuable contributions to the prevention of accidents and illnesses, the early detection of diseases, and the control of diseases are made by research in bacteriology, chemistry, epidemiology, statistics, sanitary engineering, safety engineering, and psychology. Developments of such equipment as the x-ray machine and the electrocardiograph have proved invaluable aids in early detection. The research activities of manufacturers of many types, including such diverse fields as drugs and automobiles, have propelled developments that have aided in the prevention of disease and injury. These efforts are financed by private and voluntary organizations and foundations, and by public funds, singly or in combination.

In more recent years, considerable emphasis has been placed on research in heart disease, cancer, and stroke. The report of the President's Commission on Heart Disease, Cancer, and Stroke in 1964 gave considerable impetus to research in these areas. Subsequently appreciable research activity developed in the National Institutes of Health. In 1975 the NIH, in a document "Research Advances," reported on research gains toward the prevention of cancer, heart damage, hepatitis, pneumonia, influenza, glaucoma, cataracts, meningitis, dental caries, certain environmental factors, and other aspects of health. Also noteworthy is the development in 1973 of a National Cancer Program Plan by the Institute of Medicine of the National Academy of Sciences with the objectives of:

Reducing the effectiveness of carcinogenic agents.
Modifying biological systems to decrease the likelihood of the development of cancer.
Prevention of the conversion of cells to those that can form cancer.
Preventing tumor establishment.
Assessing the presence of cancer risk in populations.
Curing as many patients as possible.
Restoring patients with residual defects.

Research into the occurrence of accidents also receives considerable attention today: their causes, specific aspects of certain types of accidents (those occurring in the home, on the highway, or at work), and the effectiveness of preventive measures.

The American National Standards Institute (ANSI) has recognized that injury and accident statistics have an important role in accident prevention, indicating areas in which safety activities should be intensified, pinpointing problems that must be solved, providing a factual base upon which a safety program can be built, and measuring the success or failure of a particular effort. To this end ANSI has established a committee to prepare standards for the collection of data on the nature and occurrence of work injuries. The classification procedure is not designed to identify accident occurrences in terms of management or worker responsibility, nor to establish primary causes in terms of hazardous conditions or unsafe acts. The standards include complete recording of the nature of the injury, type of accident, hazardous conditions involved, part of the body affected, source of the injury, and unsafe acts involved.[11]

For the same reasons ANSI has developed standards for recording data on employee off-the-job injury-frequency rates that cause death, permanent and total disability, permanent partial disability, or temporary total disability. The purpose is to determine the seriousness of the problem and the need for activities to promote off-the-job safety among employees. Included in the standards are data with which to evaluate the effectiveness of prevention activities and the progress being made in improving the experience of employees outside the work environment.[12]

REHABILITATION

The rehabilitation of disabled persons is a closely related process to preventive medicine. Since World War II there has been a growing and expanded interest in rehabilitation. It is a field in which the potential would appear to be considerable. In many respects, however, it is a relatively new field, and much remains to be learned.

The National Council on Rehabilitation has defined rehabilitation as "the process of restoring the handicapped individual to the fullest physical, mental, social, vocational, and economic usefulness of which he is capable." As used here, the term is intended to encompass all efforts (other than those inherent in routine medical care and surgery) to reduce the duration or episode of treatment or the period of disability resulting from disease or accident.

Rehabilitation efforts might be of a general nature, or they might be limited to specific aspects of the subject, such as vocational rehabilitation, reeducation and training, physical therapy, speech therapy, psychiatry, social service, or surgical convalescence. Some might be limited to persons with certain types of disabilities, such as amputees, arthritics, the blind, deaf, aged, crippled children, the mentally ill or deficient, alcoholics or drug addicts, or victims of such diseases as polio, cerebral palsy, and heart disease. Some might be limited to certain categories of persons, such as veterans, the members of a specific labor union, the employees of a specific employer, or persons covered by a particular insurer.

Many factors unquestionably can serve to avoid, deter, or retard the rehabilitation process. One is the shortage of professional personnel and facilities. Another can be the absence of the will of the disabled or handicapped person to recover and assume to the maximum extent possible a normal and full role in life. Yet a further inhibiting factor can be economic in nature. The person whose business or professional practice has suffered severe reversal as a result of disability, the person who is too old to be employed, the person who becomes disabled from an industry in which there is above-average unemployment, the person whose disability requires a marked change in the manner in which he will earn a living, with perhaps a clear indication that earnings will be sharply reduced—all these cases present an economic situation that can be discouraging and thereby inhibit the rehabilitation process. There also is the reverse economic situation: those cases in which it is more profitable economically to continue in a disabled state. In such cases the rehabilitative process can be rejected, resisted, or delayed. This situation can be accentuated if any of the preceding factors are present. Usually it can occur if some form of cash disability benefits is present.

That insurance companies have an active interest in the process of rehabilitation is evident from a statement made by the Association of Casualty and Surety Companies (now merged into the American Insurance Association) two decades ago:

The insurance industry accepts the premise that physical rehabilitation is part and parcel of medical care under Workmen's Compensation. It maintains a deep interest in the subject and will continue to cooperate with all other interested parties in an attempt to make further progress in this field. . . . The insurance industry has a well defined obligation in the field of rehabilitation. . . . The carrier's program will provide the best in realistic rehabilitation, allowing the injured man to understand and reach his potentialities and goals physically, mentally, socially, and economically.[13]

To make such principles effective, the Association recommended that insurers should:

Establish a definite plan for a well-organized rehabilitation program.

Organize a rehabilitation division or unit.

Promote a consciousness on the part of claims personnel of rehabilitation potentialities.

Develop a knowledge and understanding of all rehabilitation services available.

Maintain accurate records of cases reviewed as potential rehabilitation prospects and those referred for treatment and training.

Establish close affiliation with the insurer's medical department or advisers.

Receive from the attending physician an early report covering rehabilitation opinions and plans.

One of the forerunners among the many insurance companies that have an active interest in the field of rehabilitation is the Liberty Mutual Insurance Company. Since World War II this company has been conducting research in the subject and is convinced that periods of disability can be shortened by some of the newer developments in medicine. It has concluded that much can be accomplished through a planned process of rehabilitation. Starting with an interest in workmen's compensation cases, the company became convinced that the same basic purpose and many of the same procedures applied equally to liability and group disability cases. As a result, the company gradually developed an integrated program for the handling of serious injury cases. Since the services of its own centers are designed to provide treatment for injured employees of policyholders, there is no charge to the employee or the referring physician. The actual cost of treatment is charged to those insured employers whose employees are treated at the centers.

The effectiveness of such a program is evidenced by the fact that some 80 percent of those cases handled by the Liberty Mutual program returned to work. Savings on workmen's compensation benefits have been estimated to range from $1,000 to $50,000 per case. The total savings after the cost of the program are several millions of dollars.[14] Similarly, Aetna Life and Casualty has quite recently stated that for every $1 invested in physical and vocational rehabilitation, there is a return of at least $5 on workmen's compensation claims.[15] The rehabilitation of disabled workers, then, is an economically viable as well as a highly socially desirable endeavor.

Today a great many of the large employers directly or indirectly have developed rehabilitation programs for their disabled employees. Rehabilitation services are also provided by certain labor union health programs such as the Sidney Hillman Health Center of New York for the Amalgamated Clothing Workers of America.[16]

In relatively recent years the President's Committee on Employment of the Handicapped has done much to better this situation but much continues to remain to be done. The effort has been supported by the state vocational rehabilitation agencies. Also supporting the effort is the practice of disability-income insurance, which permits a disabled employee to attempt part-time employment without having the disability-income benefits terminated immediately. In such cases the insurance benefits are reduced by an amount equal to the earnings received or adjusted on some other preestablished basis. It is a hopeful sign that Dr. Salvatore G. DiMichael recently wrote that there is "no evidence that employed handicapped people with enough seniority were being dismissed . . . employers have shown a sense of concern for those who have served their business."[17]

BIBLIOGRAPHY

J. F. Follmann, Jr., *Medical Care and Health Insurance* (Homewood, Ill.: Richard D. Irwin, Inc., 1963); *Alcoholics and Business*, AMACOM, 1976.
George Rosen, *Preventive Medicine in the United States, 1900-1975* (Boston: Science History Publications, 1975).

Material from the following organizations was also used in the preparation of this chapter:
American Medical Association
American Public Health Association
California Medical Association
Metropolitan Life Insurance Company
National Health Council
National Safety Council

Office of Health Economics (London)
Social Security Administration
U.S. Environmental Protection Agency
U.S. Public Health Service

REFERENCES

1. American Public Health Association, "Heritable Disease: A Proposal for Public Health Action," Washington, D.C., 1962.
2. Norvin C. Kiefer, M.D., "Accidents: The Foremost Problem in Preventive Medicine," *Preventive Medicine*, March 1973.
3. Walter A. Cutter, *Accident Prevention: The Role of Physicians and Public Health Workers* (New York: McGraw-Hill Book Company, Inc., 1961).
4. H. Doyl Taylor, "Quackery: Special Problems," *Proceedings of the Second Annual Conference on Health Education of the Public*, American Medical Association, 1966.
5. Victor Fuchs, keynote address, 1967 National Conference on Medical Care, U.S. Department of Health, Education, and Welfare.
6. J. C. Duffy, M.D., "Health Screening on an International Basis," paper presented before the Kentucky Medical Association Annual Meeting, September 25, 1969.
7. "Group Health Surveys in a Diagnostic Center," *New England Journal of Medicine*, January 29, 1953.
8. Editorial, "Union's Health Circuit," *Public Health Reports*, November 1952.
9. H. Heimann, "The Health of Ferrous Foundry Workers," *Public Health Reports*, February 23, 1951.
10. David L. Sackett, M. D., "Screening for Early Detection of Disease: To What Purpose?" *Bulletin of the New York Academy of Medicine*, January 1975.
11. American National Standards Institute, "Method of Recording Basic Facts Relating to the Nature and Occurrence of Work Injuries," New York, 1969.
12. American National Standards Institute, "Method of Recording and Measuring the Off-the-Job Disabling Accidental Injury Experience of Employees," New York, 1973.
13. Association of Casualty and Surety Companies, "The Physically Impaired: A Guideline to Their Employment," New York, 1957.
14. W. Scott Allan, "Rehabilitation as a Practical Insurance Program," *Journal of Insurance*, November 1958.
15. V. S. Davidson, "Rehabilitation of the Workmen's Compensation Claimant" (New York: Aetna Life and Casualty, circa 1973).
16. "Demand for Rehabilitation in a Labor Union Population" (Sidney Hillman Health Center of New York, 1964).
17. Salvatore G. Di Michael, "The Right to Work for the Handicapped," *ICD Rehabilitation and Research News*, Volume 1, 1976.

9

Those Engaged in Prevention

Those engaged in the prevention of diseases and accidents are legion. The necessity for this is apparent from the broad and complex nature of preventive medicine shown in the preceding chapter. Preventive medicine brings into play a host of disciplines, pursuits, and agencies. Efforts can be financed with private funds, public funds, or both, and by individuals.

It is not feasible here, nor would it serve any useful purpose, to present a categorical list of all of those who play a significant role in the prevention of diseases and accidents. Foremost, of course, is the individual and the family unit. Each plays a basic role, despite the fact that each should do much more than is presently the case. The following discussion will give some indication of the forces that come into play in addition to the individual and the family.

THE PRIVATE SECTOR

Within the private sector there are a variety of people, organizations, and agencies that, to a greater or lesser degree, contribute to the prevention of diseases and accidents. In many instances their efforts are solely devoted to prevention.

The health professions and allied disciplines, individually or collectively, make a very considerable contribution. Among the professionals involved are physicians, dentists, nurses, psychiatrists, psychologists, physical therapists, nutritionists, health educators, industrial physicians and nurses, social workers, laboratory technicians, and other allied health professionals. Their organizations, which play a significant role, include the American Dental Association, American Industrial Hygiene Association, American Medical Association and state and county medical societies, American Optometric Association, American Nurses Association, American Physical Therapy Association, American Psychiatric Association, American Psychological Association, Industrial Medical Association, National Association of Social Workers, National League for Nursing, and Occupational Health Institute.

Hospitals, clinics, and other health institutions also make important contributions. Some clinics and neighborhood health centers, such as those devoted to well-child care, child guidance, mental illness, alcoholism, or drug addition, have prevention as one of their designated purposes. Organizations such as the American Hospital Association help significantly.

A great many voluntary organizations and agencies have a greater or lesser degree of interest in prevention. Included are the American Conference of Governmental Hygienists, the American National Council on Health Education of the Public, the American Red Cross, the Boy Scouts, the Girl Scouts, the Salvation Army, and many churches and religious organizations of all faiths. Many organizations are devoted entirely to the prevention of specific conditions. By way of illustration, a few of these are: Alcoholics Anonymous, Alanon, and Alateen, the American Cancer Society, American Social Health Association, Arthritis Foundation, International Society for the Rehabilitation of the Disabled, National Association of Retarded Children, National Council on Alcoholism, National Safety Council, National Society for the Prevention of Blindness, National Tuberculosis and Respiratory Diseases Association, and United Cerebral Palsy Association, Inc.

Many of the privately endowed foundations are also interested in the prevention of diseases and accidents, and most usually provide research financing or demonstration projects. Included are the Rockefeller, Johnson, Kellogg, Milbank, Carnegie, and Ford foundations.

Colleges and universities contribute essentially in two ways: in the training of those who will work in prevention and in the conduct of research. The schools of medicine, nursing, hospital administration, industrial health, public health, hygiene, and industrial and labor relations all have significant roles to play. Courses are also made available in some instances for employers and union leaders, usually in specific aspects of industrial health.

The business community makes a considerable contribution in many ways: through the efforts of sanitary and safety engineers, chemists, and architects, as well as through the manufacture of many products that help prevent diseases and accidents, such as refrigeration, heating and ventilation systems, and safety devices. Organizations that assist and encourage the process include the American National Standards Institute; Blue Cross Association; federal, state, and local chambers of commerce; Engineers Joint Council; insurance companies and their organizations; National Association of Blue Shield Plans; and National Association of Manufacturers.

Labor unions have also played a very considerable role through the years, not alone in their direct concern with prevention but also in their encouragement of legislation for the regulation of child labor, minimum wages, working hours, and working conditions.

Other forces constantly brought into play include the mass media, consumer organizations, and many nonmedical disciplines, such as engineers, chemists, and architects.

THE PUBLIC SECTOR

The many forces of the public sector, at the federal, state, and local levels, are invaluable in developing the concepts of prevention, in fact gathering, in supporting research, and in bringing to action the forces of law and regulation. The range of roles played is extremely wide—from the establishment of building codes and inspection, to police and fire protection, to highway design and traffic control, to the maintenance of hospitals and clinics, to the provision of immunization, to the control of water and air pollution, to health and driver education in the school system, to attacks on specific disease conditions, to the prevention of accidents, and to the provision of guidance and assistance to places of employment. Certain of the agencies involved have been identified in the previous chapter, others are noted in subsequent chapters and in the Appendix. Some of the federal agencies that make important contributions are the Bureau of Mines, Environmental Protection Agency, Food and Drug Administration, Department of Agriculture, Department of Health, Education and Welfare, Department of Labor, National Center for Health Statistics, National Institute on Alcohol Abuse and Alcoholism, National Institutes of Health, and Occupational Safety and Health Administration. Important counterparts to such agencies exist at the state and local levels of government.

While this diversification of activities indicates a considerable effort in the interest of prevention and brings into play a very considerable degree of expertise, there are probably few who would not argue that the effort is not nearly as effectual as it should be. One reason is that with all these interests at work, the total effort becomes seriously fragmented, to the point where it is even doubtful if any one person knows all that is going on at any given point in time. Another reason is that efforts are frequently spasmodic in nature, even subject to fads. One more reason is that efforts in a particular area of concern are often duplicated, with resultant waste and dilution of effort. Within any particular area, furthermore, efforts themselves can at times be ill conceived or ineffectively executed.

It would seem that a greater degree of coordination, with less inclination to consider a particular area of interest an entity unto oneself, could serve to improve and strengthen the present situation. A meaningful, hardheaded evaluation of the entire prevention effort might very well produce good results, which for one thing would mean the most efficacious use of the available funds. The need for progress is far too important to the public welfare to permit secondary impediments to stand in the way.

10

Some Deterrents

Despite the appreciable gains made through preventive medicine, and the conceivable potential, a number of inhibiting factors deter the realization of fuller accomplishment. In the main, these would appear to increase in intensity as the degree of individual motivation, initiative, influence, cooperation, exertion of effort or willpower, and expenditures of time and money increase.

Inhibiting factors might be categorized as (1) public attitudes and behaviors, (2) the supply of professional personnel, (3) the question of economic deterrents, and (4) the question of financial incentives.

PUBLIC ATTITUDES AND BEHAVIORS

The predominant deterrent to the full accomplishment of preventive medicine today in the United States appears to lie in public attitudes and behaviors as exemplified by the acts, or omissions, of individuals and families. This has been noted in many sources.

In 1954 E. L. Koos commented: "In the last analysis the health of the community is based upon the ideas, ideals, attitudes, and behavior patterns of the individual and his family."[1] He noted that attitudes toward illness can differ rather sharply among social groups, such differences including the perception of illness, the use of medical care, and the use of medications.

The following year a study made in Syracuse, New York, of the utilization of health resources by recipients of public assistance revealed that while there were "no basic differences in the receipt of health information and health literature" among the various income groups studied:

> (1) Fewer medical recommendations were made [by the school health services] for the middle-income group as compared with the public assistance and low-rent-housing children, but a higher percentage of recommendations were followed when made [to the middle-income group]; (2) there appeared a definite deficiency in the extent to which community mass X-ray programs were utilized by the lower-income groups; and (3) while public assistance women infrequently failed to receive any antepartum care they applied for such care much later during pregnancy and made fewer visits [than higher-income groups].[2]

Two years later the Commission on Chronic Illness, in discussing the prevention aspects of chronic illnesses, commented:

> Individual initiative is vital in the prevention of chronic disease. The public—all of us—should incorporate in our daily lives the recognized precepts of preventive medicine and should cooperate fully in practical programs of preventive medicine once they have been worked out by the professionals concerned.[3]

In 1961 Gordon Macgregor, in reporting on a study made in the Great Plains, said: "We [found that] . . . unless encouragement of individuals to adopt preventive health measures is based upon the effect of illness in disrupting occupational and social roles, such encouragement will have little influence."[4]

About this time J. Douglas Colman, a biostatistician and later Blue Cross executive, observed that "positive health is not something that one human being can hand to or require of another, [it] can be achieved only through intelligent effort on the part of each individual."

Guy Stevart has recently observed that there "is a deeply entrenched and simplistic notion that health-related behavior is susceptible to change primarily via the medium of health information accompanied by exhortation to change." He considers this notion to be the one major obstacle to progress. He goes on: "What we have tended to do in health education is with moralistic overtones to intervene in the round of ordinary daily behavior." He suggests the involvement of people in activities and services that benefit not only themselves but the larger social system.[5]

Victor Fuchs notes that "one of the central choices of our time . . . is finding the proper balance between individual (personal) and collective (social) responsibility."[6] He quotes Henry Sigerist, an advocate of socialized medicine, as observing that while the state can protect society from many dangers, "the cultivation of health, which requires a definite mode of living, remains, to a large extent, an individual matter."

It is observed, then, that while it is generally assumed that good health is something everyone desires and seeks above everything else, it can hardly be said that beyond a certain point individuals are willing to do very much to achieve that end. They are careless about, or in some cases ignorant of, such basics to good health as a well balanced diet and sufficient rest and exercise; they ignore obesity, deleterious lifestyles, and ordinary safety precautions; and, once under the care of a physician, they ignore counsel and refuse to follow the prescribed regimen for treatment. In establishing their financial priorities, the new car, entertainment, or even gambling can easily take precedence over health considerations. The observation has been made many times that people find it easier to place dependency in their healers than to engage in the more difficult task of living wisely. This, in turn, has led some to speculate whether it should be an object of public policy to attempt to change the tastes of people or the choices they make.

Recently, Dr. Laurence E. Hinkle, Jr., of Cornell University Medical College, reported to a national conference on preventive medicine that "the problems of understanding and changing motivation [with respect to such matters as cigarette smoking, alcohol abuse, obesity, and reckless driving] are crucial to many efforts to prevent disease."[7] At the same conference, Dr. Norton Nelson noted that efforts to curb such deleterious health practices had been totally ineffective and that "far more effective and reliable techniques for changing the health-related behavior of the public are required."[8]

The importance of lifestyles to health has been shown by Victor Fuchs when he draws a comparison between Utah and Nevada, two states very much alike in income, schooling, degree of urbanization, climate, and other variables, including the supply of physicians and hospitals per capita.[9] Yet the death rate in Nevada is considerably higher than that in Utah for all age groups. The differential, in Dr. Fuchs' opinion, lies in the fact that Utah is inhabited largely by Mormons, who generally lead stable, quiet lives, have a high degree of marital and geographic stability, and do not use tobacco or alcohol. These two latter factors are reflected in the excess of death rates in Nevada for malignant neoplasms of the respiratory system and cirrhosis of the liver in comparison with the rates in Utah (deaths from cirrhosis of the liver are from two to almost six times greater in Nevada, depending on age and sex differentials). It also cannot pass unnoted that Utah reportedly has the lowest rate of absenteeism from work and the lowest rate of labor turnover in the United States. Further, it might be recalled in the examination of the health status in the United States that time and time again mortality from a specific disease is lowest in Utah.

Another deterrent to good health, made amply evident by any number of studies, and regardless of the indices used, is an inferior level of general education or schooling. Granted this frequently affects income levels, social status, and other matters that in turn affect health status. But schooling is the central cord that predominates the picture. The reason is not yet clearly understood. But the fact remains that premature deaths, disabilities, illnesses, accidents, days lost from work, and infant mortality are all in greater proportion among the lesser educated. When it is recognized that 8 percent of the United States population is considered to be functionally illiterate, the magnitude of the problem is readily apparent. Victor Fuchs is impressed with the correlation of schooling and health: "So far all research on the relationship between health and schooling has utilized retrospective statistical analysis . . . such research has nevertheless suggested an important connection between an individual's behavior and his health."[10]

With respect to periodic physical examinations, other inhibiting factors present themselves, at least in the mind of the individual. There may be a lack of confidence in the techniques employed, or in the findings. Time off from work may be required. There may be concern over resultant unemployment, possible deterrents to advancement on the job, or reemployability. There may be a distinct fear of the findings, a lack of knowledge concerning the facilities for such examinations, or a lack of encouragement from physicians. Social mores, religious beliefs, ethnic backgrounds, long-entrenched habits, and natural inclinations to take good health and the indestructibility of the human being for granted can serve as inhibiting factors.

Also of interest is a 1955 survey of public attitudes, which showed that while 80 percent of those interviewed believed in the value of regular physical checkups, only 29 percent actually had such examinations. Of the physicians and the public interviewed, 20 percent saw no value in physical examinations.[11] A study by Helen Avnet in 1967 reported that only 17.6 percent of Group Health Insurance, Inc. members used preventive services. The greatest use was for young children. Women used such services 60 percent more than men. Blue-collar workers used such services least.[12] Other studies show similar lack of participation despite availability and the absence of cost.

In 1965 the U.S. National Health Survey reported on the problem of persuading people to cooperate in a health examination survey. It found that 71 percent might be willing to come to a health examination if the time and place were convenient; 25 percent would not; and 4 percent did not know. Those least inclined to participate were those living in the Northeast, people over age 45, white persons, males, and those at both extremes of income levels.[13]

Thus J. M. Mackintosh, professor of public health at the University of London, has commented: "Everyone says that prevention is better than cure and hardly anyone acts as if he believes it. . . . Palliatives nearly always take precedence over prevention, and our health services [the National Health Service] today are heavily loaded with salvage. Treatment . . . is more tangible, more exciting, and more immediately rewarding than prevention."[14]

THE SUPPLY OF PROFESSIONAL PERSONNEL

A great number of professional disciplines play significant roles in preventive medicine. Yet the supply of physicians, as well as other professionals, is by no means unlimited. Contra, in all the related disciplines shortages are reported and, as a consequence, personnel training loans or funds are demanded. These shortages present a serious handicap to preventive medicine and, in turn, stimulate an active interest in multiphasic screening. This has led Seymour Harris, an economist, in examining the economics of American medicine, to comment: "If each member of the population had but one examination per year . . . the cost in time would be about 350 million hours, or about 40 to 45 hours a week of all physicians in private practice."[15] That, obviously, would not leave much time for treating those who are already sick.

The shortage of professional personnel, then, cannot be overlooked, particularly in attempts to accelerate the early detection of disease. The extent to which multiphasic screening would relieve the present situation, assuming it could be made generally available, is not known, nor is there enough knowledge upon which an answer could be hypothesised with any reasonable degree of certainty. In view of the present situation, study of this aspect of preventive medicine would be worthwhile.

THE QUESTION OF ECONOMIC DETERRENTS

Preventive measures cost money, and it is not yet clear in certain respects whether such monies spent would be cost-effective. Certainly when public funds are needed to advance the concepts of preventive medicine, inhibitions can be created by restrictions on the funds the taxpayer is willing to make available.

The same process can come about as respects the individual or family unit. As Victor Fuchs has commented, health must compete for resources (money) along with choices for creature comforts, and "if we are honest with ourselves there can be little doubt that other goals often take precedence over health."[16]

Certainly when an individual considers periodic physical examinations that are not provided by the employer, labor union, or through some type of prepayment arrangement, it is logical to assume that he would be concerned with the costs. Physical examinations are costly, and it seems logical that for lower-income people cost could be a deterrent. This would be particularly so in the absence of symptoms of disease, when loss of time from work can be involved (with a possible loss of pay), and if negative findings will mean additional expenses for treatment and further work loss. Just how important this factor is and whether it is a primary factor standing alone remains a subject for question. There is evidence, however, that casts reasonable doubt upon its importance as a single or all-pervading influence.

In 1951 a multitest clinic was made available in Richmond, Virginia. Over one-third of those approached did not bother to take advantage of the tests, over one-fifth said they could not arrange to get to the clinic, one-seventh said they did not know about the tests, and one-tenth said they took care of their own needs.[17]

Several years later Johns Hopkins Hospital in Baltimore invited a representative sampling of the community population to have a complete physical examination at no personal expense. Appointments were offered at the convenience of the individual and free transportation was provided. In instances where no response was received, follow-ups were made, including personal calls or visits. Nevertheless, 41 percent of the sample refused to be examined. The proportion of refusals was higher in the middle-income groups, among older people, and among the white population. From this experience, the Commission on Chronic Illness concluded that the attitudes of people were "major obstacles to the application of available therapeutic measures."[18]

Under the Kaiser Foundation Health Plan there is no serious dollar deterrent to the multiphasic tests that are available. Yet utilization over the period 1953 to 1960 for the membership of one large union group was annually only 11 to 12.5 percent of eligible persons. (There were specific reasons, in part, for this degree of utilization, including loss of pay in some instances, nonavailability of services in some areas, and examinations obtained elsewhere. On the other hand, there was active encouragement among the group to have such screening examinations.)[19] Fortunately, recent Kaiser experience for its entire membership shows a much higher rate of utilization.

In the Health Insurance Plan of Greater New York, where one of the purposes is to bring about the early detection of disease by removing any cost barrier that might exist, a routine health examination program at no added cost is promoted through periodic bulletins and pamphlets and at subscribers meetings. Yet it has been reported that the same reluctance to get such examinations exists in groups as it does in individuals. Patients were scheduled by the leaders in one particular union to appear for examinations, but only 30 percent of the group appeared for examination.[20] One experiment, the Montefiore Family Health Maintenance Demonstration, conducted by the Montefiore Medical Group of HIP (which had a 22 percent utilization of health examinations), is worth noting. People were preselected on the basis of their interest in the objectives of the demonstration and their agreement to cooperate. They experienced 140 examinations per 100 persons the first year (compared to 40 percent for other families). On the basis of this experiment Dr. George Silver, who directed the demonstration, recommended against the categorical promotion of periodic physical examinations for all families. Instead he concluded that routine health conferences would be more efficacious and economical in the implementation of a preventive service program.

These experiences evidence that neither the original cost nor concern over the possibility of subsequent costs was the predominant deterrent to physical examinations for adults.

THE QUESTION OF FINANCIAL INCENTIVES

In literature advocating preventive medicine the point is frequently made that the absence of health insurance coverage for physical examinations acts as a disincentive to the early detection of disease. Paralleling this is a rationale that if the individual has to share in the cost of such services, that also acts as a disincentive. Such arguments can appear reasonable, but are certainly an oversimplification of the problem.

Yet the concept prevails. In 1955 A. Hays, a labor leader, said: "It is high time, I believe, that we stopped emphasizing insurance against the high costs of neglected health, and devoted more of our efforts to developing a system of insurance or prepayment which will give the American people a greater access to the kind of health care which prevents illness, or nips it in the bud." In 1959 Edward W. Chase maintained that health insurance "by its very nature has inhibited the most promising development in health care—preventive and restorative medicine." In 1960

Jerome Pollock, then a spokesman for the United Automobile Workers, said: "There is no fundamental reason why the diagnostic examination, so valuable for the early detection of disease, must continue to be dismissed as something people ought to save for."

The following year Thomas Parran, then surgeon general of the U.S. Public Health Service, said: "A prepayment scheme for medical care . . . should contain an incentive for prevention." That same year John D. Porterfield, a deputy surgeon general in the U.S. Public Health Service, said: "The character of most health insurance programs . . . reinforces the idea in the community's mind that the smart thing to do is wait until you get sick and then let someone else pay the bill, particularly since no one seems ready to pay your bill for 'merely' preventive services." More recently, John H. Knowles said that "dramatic changes in health insurance are needed to favor lower-cost services, health education, and preventive-medicine programs."

That insurance coverages for certain preventive services, including periodic physical examinations, are available for employed groups but infrequently purchased will be shown subsequently in later chapters.

Yet another rationale warrants comment. It is not uncommon to find proposals, even demands, that insurance programs, private or public, should provide financial incentives for preventive measures, failing to recognize, of course, that the converse of an incentive is a penalty. Unfortunately, no one has indicated in pragmatic terms how this is to be done or on what basis calculations would be made.

The concept, of course, is intriguing. The insurance business is not without a long history of precedence. In fire insurance, reductions in premiums are made for improved construction and for the use of fire-resistant measures. In automobile insurance, safe driver awards are given, but conversely persons in certain age brackets are penalized for the deleterious accident record of their peers. In workmen's compensation insurance, credit is given for improved work health and safety conditions. In life insurance, persons with a good medical history and health record pay less than others. In health insurance, experience rating is applied to larger insured groups, so that their premiums eventually reflect the loss experience of their particular employees. This most certainly should provide an incentive for preventive measures, and insurers will so argue. The degree to which it does act as an incentive, however, simply is not clear or proved, although the logic of the situation would be toward a positive influence.

A contrary influence is to be noted in the health insurance field, however. Through the years, and extending to the present (see, for example, the Congressional and National Association of Insurance Commissioners' debates on HMOs), there has been strong insistence that health insurance should be community-rated (that is, that each individual in a community should pay the same premium regardless of his personal risk). The argument is made on the grounds of social equity. While in practice pure examples of community rating are extremely rare or impossible to find even among the advocates, and while the trend is distinctly toward experience rating, the voices demanding community rating continue. Certainly, the question is not one to be resolved here. The debate and the concept are important to note, however, since community rating would per se rule out the possibility of incentives for preventive measures. The concept of incentives for loss prevention, then, is not foreign to insurers.

A strong question is presented with respect to the practicalness of certain types of proposed incentives, however. For example, one who is particularly concerned with the adverse effects of smoking on health would advocate an incentive for nonsmokers. Another, concerned with alcoholism, would lower the premiums for nondrinkers. And so forth. Thus the smoker and the drinker would be penalized. This raises difficult questions. Are ethnic groups with abnormal rates of certain diseases or accidents to be penalized? The trend in the last few decades has been decidedly away from such methods of insurance rating. At a practical level a further question is raised: How can it be proved that a person is a nonsmoker? This would require periodic investigation, and investigations cost money. What, then, would be the amount of the reduction in premium for the nonsmoker, and what would the administrative cost of the system be if the veracity of the nonsmoker is to be substantiated? It could well be that the latter could more than consume the former, with the result that the nonsmoker would have to pay a greater premium.

There is a strong question, then, whether this approach to incentives is practical. Added to this is the fact that the vast majority of health insurance is written on groups of people, and an incentive approach like this, therefore, would be inapplicable to most people. Hence the desired objective would not be attained.

Those who advocate incentives would do well to go beneath generalities to explore the potential of a practical and workable system that might have reasonable possibility of accomplishing the desired objective. Joint discussions with insurers might be constructive.

A still further rationale that pervades the subject of preventive medicine, particularly as respects periodic health examinations, is that it will save money presently spent for health services, will reduce the incidence and duration of disability, and will reduce the number of premature deaths. Contra, there is a concept that it will result in increased expenditures for health care by revealing conditions not otherwise known to exist and requiring treatment. The matter of the cost-effectiveness of preventive medicine will be discussed subsequently.

It is not a simple matter to evaluate in concrete terms what the role of health insurers should be with respect to the prevention of illness and accidents and the early detection of disease. There appears to be a persuasive element of inherent logic that would argue that any form of insuring process should, in its own self-interest, as well as in the public interest, prevent the occurrence of loss whenever possible. It is not an easy subject to come to grips with, however.

Insurers cannot ignore the several inhibiting factors that have been noted. To what degree is it within their province to overcome them? To what degree can they, through the design of their coverages, the structuring of their premium rates, their educational endeavors, or their research activities, alter the inclinations of human nature? Who can say, for example, whether more loss ultimately would be prevented, and the public interest better served, if a given amount of resources were directed into research channels rather than into providing or covering periodic physical examinations? Is health education, actively pursued, more effective than either? Can the influences of experience rating be more directly brought to bear on the subject? Would the inclusion of periodic physical examinations in health insurance coverages actually result in an appreciable increase in the early detection of chronic illnesses and in a reduction in disability? If so, would it produce savings or increases in medical care expenditures? If such matters could be resolved, considerations by insurers might be more concentrated and interest in the subject appreciably heightened.

What is said here is not intended to argue for complacency in the status quo; it is presented to acknowledge that the subject is complex and inhibited by many factors not easily overcome.

Concern over the prevention of economic losses resulting from injury, illnesses, and premature death is of greater potential concern to insurance companies than to other types of health-insuring mechanisms, since insurance companies have much more at stake than the expenditures for medical care. Because they also insure against loss of income, workmen's compensation, public liability, accidental dismemberment and death, and natural death, any effective means of reducing the incidence or the severity of injury and illness can be a matter of consequence. When these can be brought about by the insurance company directly, insurance purposes are served by containing or reducing the costs of the insurance, and the public interest is served by reducing both economic and personal loss.

The question of incentives for improved health through the functioning of insurance programs remains, then, a moot question. A long-term study under way by the Rand Corporation on a grant from the U.S. Department of Health, Education, and Welfare to determine the effects of different types of financing (including the use of deductibles and coinsurance) on the receipt of health care should throw some light on the relationship between health care and economic deterrents.

The question of economic deterrents or incentives to the progress of preventive medicine, then, is a difficult one. Despite their apparent logic, available facts by no means bear out the rationale. Thus Rashi Fein, an economist, has said: "The issue involved in breaking the barriers of preventive services is not really one of fiscal constraints but one of lack of interest and commitment . . . the issue, I submit, is not economic; the issue rather is embodied in an examination of our values." Similarly, the Office of Health Economics in London has stated that the concept of a price barrier to health is a naive assumption, shown to be no longer valid with the existence of the National Health Service. It points out that studies by Shuval in Israel and Robinson in Swansea in the early 1970s have shown that health decisions are often determined more by cultural, social, or psychological factors than by purely medical ones.

An international comparison of medical care reported by Kerr L. White is also of interest. He points out that health care for entire populations is one of the most difficult of the social services to plan because of the large variation in the demand for services, the difficulty in achieving a rational equilibrium between demand and resources, the increasing variety of medical services, and the rise in the amount of public money spent to support the system. His study found that the indices used provide no evidence for the generally accepted view that the presence or absence of financial barriers significantly influences the use of medical services. Differences in the supply of physicians were also found not to be a factor in the use of services. With respect to physical examinations generated by administrative requirements in schools and places of employment, or for insurance purposes, the study concludes that this form of care can place a substantial burden on health resources. He found marked variation in utilization among the geographic areas studied (municipalities in Canada, Argentina, England, Finland, Yugoslavia, the U.S.S.R., and the United States) and reflects that this creates a large demand for a procedure "that is of doubtful efficacy in many instances" and thus a questionable use of scarce medical resources. He recognizes that this then presents the question of where policymakers should make their next investment "in order to reduce the relatively high levels of perceived need. . . ."[21]

It is possible that employers and labor unions, functioning in relation to specific groups of active people with whom they have close and meaningful contact, can provide a cohesive force than can overcome at least some of the deterrents noted in this chapter. Such possibilities are discussed subsequently.

REFERENCES

1. E. L. Koos, *The Health of Regionville* (New York: Columbia University Press, 1954).
2. Knowledge and Utilization of Health Resources by Public Assistance Recipients," *American Journal of Public Health,* February 1958.
3. Commission on Chronic Illness, *Chronic Illness in the United States: Prevention of Chronic Illness, Volume I* (Cambridge, Mass.: Harvard University Press, 1957).
4. Gordon Macgregor, "Social Determinants of Health Practices," *American Journal of Public Health,* November 1961.
5. Guy W. Stevart, "Breaking the Barriers of Prevention," paper presented before the New York Academy of Medicine, April 25, 1974.
6. Victor Fuchs, *Who Shall Live?* (New York: Basic Books, 1975).
7. Laurence E. Hinkle, Jr., M.D., paper presented before the American College of Preventive Medicine, June 1975.
8. Norton Nelson, M.D., paper presented before the American College of Preventive Medicine, June 1975.
9. Fuchs, op. cit.
10. Fuchs, op. cit.
11. Odin W. Anderson and George Rosen, M.D., "An Examination of the Concept of Preventive Medicine," Health Information Foundation, Research Series 12, 1955.
12. Helen Hershfield Avnet, "Physician Service Patterns and Illness Rates," Group Health Insurance, Inc., New York, 1967.
13. HEW, National Center for Health Statistics, "Cooperation in Health Examination Surveys," 1965.
14. J. M. Mackintosh, *Trends in Opinion About the Public Health* (London: Oxford University Press, 1953).
15. Seymour E. Harris, *The Economics of American Medicine* (New York: The Macmillan Company, 1964).
16. Fuchs, op. cit.
17. Walter E. Book, *An Analysis of the Multi-Test Clinic of Richmond, Virginia,* Health Information Foundation, 1951.
18. Commission on Chronic Illness, op. cit.
19. Goldie Krantz, "A Multiphasic Health Program: A Supplement to Health and Welfare Plans," paper presented before the American Public Health Association, November 15, 1961.
20. Karl Pickard, "Quality of Medical Care in HIP," paper presented before the Group Health Institute, May 27, 1959.
21. Kerr L. White in *Scientific American,* August 1975.

11

The Potential

Today, when health expenditures in the United States exceed $100 billion a year, and when there is such public concern over the constant escalation in the costs of medical care, it is not surprising that interest in preventive care should accelerate, particularly in view of the assumptions of cost savings.

That these assumptions, on a broad population basis, can be subject to question is perhaps evident from the fact that such a hypothesis played a role in the formulation of the British National Health Service (NHS). Following a series of developments over several years, it was the Beveridge Report in 1942 that is often considered as having paved the way for the enactment of the British National Health Service Act in 1946. In that report it was stated that the "ideal plan" would be "a health service providing full preventive and curative treatment of every kind to every citizen without exception, without remuneration limit and without an economic barrier at any point."[1] The report estimated the cost of such a comprehensive service to be provided to everyone, practically without charge (other than taxes). Interestingly, the report reasoned that the annual costs would be unchanged by 1965 because the increases in expenditures due to the development of the service would be offset by the improvement in general health resulting from the operation of the service. In

1943 the Churchill Coalition Government accepted the Beveridge Report in principle. After the Labour Party came into power in 1945, its Minister of Health, Aneurin Bevan, caused to be introduced what was known as the Bevan Act. The original principles, and the original cost assumptions, went unchallenged. The proposal was passed by Parliament in 1946 as the National Health Service Act, which became effective in 1948.

Experience has borne out the unpleasant fact that the original cost assumptions were grossly erroneous. Costs were immediately twice those anticipated and have continued to increase precipitously. As Odin W. Anderson has recently noted in his study of the health systems in England, Sweden, and the United States, under each system, having marked differences from the others, the costs of health services are increasing faster than the expenditures for other goods and the gross national product.[2]

In recent years the NHS has been bogged down by financial and other difficulties. It is evident, then, that making universally available under a national program the services for preventive care and curative treatment without charge at the time of use does not bear out the hypothesis that such availability will reduce, or even stabilize, health care expenditures.

Also of interest is the statement made in June 1975 by the U.S. Department of Health, Education, and Welfare in preparing its "Forward Plan for Health" for the fiscal years 1977 to 1981: "It has become clear that only by preventing disease from occurring, rather than treating it later, can we hope to achieve any major improvement in the Nation's health." The HEW report notes, however: "While there is much that can be done to prevent diseases and death, there is still much about which we are uncertain. Some preventive approaches are effective and others are either too costly or the benefits obtained are unclear."

In passing, and to place the subject in fuller perspective without by any means serving to sell short the possibilities that should be inherent in preventive medicine, a statement by the World Bank is worth noting: "Secular increases in health standards in Western Europe and North America were brought about much more by rising living standards and improving socioeconomic conditions than by medical care per se."[3]

Regardless of any caveats, however, important gains in health and welfare can be made through preventive medicine and are not to be negated simply because their reported potential might, in some instances, appear to overstate the case. The dimensions of the potentials of preventive medicine in American society today are very considerable. The following material will serve to illustrate the point, as well as to suggest areas where future progress might be made.

Heart disease. The diagnostic techniques of heart disease have improved very considerably over the past 30 years. With newer techniques and the early identification of these diseases, the prospects of reducing mortality and morbidity appear to be good. High-risk persons can now be identified, principally those with deleterious lifestyles, high blood pressure, diabetes, diseases associated with certain ethnic backgrounds, and hereditary conditions. Most of these conditions can be controlled in the interest of prevention. Accordingly, it has been estimated that the reduction of these risk factors could reduce the incidence of heart diseases by 50 percent.

The President's Commission on Heart Disease, Cancer, and Stroke said over a decade ago that "the prospects are excellent for reducing the toll of heart disease in the years immediately ahead." Noting the advancements made in recent years, the report nonetheless recognized that "the challenges are many and formidable . . . substantial reduction of the toll of heart disease awaits a major nationwide effort to apply what is already known."[4]

One study of the subject, however, stated: "The current position can be summarized succinctly by stating that there is no conclusive evidence yet available that control of any of the risk factors has been followed by reduction in the incidence of coronary heart diseases whether on a secondary or primary preventive basis." The study states elsewhere that "screening tests and more complex investigations for the early detection of ischaemic heart disease . . . can be relied upon to identify less than half of those who will develop the disease in the foreseeable future," and concludes that "these limitations, and the fact that it has not yet been shown that myocardial infarction can be prevented from developing in healthy people or from recurring . . . , indicate that surveys to uncover occult ischaemic heart disease should be deferred."[5]

Hypertension. Today much can be done to control hypertension through early detection, therapy, relaxation, control of diet, obesity, and smoking, and the prescription of a wide array of drugs. Significant reductions in premature death and disability appear to be possible. In fact, the death rate from hypertension has already been reduced 50 percent in the past decade.

The risk of developing hypertension is related to family history. It can cause stroke, heart disease, and kidney failure. Unfortunately many, perhaps half, of the hypertensives are not aware of their condition. Programs are needed to correct this situation by bringing about the early detection of the condition, making necessary diet changes, and prescribing drug treatment. Through such an approach it should be possible to further reduce the mortality and morbidity toll of hypertension quite considerably.

Stroke. Stroke-prone people can usually be detected by the presence of certain apparent symptoms. This points to the importance of early detection and to a planned course of treatment. About 80 percent of deaths from stroke occur in persons over age 65, and about 40 percent of these are over age 75. About 80 percent of stroke victims survive the acute initial phase of the disease but most live in a seriously disabled state. It is said that most of this toll could have been obviated by the timely application of preventive or rehabilitative treatment. Stroke is proving to be neither inevitable nor irremedial.

Cancer. Progress has been made in reducing mortality from cancer in certain sites: the stomach, prostate, digestive organs, and uterus, and from leukemia. On the other hand, the mortality rate from lung cancer has been increasing, taking 84,000 lives in 1976. In some instances it is estimated that the death toll from cancer could be considerably reduced through early detection and treatment, particularly in cases of cancer of the breast, the uterus, and the prostate. The same might be so with respect to cancer of the colon and rectum. The survival rate today for those stricken with cancer of all types is one in three. It is estimated that this could be increased to one in two by early detection and treatment. It has also been estimated that if smoking were eliminated from our mores, lung cancer mortality could be reduced at least 60 percent, saving 20,000 lives a year. Lung cancer is difficult to diagnose in time for cure, however, with the result that only about 10 percent of the detected cases are saved.

The American Cancer Society has said that some cancer can be prevented, but not all. According to the Society, most lung cancers are caused by cigarette smoking, and most skin cancers by frequent overexposure to sunlight. These can be prevented by avoiding their causes. Also, certain cancers caused by occupational factors—particularly bladder cancer in the dye industry—have been prevented by eliminating the causative agents. The immediate goal of cancer control in this country is the annual saving of 330,000 lives, or half of those who develop cancer each year.

Cancer of the colon-rectum, which strikes almost 100,000 Americans each year and kills about 50,000, can be detected early as part of a health checkup. When detected early it can be cured. It is estimated by the American Cancer Society that a procedure undergone regularly by all persons over age 40 might help save more lives from cancer than any other step in the health checkup. Yet only about 15 percent of people have this test (proctosigmoidoscopy) each year.

Deaths from uterine cancer have decreased 65 percent over the past 40 years because of the Pap test. With early detection the chances of cure are just about 100 percent. However, each year only about 15 percent of women over age 21 have this test. The American Cancer Society says that deaths from uterine cancer could be dramatically reduced.

Current therapy is highly effective when breast cancer is discovered early and in its localized stages. Today the five-year survival rate is 84 percent when this is done through physical examination or self-examination; yet in 1976 deaths from this cause were 33,000.

Each year some 8,000 Americans die as a result of oral cancer. Yet this form of cancer is easily observable. Unfortunately by the time of diagnosis 50 percent of oral cancers have metastasized as a result of neglect, and of these the survival rate is only 30 percent. Progress would appear to lie in public education leading to early detection and treatment.

There are some perplexing aspects, however, as respects the prevention of cancer or of reducing its mortality or morbidity toll. For example, the incidence of stomach cancer has declined markedly in the United States, but no one knows why. The question of multiple causative factors also calls for clarification. One such factor is excessive cigarette smoking, yet the question remains whether this alone is the causative factor when some 20 percent of cases of lung cancer occur among nonsmokers or whether other factors such as air pollution, occupational hazards, automobile driving, urban living, or lack of fresh air and exercise combine with excessive smoking to produce the cancer. Why cancer of the liver is common in Indonesia and parts of Africa and Asia and rare in other regions, why cancer of the stomach is more frequent in Iceland, Scandinavia, and Japan, and why breast cancer is much less frequent in Japan are questions to which the answers are not known.

Respiratory diseases. Tremendous gains have been made in controlling the incidence and the effects of the more serious respiratory diseases, particularly tuberculosis and pneumonia. Influenza remains a problem, principally because of the changing nature of the virus. The maintenance of good general health, the avoidance of contagion, more healthful lifestyles, improved working conditions, and the control of harmful air pollution could all play a role in further reducing the toll taken by respiratory diseases. For the chronic respiratory diseases, primary prevention could be accomplished by reduction in exposure to air pollutants; secondary prevention could be accomplished through the application of screening tests and other forms of early detection.

Deaths from bronchial diseases have been rising sharply in the United States, today taking some 10,000 lives each year. Of chronic bronchitis it has been said that the most important means for prevention and treatment is the avoidance of tobacco smoke. Of emphysema, which has been increasing in incidence, it is said that the etiology of the disease is unknown but that a large majority of the cases also have chronic bronchitis and that this points to a contribution by excessive cigarette smoking. Of bronchiectasis it is said that while there are many causes, not all of them are known, and that therefore there is no simple way of preventing the condition.

The relationship of excessive cigarette smoking to chronic respiratory diseases has also been noted in England when it is said of chronic bronchitis and emphysema that after diagnosis "the only method likely to arrest development of disability is to persuade subjects with early evidence of airways obstruction to stop smoking."[6]

Mortality from pneumococcal pneumonia has been reduced, especially among people in their productive years, by early treatment with antimicrobials. Prevention lies in the promotion of better general health, avoidance of fatigue and overindulgence, and good housing. Bronchopneumonia is treated with penicillin and other approaches to the basic cause. Prevention is through the avoidance of infection. Pulmonary tuberculosis, today uncommon, is prevented through healthful living conditions and avoidance of infection, as well as early detection through chest X-rays. It has been estimated that accelerated research should be able to avert 7,000 deaths yearly. Prevention of influenza is said to be best accomplished through the avoidance of infection and, to a certain extent, with vaccines. The common cold can only be prevented by avoidance of infection. Fungus infections of the respiratory system are exceedingly common and one form of such infections, histoplasmosis, is usually not subject to prevention on the basis of present knowledge.

Pneumoconiosis is a general term for respiratory diseases caused by dust. One such disease is silicosis, which usually results from the inhalation of silica dust. Anthracosilicosis results from coal dust. Asbestosis results from inhalation of asbestos fibers. Berylliosis is caused by beryllium dust. All these diseases are essentially the result of uncontrolled occupational hazards. Their prevention lies in sound industrial health measures.

Diabetes. While there is no cure for diabetes, the disease can easily be detected and brought under control. A great many diabetics, perhaps 50 percent, do not know they have the disease. This points to the need for early detection and treatment. The American Diabetes Association maintains that if either the hereditary factor or obesity could be controlled, diabetes in many people could be prevented or postponed. In any instance, public information should make possible a much sharper reduction in the deaths and disabilities that continue to result from this disease.

Urinary tract infections. Urinary tract infections are probably the commonest bacterial infection. These infections have a high spontaneous cure rate, perhaps in a magnitude of 40 percent, and treatment results in cure in some 80 percent of the remaining cases. There has been, however, an increase in the number of deaths resulting from infections of the kidney. Diagnosis of urinary tract infections in a latent stage is now possible at reasonable cost, and screening is relatively simple. While prevention is preferable to early diagnosis, the only practical means is the avoidance of unnecessary instrumentation of the urinary tract. Research, however, may widen the scope of primary prevention.

Arthritis and rheumatic diseases. These diseases are not curable except that many cases result in a spontaneous cure. Early detection and treatment can at times avoid serious crippling effects, and it has been estimated by the Arthritis Foundation that there is a potential of adding from one to five years of productivity for 936,000 arthritics.

Communicable diseases. Constant vigilance is required to bring about the total eradication of many of the communicable diseases. The problem is essentially one of public complacence and apathy. This is evidenced by the decline in immunization in recent years. The means of prevention, even eradication, are at hand. What is needed is public education on a continuing basis, with emphasis on poverty, low-income, and ethnic groups.

Venereal disease. It has been shown that the incidence of venereal disease has been increasing in the United States in recent years. While some progress has been made in reducing the toll of venereal disease, eradication of the disease should, at least theoretically, be possible. Public awareness is obviously lacking.

The first attempt at controlling venereal disease was in 1918 when Congress enacted a law creating the Division of Venereal Disease Control of the U.S. Public Health Service. Later, premarital and prenatal examinations were made compulsory in some of the states. It was during World War II, however, that real progress was made, with therapy available on an outpatient basis by physicians and clinics. Social taboos, however, and reluctance on the part of those infected to seek treatment, provide real drawbacks to what should be a preventable disease.

The following steps to bring about early detection and prevention have been proposed:

> Routine serological testing in high schools, trade schools, and colleges.
> Routine serological tests as a condition of employment.
> Routine serological tests for applicants for marriage licenses.
> Routine serological tests for persons admitted to hospitals.

Mental retardation. It has been estimated that the incidence of mental retardation could be reduced 30 to 50 percent, and that mongolism could be eliminated by judicious use of the rubella vaccine. Other approaches to the prevention of mental retardation include sound maternal care, the control of diabetes, accident prevention, and the control of venereal disease. Total elimination of retardation, however, is not possible with available knowledge. More research is quite obviously dictated.[7]

Alcoholism. The means for controlling alcoholism are becoming increasingly available, particularly in places of employment. The means for preventing the development of alcoholism also exist if they are conscientiously employed by social groups, business groups, and by the individual. A sine qua non would seem to be public education on a continuing basis. In theory, it should be possible to eliminate the disease. In practice the toll taken by alcoholism can be reduced very considerably, and distinct evidences of this are now readily apparent.

Accidents. Accidents, almost by definition, should be reducible to a bare minimum. When the toll in premature deaths, disability, and lost work is noted, it becomes evident that the gains could be tremendous. As has been shown, there has been a considerable reduction in accidental deaths and disabilities in relatively recent years, particularly those occurring in the workplace and very recently in motor vehicle accidents. The progress made, however, has by no means approached the maximum, especially when one considers that about 95 percent of all accidents are preventable.

Prevention of accidents would appear to require different approaches to different kinds of people, giving due consideration to such differences as age, sex, educational level, and socioeconomic status. Different approaches are also called for in relation to the places where accidents occur. In many cases mechanical devices of all types require constant vigilance and research to make certain that engineering design, standards, construction, and safety devices eradicate all possible hazards and that the surrounding environmental factors (lighting, vibrations, ventilation, temperature, moisture, and noise control) are regulated to provide maximum safety. In other instances more stringent enforcement of laws and regulations can serve a valuable purpose. Overall, however, increased public awareness to hazards is a basic requirement.

Accidents that occur in or around the home call for special attention since they are so difficult to control, except on the part of the individual, since they can result from so many different causes, and since they are so numerous. They result from faulty construction, neglect in making repairs, neglect of all types of fire, burning, and explosion hazards, electrical shock, gas leakage, slippery floors, mechanical failures, glass doors, power mowers, ladders, firearms, sports participation, improperly protected poisons and medicines, swimming pools, toys, misplaced furniture, carelessness in the bathtub, or insect and animal bites.

Because of their high toll of human life and limb, motor vehicle accidents also warrant special attention. It has been speculated that deaths and serious injuries sustained in highway accidents could be reduced one-half in a decade provided that a scientific program to that end received the full cooperation of automobile manufacturers, legislators, law enforcement agencies, safety groups, scientists, and the general public. Since this speculation was made in 1960 it is all too evident from the available statistics that such a program did not come to pass.

There probably is no reason why, by the year 2000, the prevention of accidents cannot be as historically successful as the prevention of many diseases today. It is necessary to recognize, however, that such a transition will not happen by itself. It will require an intensive, multifaceted approach on a continuing basis.

Lifestyles. The correction of harmful lifestyles is a difficult area in which to produce positive results. Some gains have been made in relatively recent years, but a great deal remains to be known of the subject. That the potential for reducing the incidence of disease, accidents, and premature mortality appears to be great is readily evident. The question, basically, is how to accomplish the results.

One frequent fault in our lifestyles is our dietary habits. Adequately balanced diets not only engender better health but also are important in warding off disease. The U.S. Department of Agriculture estimates that improved nutrition could result in a 20 to 25 percent reduction in heart disease, a 10 percent reduction in mental health disabilities, 3 million fewer children with birth defects, a 50 percent reduction in infant deaths at birth, a 50 percent reduction in the incidence and severity of dental defects, and an improvement in the survival rate to age 65 from approximately 65 percent to 90 percent.

Inadequate diets, of course, can result from insufficient income, but they can be found in all economic strata of society as a result of ignorance of their importance, self-indulgence, and simple carelessness. One result can be obesity, with all of its harmful effects on health. Prevention would appear to be in continuous public education, in an awareness to the importance of diets, and in a wholesome skepticism of food fads, crash diets, and highly advertised solutions to obesity that approach quackery. The counsel of a physician is always helpful. For those with inadequate incomes, such programs as school lunches and food stamps can be helpful, but the program remains essentially one of education.

Insufficient exercise is another manifestation of an unhealthy lifestyle. Consistent, moderate exercise, principally walking, swimming, and mild gymnastics, can be a control against obesity, hardening of the arteries, stress, fatigue, and improper posture.

Functioning under stress may have a harmful effect on the heart and on blood pressure. It can result in accidents of all types. Prevention against such consequences can be found in adequate relaxation and rest, moderate exercise, vacations, and balancing work and play. Too many people function under undue stress. Most certainly this can be found at the managerial level in places of employment. It can be deadly. It can be controlled.

While excessive cigarette smoking might not of itself produce disease, it most certainly, as has been shown, is a contributing factor in the development of many diseases. Obviously, this matter can be prevented by reducing or stopping smoking. Yet the habit is difficult to break.

Many efforts are devoted to reduction of excessive smoking, including those of the American Heart Association, the American Cancer Society, the Consumers Union, "Smokenders," the American Health Foundation, the American Public Health Association, the American Lung Association, the National Tuberculosis and Respiratory Disease Association, state and county departments of health, the U.S. Public Health Service, and the National Clearinghouse for Smoking and Health. As a result of such health education activities some progress has been made in reducing the amount of excessive cigarette smoking. Parental example is also considered an important factor in its effect on younger people.

Other approaches have also been proposed in an effort to reduce excessive smoking. Included are regulation of the content of cigarettes and requirements that filters be used, the prohibition of smoking in public places, control of cigarette advertising, cautionary notices on packages, increased taxation of tobacco products, group therapy, and public education in the schools. It has not yet been proved how effective any of these approaches have been in the improvement of health levels.

Age-old enemies of health. One final potential for preventing disease and accidents lies in the correction of the age-old enemies of good health and well-being. The approaches to prevention here lie in the eradication of poverty, in improved housing, in bringing all educational levels to at least a reasonable minimum, in the eradication of pests and rodents, in the conquest of air and water pollution, and in teaching everyone the fundamentals of hygienic living. This, quite obviously, is an extremely broad area involving a great many different approaches. Yet its basic importance is evident from what has already been accomplished, as evidenced by the health gains shown earlier. In many of its aspects it ranges far afield from the health field per se, but it is fundamental to improvement in the health status of the nation. Involved are the combined forces of the health professions, civic leaders, voluntary agencies, foundations, educators, the churches, private business, and government at all levels. It is a formidable but worthy and humane task.

REFERENCES

1. Sir William Beveridge, "Report on Social Insurance and Allied Services" (London: His Majesty's Stationery Office, 1942).
2. Odin W. Anderson, *Health Care: Can There Be Equity?* (New York: John Wiley and Sons, 1972).
3. World Bank, *Health,* March 1975.
4. The President's Commission on Heart Disease, Cancer, and Stroke, "A National Program to Conquer Heart Disease, Cancer, and Stroke," 1964.
5. M. F. Oliver, M.D., "The Early Diagnosis of Ischaemic Heart Disease" (London: Office of Health Economics, 1965).
6. Cochrane and Fletcher, "The Early Diagnosis of Some Diseases of the Lung" (London: Office of Health Economics, 1965).
7. President's Committee on Mental Retardation, 1976.

Part IV

Industrial Health Programs

Industrial health programs are a significant aspect of preventive medicine. In a certain sense—and if some day they were to serve all employed people regardless of the type or size of the place of employment, and if the fullest concept of what an industrial health program can achieve were to be realized—industrial health programs could probably be the single most important aspect of preventive medicine.

Many would disagree with this statement, arguing that what will remain paramount is the broad scope of public health measures, or an approach to good health such as care of the well child. Certainly there is no intent here to detract from the important contributions of such efforts. But industrial health programs, in their broadest application, have two distinct and unique aspects. They are concerned, first, with adults in their productive years. And second, they function within a controlled setting, the word *controlled* being used in the scientific sense, in that they permit the creation of a healthful environment, they permit observation on a continuing basis, and they make possible an influence over an individual, or a group of individuals, that most usually does not occur in a more general setting. Thus it has been said: "American industrial plants represent a natural point of first contact with the health care system for the employed population. A health center at the workplace would have a very high potential for being used as the multipurpose primary care center for emphasizing health promotion and preventive medicine."[1] Dr. William Jend, Jr., of the Michigan Bell Telephone Company, also maintains that "the only concerted, planned, large-scale, personal, adult health maintenance programs in this country today are in the occupational health programs of American industry. The workplace is probably the ideal locale to practice real preventive medicine on a wide and effective scale."[2]

As a result of the tremendous progress made in this century, preventive medicine is now in a quite sophisticated stage. The next era of progress in health depends, perhaps more than anything else, on whether we can come to grips with the deleterious health habits of individuals. As has been shown, this is a far different matter from purifying a water system: It involves motivation and willpower on the part of people as separate entities in controlling their own destiny. No marked successes are yet evident in breaking through this difficult barrier. For example, although some 10 million Americans stopped smoking as a result of the surgeon-general's widely advertised campaign against it as carcinogenic, excessive smoking has increased worldwide, and starting at lower ages. It is not beyond the realm of possibility that the necessary breakthrough will come in places of employment. Certainly the progress being made today in some places of employment in coping with and controlling problem drinking and alcoholism gives evidence that such a possibility is not improbable.

It can be argued that no matter how effective industrial health programs are, it will be a very long time before all employed people, on a broad basis, will be reached. This, of course, is a perfectly valid point. One can only respond by noting that the trend is in that direction, that the broad view of industrial health programs is relatively very recent, and that such a goal should not be debunked. Many other attempts at preventive medicine have fallen far short of achieving their ultimate potential, and we must look in other directions for solutions.

Whether or not industrial health programs will ever be sufficiently effective to make a real difference rests essentially with the attitudes of both employers and employees. Today there are many forces at work that make it eminently possible and that should favorably influence the thinking of those who can bring it about. There are, first, the growing interest of the business community and organized labor, and the expanded views taken by the health and other professions. There are also support of many types of voluntary organizations and foundations, increased studies of the subject by universities and research agencies, the active interest of insurance or prepayment organizations, the helpful assistance of government agencies at all levels of government, and the considerations of legislators. Not to be overlooked are both the direct and the indirect effects that can flow from the Occupational Safety and Health Act of 1970.

REFERENCES

1. George James, "The Teaching of Prevention in Medical and Paramedical Education," *Inquiry,* vol. 8 (1970), no. 1.
2. William Jend, Jr., M.D., "Where Do We Want to Be in Occupational Medicine?" *Journal of Occupational Medicine,* July 1973.

12

The Concept

Originally, the concept of industrial medicine was limited essentially to occupational accidents and diseases. Unquestionably a compelling force was the enactment of workmen's compensation laws commencing in 1911. Prior to this, however, certain employers had made general provision for the health care of their employees and in some instances for their families, usually in such industries as railroading, lumbering, and mining where there was no ready access to medical care services or facilities.

Effective concern with industrial health in the United States, then, is of relatively recent origin. For example, the First National Conference on Industrial Diseases was not held until 1910. Between 1910 and 1920, however, industrial health became a more active subject of concern as employers, union leaders, insurers, economists, physicians, nurses, engineers, social scientists, and government began to take a greater interest. In part, these developments reflected the revolution that was taking place in the technology of industrial production and in the larger-scale organization of business and industry.

At the end of World War II, Bernhard J. Stern expressed apprehension that the specialization and refinement of industrial medicine, rather than evolving as an integrated approach to include the complete health status of workers, "has tended to sever industrial medicine from the mainstream of medicine and public health by dealing exclusively with hazards in the factory, shop, mill and mine, isolated from the worker's health in the community."[1] In 1960 a definition of occupational or industrial health by the American Medical Association was: "A program that embraces the principles of preventive medical care as provided by management to deal constructively with the health and safety of employees in accordance with their work environment and job description."

In between, many other developments were in progress that could only serve to improve the health status of working people. Child labor laws, shorter workdays and workweeks, a minimum wage, and improved working conditions all played a role. So, also, did the raised standard of living and improved educational levels.

In the course of these developments an awareness began to emerge that health in the workplace and health in the community at large were not separable entities, but rather that they reacted one upon the other. That very considerable gains could result from such a broadened concept came to be a matter beyond question. Today it is increasingly being recognized that it is difficult, it not impossible, to disassociate occupational health from the total environmental health of employed people. The employee spends some 40 hours a week on the job and 128 hours plus vacations and holidays away from the job. It was shown earlier that the hours spent away from work are, with relatively few exceptions, more hazardous to health than the hours spent on the job. Accidents that occur off the job are just as likely to result in disability and premature death as those that occur on the job, perhaps more so. Over and above the worry, anguish, and sorrow suffered by the individual and his family, there is the loss of well-being, temporarily or for a long period of time, and fear of the risk of job stagnation or the threat of unemployability. Economically, the loss of wages and medical expenses can be considerable, even disastrous. For the employer, the result is absenteeism, lost production, inefficiency, labor turnover, increased insurance costs, and frequently a lowering of employee morale. Employers have a very considerable investment in nonoccupational disability insurances and salary-continuance programs for their employees, and health insurance for their employees and their dependents. The costs of these programs unquestionably reflect the illnesses and accidents that occur away from the place of employment.

Such costs, even apart from a general concern for the health and welfare of employees, must broaden the horizon of industrial health programs. They make it evident that health is not divisible, that health hazards cannot be compartmentalized into occupational and nonoccupational hazards.

It is rather logical, then, to find today a trend in industrial health toward an expanded concept of total concern for the health of employed people. This trend is leading to companies' assuming some responsibility for employees' well-being, through counseling and the application of the techniques of preventive medicine regardless of the source of illnesses and accidents.

Additional weight and importance are given to this broad concept of industrial health programs by the fact that places of employment, both private and public, are increasingly called upon to employ the socially, economically, and educationally underprivileged, as well as the physically handicapped. This development, given impetus by the Equal Employment Opportunity Act, places new and at times difficult demands on employers. It may mean coping with employees who have a low level of health awareness and little knowledge of the health care complex and what it entails. It means that increasingly there will be employees with a backlog of unmet medical needs. It calls for the development of a familiarity with different ethnic cultures and subcultures. It requires a different approach to on-the-job education and training, and to the development of an awareness of the value of good health.

This changing concept, it might be noted, is not unique to the American scene. In the Scandinavian countries, and particularly Sweden, sharp distinctions are no longer made between occupational health and personal health.[2] The goal is to reduce the occurrence of premature death, the incidence of disability, absenteeism and wage loss, lost production, the utilization of and expenditures for health services, the need for early retirement, employee turnover, and the burden placed upon many types of publicly provided programs. The ultimate purpose is the improved well-being of individuals, families, and society as a whole.

The broadening concept in industrial health has been noted in very recent years by many persons outstanding in the field of preventive medicine. One of these, Dr. Lorin Kerr of the United Mine Workers, and a recent president of the American Public Health Association, has commented:

> The prevention and control of all job-related illnesses, injuries, and hazards, including those which are socially related to the worker and his family, can be achieved by expanding existing programs and developing and implementing new ones. First and foremost is the need to develop a personal health services delivery system in which the prevention, diagnosis, treatment, and rehabilitation of occupational illness and injury will be coordinated and integrated with all of the health services provided for the worker and his family.... The end of the isolation and fragmentation of occupational health is close at hand [and] for the first time "prevention," the hallmark of occupational health, becomes feasible.[3]

Dr. William Jend of the Michigan Bell Telephone Company has said: "Our responsibilities under the National Occupational Safety and Health Act add a vast new dimension to our medical responsibilities. The physical and mental health of our employees is a major consideration in the ability of any business to provide good service or a good product."[4]

Dr. John Foulger of Du Pont has said: "Management must be concerned with total health — there is no such thing as occupational or nonoccupational health . . . industrial medicine has the greatest potential for supervising total health."[5]

Dr. G. H. Collings, Jr. of the New York Telephone Company stated: "We, in occupational or industrial medicine, are involved with and to some degree responsible for the health of large semi-captive populations. In addition, our companies pay hundreds of millions of dollars annually to protect the worker against adverse health efforts on the job and to purchase medical care for employees and their families. We are, therefore, likely to visualize health care in its broadest connotations as it affects such populations." He then notes such in-house supplements to primary health care as the health education of employees, early diagnosis in the form of periodic health inventories or multiphasic screening programs, and health counseling of employees.[6]

That there is an absence of unanimity concerning this newer concept goes without saying. To wit, the American Medical Association, while stating that "organized medicine should exercise leadership in improving occupational health programs," maintains that "definitive diagnosis and therapy of nonoccupational injury or illness are not responsibilities of the employer, but he may provide certain preventive health measures. . . . Treatment of nonoccupational injury or illness never has been and is not ordinarily considered to be a routine responsibility of an occupational health program." Exceptions to this statement are (1) the care of emergency cases, (2) the provision of first aid or palliative treatment for minor disorders, and (3) the provision of care in cases where the employee cannot locate a personal physician or health services, and if requested by the employee. The AMA statement does take cognizance of the fact, however, that an industrial health program should have as one of its purposes the protection of "the general environment of the community."[7]

Regardless, it might be accepted as a trend of the future that, given a realization by employers and unions of the need for consideration of the total health of employed people, and given a sufficient supply of trained professional occupational health personnel, industrial health programs will move in the direction of a growing concern for the total health and safety of employed people both on and off their jobs, as well as for the health of the families of employees. This is a responsibility that must be accepted, including a concern for the newer processes of modern industry and the multiplicity of environmental disturbances that have created new health problems and that affect the worker not only on the job but as a member of his community. Many of these hazards, or potential hazards, have not been adequately evaluated to date. These developments and changes, furthermore, in such a dynamic society as the United States, occur with such rapidity that in some instances there is not even time to make a proper evaluation, which sometimes can require multiple decades of observation. Just as society expects industry to accept responsibility for pollution and its concomitant ills, possibly we can expect industry itself to expand the concept of industrial health, relating it to the community at large.

REFERENCES

1. B. J. Stern, *Medicine in Industry* (New York: The Commonwealth Fund, 1946).
2. Morris M. Joselow, "Occupational Health in the U.S.: The Next Decades," *American Journal of Public Health*, November 1973.
3. Lorin E. Kerr, M.D., "Occupational Health: A Discipline in Search of a Mission," *American Journal of Public Health*, May 1973.
4. William Jend, Jr., M.D., "Where Do We Want to Be in Occupational Medicine?" *Journal of Occupational Medicine*, July 1973.
5. National Health Council, *The Health of People Who Work*, New York, 1960.
6. G. H. Collings, Jr., M.D., "What Is Primary Care: An Occupational Health Perspective," paper presented before the New York Academy of Medicine, April 22, 1976.
7. American Medical Association, "Scope, Objectives, and Functions of Occupational Health Programs," Chicago, 1971.

13

The Aspects of Industrial Health Programs

If the industrial health programs of the future are to be broad enough to encompass the wide range of health problems that affect people both as workers and as public citizens, there will be four basic aspects, separate yet interwoven and interdependent and having a common purpose and goal.

1. The basic, focal concern is the prevention of occupational accidents and diseases to the greatest extent possible, and the prompt treatment of those that do occur.
2. A closely related concern is nonoccupational diseases and accidents among both employees and their dependents. Again, the purpose would be the prevention of diseases and accidents to the extent possible and, failing this, an assurance of prompt and adequate treatment.
3. To the extent that the health of the community at large is affected by the processes, practices, and products of industry itself, a responsibility rests with the industrial health program to reduce or eliminate any harmful effects on the health of the community.
4. A final important aspect of an industrial health program is the rehabilitation of a disabled employee to the maximum extent possible and, corollary to this, the employment of handicapped people, whether or not former employees.

THE ASPECTS OF INDUSTRIAL HEALTH PROGRAMS 161

A few words on each of these four aspects of an industrial health program are warranted as background for the discussions in later chapters.

OCCUPATIONAL ACCIDENTS AND DISEASES

The sine qua non of an industrial health program is, of course, the prevention and treatment of occupational accidents and diseases. Quite naturally, occupational hazards vary, at times very considerably, among different types and sizes of places of employment. The hazards in a steel mill, for example, differ appreciably from those in a bank. What is said here and subsequently, therefore, can only be couched in general terms, since each place of employment would present its own problems and its own solutions and approaches to those problems.

The dimensions, toll, and costs of occupational diseases and accidents have already been described. That these statistics are important to workers and to places of employment is readily evident. The resultant lost production, absenteeism, wage loss, labor turnover, insurance costs, and premature mortality are matters of very considerable economic significance to both workers and business management.

The National Safety Council has defined occupational injury as "any injury, such as a cut, fracture, sprain, or amputation, which results from a work accident or from exposure in the work environment." Occupational disease is defined as "any abnormal condition or disorder, other than one resulting from an occupational injury, caused by exposure to environmental factors associated with . . . employment. It includes acute and chronic illnesses or diseases which may be caused by inhalation, absorption, ingestion, or direct contact." Included are skin diseases and disorders, dust diseases of the lung (pneumoconioses), respiratory conditions due to toxic agents, poisoning (systemic effects of toxic materials), disorders due to physical agents, disorders due to repeated trauma (such as noise, motion, vibration, or pressure changes), and such occupational illnesses as anthrax, brucellosis, infectious hepatitis, malignant and benign tumors, and food poisoning.

The NSC has categorized occupational safety and health hazards in places of employment as:

- Workplace hazards, including faulty construction and layout, inadequate illumination and ventilation, unsanitary conditions and poor housekeeping, inadequate maintenance, fire and explosion hazards, and faulty storage.
- Machine and equipment hazards, including inadequate space, poor maintenance, inadequate lubrication, and absence of protective devices.

- Materials hazards, including not alone their use but also their handling and their storage.
- Power source hazards, which can differ among such sources as electric, pneumatic, hydraulic, and steam power.
- Operation hazards, involving such operations as welding, cutting, blasting, and trenching.
- Employee hazards, including the failure to use protective devices, improper use of machinery, incorrect methods of lifting and handling, and failure to use sound judgment.

Accidents result most generally from machine use, falls and slips, and materials handling. They also result from such things as cave-ins, inadequate tunneling and quarrying, fires, explosions, the use of motor vehicles, faulty machine or equipment maintenance, poor housekeeping, improper storage, or inadequate space or clearance. Contributing factors to the occurrence of accidents may be noise, faulty lighting, inadequate ventilation, plant construction and layout, inadequate employee selection and training, and general neglect on the part of management. Other contributing factors rest in the employee himself: age, ill health, fatigue, carelessness, improper use of machinery, failure to use protective devices, incorrect use of the body in doing the job, failure to use sound judgment, horseplay, or the use of alcohol or drugs.

Accidents happen to employees off the premises as well, particularly to drivers, delivery personnel, service personnel, salesmen, and employees traveling on business.

Regardless of the direct or indirect cause, it is the purpose and responsibility of an industrial health program to prevent such accidents to the greatest degree possible. The approaches to be taken are many, again depending upon the type of place of employment and the nature of the operations.

At the outset, it should be made certain that the construction and layout of the workplace cannot themselves become a cause of accidents and cannot induce accidents by any inherent faults in their planning. Lighting should be adequate and not glare-producing, ventilation should be sufficient to keep the air fresh, and temperatures should at all times be moderate. Machines, tools, and other work equipment should be designed, among other things, to protect the safety of the worker, should have all necessary safety guards and devices, and should at all times be maintained in good working order. Materials-handling accidents, which can result from the use of cranes, slings, ropes, chains, pulleys, hoists, conveyors, hand trucks, or manual operations, should be guarded against to the fullest extent possible, including instruction of the worker in safe body position and footage and correct methods for lifting, carrying, and gripping, as well as the provision of adequate space and clearance for such operations.

All due protection should be taken, also, against fires and explosions. The building itself should be fire-resistant and should be equipped with an automatic sprinkler system, sufficient protected fire exits and means of evacuation, an alarm system, hoses, and sufficient extinguishers. Electric wiring should be as safe as possible. The handling, packaging, and storage of fuel and all flammable materials, solvents, and chemicals should be treated with all possible precautions. Fire-inducing operations such as those involving open flames, hot surfaces, friction, portable heaters, spontaneous ignition, or static electricity should be protected in every way possible. Refuse and other flammable waste should receive adequate disposal. Smoking in the work areas should be restricted.

The indirect or even direct injuries that result from excessive noise, vibration, or repeated motion should be protected against by adequate machine design and maintenance, substituted equipment or materials, sound-absorbent construction, the use of mufflers, the isolation or enclosure of such operations to the degree possible, and the provision of properly designed and fitted ear muffs and ear plugs for the workers affected.

On a continuing basis there should be inspection of the work plant and its facilities and equipment and, where hazards have developed, corrective measures should be taken or the faulty equipment condemned. When accidents do occur they should be promptly investigated, and complete records should be maintained. The standards for such records developed by the American National Standards Institute have previously been set forth in some detail. Much can be learned about accident prevention in a particular place of employment from a study of such records.

It goes without saying that when accidents do occur first aid, emergency treatment, and needed medical care should be provided promptly, including immunization against tetanus where warranted.

Beyond this, an important part of accident prevention is the establishment of job standards and the application of these in the selection of new employees, careful training of the new employee in the use of machines, tools, and other equipment, and constant supervision of the employees' work methods. Above all, the employee should be taught an awareness of the potential hazards and of the necessity to observe all established safety precautions.

Occupational diseases can result from any number of industrial processes, again depending upon the potential hazards of a particular type of operation, unless due protective steps are taken. Included are the hazards inherent in the use of chemicals and toxic substances or materials (such as carbon monoxide, carbon disulphide, vinyl chloride, phosphorus, plutonium plastics, corrosives, solvents, and aerosols), harmful dusts, particulates, smoke, and fumes (such as coal dust, cotton dust, asbestos, silica, fibrous glass, rock wool, and various types of gases, mists, and vapors), certain metals (such as lead, beryllium, and mercury), radiation (from X-rays, lasers, and uranium), infections of various types (resulting from fungi, molds, bacteria, and insects), excessive noise and vibration (which can cause hearing impairment or loss, digestive disturbances, and nervous disorders), and extreme temperatures (causing respiratory ailments). It also cannot pass unnoted that by mid-1976 it was an open question, to be decided by the courts, whether an employer could be held responsible for the development of alcoholism in an employee, which might have resulted, implicitly or explicitly, from executing his duties.

In March 1976 a Ford Foundation study of occupational safety and health reported that "a significant proportion of heart disease, cancer and respiratory disease may stem from the industrial process. This includes white collar workers, not just blue collar workers." The study states that "the task is monumental. It involves redesigning technology in some cases, redesigning jobs in others."[1] Other diseases frequently identified as having been caused by the work process are hypertension, certain cancers, digestive disorders, skin diseases, liver ailments, nervous and mental disorders, and undue stress and tension.

A matter of particular concern today is the potential hazards of the newer materials, processes, and techniques being used by modern industrial establishments. With the growing emphasis on electronic and chemical operations, nuclear energy, and automation, it is possible that physiologic, pathologic, and psychologic health problems are developing, whose consequences are not yet apparent.

That there is considerable difference of opinion, to say nothing of open conflict, about many of the industrial disease hazards and their ultimate effects on either workers or the community at large is obvious: witness the citations and fines by government agencies, the contests of these in the courts, and the flow of reporting in the mass media. Unquestionably some of the charges made are based on flimsy evidence, if not pure speculation. Unquestionably the subject can easily become charged with emotion and lends itself to overstatement. And unquestionably the subject is, or is thought to be, newsworthy. To illustrate the degree of public interest in the subject, here are some headlines chosen at random from *The New York Times*.

"Black-Lung Benefits' Cost Rising: Further Aid Sought"—January 27, 1976
"As G.E. Goes, So Goes Economy of 2 Counties"—February 15, 1976

THE ASPECTS OF INDUSTRIAL HEALTH PROGRAMS

"U.S. Agency Urges a Drive to Bar Cancer by Screening Chemicals" — February 28, 1976
"Health Problems Traced to Jobs" — March 17, 1976
"Court Backs Curb on Gasoline Lead" — March 20, 1976
"Unions and Physicians Setting Up Detection Systems for Occupational Diseases" — March 27, 1976
"PCB Cleanup Cost Put at $20 Million" — April 26, 1976
"Factory Noise Is Union's Target" — May 2, 1976
"A Battery Plant and Lead Poisoning" — June 6, 1976
"New York City Defends Way It Disposes of Waste" — June 25, 1976
"Utilities Pressed by EPA to Burn More Coal as Fuel" — July 2, 1976
"Major Nuclear Power Producers Agree to Widen Exchange of Data on Wastes" — July 13, 1976
"EPA Tests Show Ill-Set Carburetors Are Polluting Air" — July 22, 1976
"Companies Use of Annual Physicals Grows" — July 31, 1976
"Jersey Moves to Restrict Chemical Waste Disposal" — August 14, 1976
"Fuel Evaporation of Cars to Be Cut" — August 23, 1976
"House Votes Ban on Output of PCBs Within 3 Years" — August 24, 1976
"Two Executives Acquitted in Kepone Case" — September 3, 1976

The frequency of these news items and the diversity of the areas of concern speak for themselves.

Such concerns on the part of public agencies, the news media, certain labor unions, and many employers should, over time, produce good results. Among other things, they should lead to much-needed additional study and research to more clearly identify the various potential hazards and determine the full extent of the suspected hazards. Until this has been accomplished, many aspects of the subject will continue to lack the sound scientific base that is necessary for the development of protective steps on which reasonable minds could come to agree. In the interim, unless care, good judgment, and freedom from emotionalism and loose talk prevail, society as a whole will suffer from plant shutdowns, worker layoffs, and the costly movement of industrial operations to another site, with economic and tax losses in the original community, damaged public relations for specific businesses, and in some instances added costs to the consumer in the marketplace.

A primary step in fulfilling the responsibility of industrial health programs to prevent occupationally induced illnesses and diseases is the identification of the hazard, its nature, and its effects. This can be done by empiric investigation and observation of the procedures, processes, and materials involved in a particular instance. It can be done by the maintenance, study, and evaluation of complete health reports for the workforce. It can be aided by consultation with chemists, engineers, industrial hygienists, professional organizations, and public agencies. The result should lead to the evaluation, control, correction, or elimination of any detected hazards to the health of workers. Where suspected hazards exist, careful surveillance is warranted in the form of periodic health surveys and periodic physical examinations of the workers likely to be affected. The worker can be given interim protection through the use of special clothing or protective blankets, and the operation can be isolated to the greatest extent possible.

Where disease hazards are detected, many preventive steps can and should be taken for the protection of the worker. Depending on the nature of the hazard, there may be substitution of materials, altering of the processes or engineering controls, enclosure or isolation of the process, or the issuance of protective equipment such as special clothing, gloves, or respirators. Certain hazards can be overcome by improved ventilation or by wetting or spraying. In other instances special containers for storage of materials are necessary. Where a radiation hazard is a possibility, both the equipment and the personnel should be monitored regularly, as is done by American Airlines, the Ohmart Corporation in Cincinnati, and Conam Inspection, Inc. in Tulsa, Oklahoma. Another good example of employee protection is the action of the General Motors Corporation in Canada in barring women of childbearing age from working at its battery plant because of the potential hazards of lead poisoning.

Beyond this, general considerations of health should be given the workplace. Such matters as sanitation, housekeeping, cleanliness, ventilation, water supply, waste disposal, vermin control, and food handling are important. Employee facilities (lunchrooms, washrooms, showers, toilets, change rooms, and lockers) should all be maintained free of disease-inducing influences. Personal cleanliness should be encouraged and rest periods should be given consideration. When evidence of disease does occur, prompt examination and treatment should be forthcoming.

Clearly, concern for occupational accidents and diseases alone is a formidable responsibility in many places of employment. Small wonder, then, that the other aspects of such programs not infrequently receive little, if any, attention.

NONOCCUPATIONAL DISEASES AND ACCIDENTS

An equally important part of an industrial health program is the prevention of nonoccupational diseases and accidents and, at least to a certain extent, their treatment or the encouragement of their treatment. There are differences of opinion on this subject, but its importance is increasingly becoming recognized. The effect of nonoccupational diseases on absenteeism, wage loss, lost production, labor turnover, and the costs of health insurance, salary continuance programs, disability insurance, life insurance, and health and welfare funds are today matters of very primary concern.

Nonoccupational diseases would include such chronic illnesses as diabetes, arthritis and rheumatism, gout, bronchitis, pneumonia, and heart and circulatory diseases. The common cold and digestive upsets are factors of considerable significance. A far-reaching influenza epidemic can be extremely costly. Nervous and mental conditions can take a greater toll than is often recognized. Conditions resulting from obesity or malnutrition, excessive smoking, inadequate rest and recreation, insufficient exercise, or the abuse of alcohol or drugs can have long-range, costly results. Accidents occurring in or about the home, happening in public places, involving the use of motor vehicles, or resulting from recreational activities have an effect on places of employment that is constantly apparent. The Eastman Kodak study, which found that nonoccupational causes resulted in 60 days of absenteeism each year for every one day of absenteeism due to occupational causes, clearly shows the costliness of nonoccupational illnesses and accidents.

A report by The Conference Board stated that "business has compelling reasons to be interested in developing national debate over health care. It has been paying a considerable portion of the nation's health bill through employee benefits programs. . . . The increasing inadequacy of the health services available to employees in many communities imposes an indirect burden on business costs and efficiency." The report also notes that "while in-house medical departments have generally been established by companies principally to serve occupational health objectives, many firms also have been providing employees some measure of preventive, diagnostic, or treatment services for nonoccupational conditions as well. . . . Such nonoccupational health services can reduce their medical insurance costs."[2]

The broad scope of such nonoccupational health services would include immunization against infectious and communicable diseases, health education, health screening and physical examinations, combating deleterious lifestyles, control programs for alcoholism and drug addiction, accident prevention, and such public health efforts as the control of water and air pollution, improved sanitation and waste disposal, and improved housing. That there are inhibiting factors to such efforts has already been discussed, but these are sometimes more easily overcome in places of employment than in the community as a whole.

The fulfillment of such broad industrial health responsibilities compels approaches from within the place of employment itself *and* outside, in the communities where the employees live. Within the place of employment, preemployment examinations constitute a first step to assure the employer of reasonably healthy people. Beyond that, periodic screening or physical examinations can do much to detect the onset of disease and to bring about its early treatment. Medical treatment can be made available for nonoccupational illnesses and accidents, and although the extent of such treatment is a subject of dispute, there should be no argument against the treatment of minor conditions and, for more serious conditions, health counseling and advice concerning needed treatment. Immunization can also be made available, particularly at times of threatened epidemics, but also as protection against influenza, tetanus, smallpox, rubella, or infectious hepatitis. Physical fitness can be both encouraged and promoted.

Accident hazards away from work, their nature and places of occurrence, and preventive measures should be detailed for the information of employees. A good example of what can be done is a report in the June 1975 issue of *National Safety News* on the training Union Carbide employees receive in fire prevention and the use of home extinguishers; and the action of International Harvester at its Melrose Park, Illinois, plant selling home extinguishers to its employees at less than cost.

Equally important is the inculcation of health awareness in employees and a continuing dissemination of health education information on the nature of various diseases, their symptoms, and what can be done about them. Emphasis should be placed on the consequences of such matters as obesity, inadequate rest and exercise, exessive smoking, and the abuse of alcohol and drugs. Full and adequate records of all nonoccupational illnesses and accidents should be maintained on a continuing basis. Much can be learned from such records, which can point the way to more effective approaches. The American National Standards Institute and the National Safety Council are two organizations that provide forms for the types of records that should be maintained.

Beyond the workplace itself, an industrial health program can do many things to reduce or prevent illnesses and accidents among its employees. Most certainly coordination should be established with the health professionals and facilities in the community and with the many voluntary and public agencies at all levels of government concerned with accident and disease prevention and the improvement of health conditions. In all instances full and active cooperation should be offered, since much help can be received in return. Dr. Robert Hilker of the Illinois Bell Telephone Company has aptly commented that an industrial health program should be actively concerned with the encouragement of needed health facilities and manpower, including neighborhood health centers;

that it should work cooperatively with medical societies, hospital boards, health planning agencies, and all efforts aimed at cost controls for medical care; that it should consider the establishment or support of health-maintenance organizations (HMOs) and community mental health centers; that it should, where feasible, consider the establishment of a health unit as part of its industrial park which could serve as a health center for the community; and that it should actively advance health education in the community. By so doing, he maintains, not only are the health interests of the community served, including those of employees and their families, but employee relationships are improved and good customer and public relationships attained.[3]

There are many influences within a community that have a direct, if not always obvious, effect on the health of employed people. Included are educational levels, housing conditions, sanitation and sewage disposal, the nature of the community water supply, the amount of air pollution, pest control measures, the control of infectious and communicable diseases, the availability of recreational facilities, highway construction, traffic control, highway speed limits, and most certainly the availability of medical care facilities. People concerned with industrial health programs are remiss if they do not have an active involvement in all such matters. The relationship to the health and welfare of employees is too direct to be ignored.

THE COMMUNITY AT LARGE

A third aspect of an industrial health program, and one closely related to the foregoing, is the acceptance of some degree of responsibility for the health of the community at large. This responsibility might be looked upon as having three segments: (1) ensuring the safety of persons while on the premises of the place of employment or while having contact with employees, (2) protecting the users of products or services offered by the employer, and (3) protecting the surrounding community from harmful consequences which might result from industrial processes.

Ensuring the safety of persons who come into contact with employees is largely a matter of accident prevention and involves, among other things, the cost of public liability insurance as well as public goodwill. All steps should be taken to ensure the safety of persons entering the premises of a workplace or its adjacent property, including sidewalks and parking areas. Attention must be paid to such potential hazards as glass doors, loading ramps, and cellar doors, as customers patronize restaurants, retail stores, banks, and hotels. Other persons affected include patients in doctors' offices, hospitals, and clinics; salesmen and job seekers;

guests or business associates; tenants; and disinterested pedestrians. Potential hazards away from the premises must also be guarded against, particularly the motor vehicles driven by employees in the course of their duties and those inherent in the acts of employees while providing services or making deliveries. An industrial health program should have an involvement with all safety measures, including the adequate education and training of all employees.

A second area of responsibility for an industrial health program is the *safety of the products* made available to the public: the contents, additives, and preservatives of foodstuffs, as well as their refrigeration, storage, and freshness; the contents and packaging of drugs, cosmetics, and other sundry items, including the means of protection against use by children and any necessary warning labels; the materials and hazards inherent in certain clothing and household goods; the safe and healthful performance of many kinds of machines from automobiles and power mowers to ovens and heating devices in the home; the nature of pesticides; and the safe packaging, including any necessary warnings, of such items as aerosols, detergents, and bug killers. The National Commission on Product Safety (NCPS), for example, has estimated that 20 million people are injured each year as a result of incidents connected with household consumer products, and that of these 30,000 result in loss of life and 110,000 in permanent disability, at a cost of $5.5 billion to the nation. NCPS estimates that 20 percent of the injuries could have been prevented had adequate safeguards been taken in the manufacture or packaging of these goods. With the rising interest in consumer affairs, the subject of product safety will become increasingly important to corporate interests.

A third area of concern for an industrial health program is *protecting the surrounding community* from any harmful consequences of industrial processes. As in the case of occupation health hazards, the subject is fraught with differences of opinion and open conflict, with charges sometimes based on flimsy evidence or even pure speculation and hard fact buried within emotionalism. As has been shown, the subject attracts considerable publicity, which might result in good — in the form of the correction of certain health hazards — or ill — in the form of economic penalties to places of business, employees and their families, or entire communities (such as plant shutdowns, layoffs, unemployment, or movement of a plant to another location). More study and research are needed to reduce the questions to scientific realities.

Regardless of the difficulties and disagreements, the elimination of potential hazards to the community should be an important facet of concern to an industrial health program.

The U.S. Environmental Protection Agency has said in a publication that "the increase in chronic and degenerative diseases is due in part at least, and probably in a very large part, to the environmental and behavioral changes that have resulted from industrialization and urbanization. The modern environment is dangerous on two counts: It contains elements that are outright noxious; it changes so rapidly that man cannot make fast enough the proper adaptive responses to it.[4]

In the Policy Statement on Environmental Pollution adopted by the U.S. Chamber of Commerce in 1970, it was noted that "pollution of our environment is the collective responsibility of all elements of society," including the public as individuals, industry, and government. With respect to the responsibility of industry, the policy states: "Industry should acknowledge a sense of stewardship for the natural resources upon which our environment depends—air, land, and water. . . . Industry has an obligation to recognize the impact of a growing population and its concentration. . . . Industry should assume leadership in jointly developing information from within the industrial community on which sound decisions can be based. . . . It is essential that industry commit the technical and financial resources and talent needed to implement achievable improvement in our environment, as well as to undertake the basic and applied research programs that will provide for the continuing development of new concepts, methods, and technology for managing the quality of our air, land, and water."

President Nixon, in his health message to the Congress in 1972, emphasized the environmental impact on health and noted that "the burgeoning modern industrial society threatens to tip the balance of man's normal adaptability to pollutants the other way."

There are certain aspects of these concerns that merit a brief discussion beyond what has been noted in earlier chapters.

Population and industrial expansion since the 1920s has vastly increased the amount and the dangers of *water pollution* to the point where in 1973 one industrial plant poured six million gallons of waste water, which included 600,000 gallons of sulphuric acid, into the Savannah River every day, and a paper bag manufacturer in the area dumped 37 million gallons of waste water into the river daily.[5] By 1975, 79 cities in the United States were found by the Environmental Protection Agency to have drinking water pollution with traces of organic chemicals, including some suspected of being carcinogenic (chloroform, carbon tetrachloride, pesticide deldrin, vinyl chloride).[6]

The American Medical Association has identified the following as sources of water pollution:

Municipal sewage wastes.
Agricultural wastes.
Recreation and navigation.
Air pollutants.
Industrial wastes from more than 300,000 water-using factories principally involved with the manufacture or processing of paper, organic chemicals, petroleum, and steel and involving such pollutants as chemicals, oil and grease, toxic materials, radioactive substances, and mine acids.

The Manufacturing Chemists Association has identified eight general categories of water pollutants as:

Organic—from human, animal, and plant life.
Disease-causing wastes—from humans, animals, and certain manufacturing processes, especially meat packing and tanning operations.
Plant nutrients—such as nitrogen and phosphorus compounds.
Toxic substances—in the form of synthetic organic compounds (such as solvents, cleaning agents, and bug killers), inorganic substances (such as insecticides, mercury, mineral salts, and acids), and arsenic, barium, cadmium, cyanide, lead, nitrates, selenium, and silver.
Persistent substances—which are not subject to biological breakdown and which can emanate from rainfall runoff, seepage, agricultural practices, and certain industrial operations.
Sediments—particles of soil, sand, and minerals.
Radioactive substances—resulting from mining and processing radioactive ores and the use of radioactive materials for industrial, medical, and research purposes, as well as nuclear fallout.
Heat—from power and manufacturing plants using water, which speeds up the rate of biological degradation.

The harmful effects of polluted water on humans include typhoid fever, cholera, bacillary dysentary, amebic dysentary, infectious hepatitis, polio, poisoning, hypertension, and bone neoplasm. Additionally, polluted water has adverse effects on fishes, shellfish, and other forms of marine life, which, in turn, are eaten by humans. The consequences may range from temporary discomfort, to illness, disability, and even death.

There are essentially three forms of controlling water pollution:

Primary treatment—the physical separation of undissolved solid materials through screens, followed by a settling process. This method removes only 25 to 30 percent of the contaminants and is considered inadequate for most needs today.

Secondary treatment—the removal of organic matter by making use of the bacteria in it to decompose the organic matter. Trickling filters or a process called "activated sludge" are used in this form. The results are the removal of about 90 percent of organic matter.

Tertiary treatment—involves a variety of processes based upon specific needs. These include chemical oxidation, coagulation-sedimentation, absorption, desalination, ion exchange, electrodialysis, reverse osmosis, and distillation. Such approaches are 99 percent effective in removing pollutants from the water.

Industrial establishments can determine which approach is the most effective for their particular type of waste discharges. Consideration must always be given to the water source of the community and how the treatment may affect it. The Manufacturing Chemists Association has prepared information on the disposal of hazardous waste and has recommended safe practices and procedures. It has also prepared materials on the solid-waste management of plastics.

Air contamination and pollution arises from many sources: natural decomposition, forest fires, volcanoes, trash burning, garbage disposal, incineration, motor vehicle operation, agricultural burning, fertilization, manure, insect control, boats, diesel engines, and airplanes, as well as the processes of industry. The contaminants are many, including carbon monoxide, oxides of nitrogen and sulfur, hydrocarbons, particulates, radioactive materials, arsenic, asbestos, and a variety of chemicals, vapors, gases, and metals, including berylium, cadmium, chromium, lead, mercury, nickel, fluorides, aldehydes, ethylene, inorganic fluorides, iron oxide, lead, phenol and cresol, natural gas, fuel oil, and mineral acids.

Dependent upon the type and amount of air pollution, the adverse health effects can be acute respiratory diseases, chronic bronchitis, impaired vision, heart disease, and cancer of the liver, lung, bladder, stomach, pancreas, prostate, or mouth. Accidents also occur as a result of air pollution. The effects of air pollution can be extremely serious for a person who smokes cigarettes excessively.

That there are causes for concern is made evident by several recent studies. A survey by the National Cancer Institute revealed that between 1950 and 1969 the death rates among males from cancer of the liver, lung, and bladder was higher in the 139 counties in which the chemical industry is most highly concentrated. Residents of communities where copper, lead, or zinc smelters were situated also had above-average death rates from lung cancer. Arsenic spewed into the air by the smelting process was considered a causative factor.[7] Another study showed that males living in the vicinity of an arsenical plant are four times more likely to die of lung cancer than males living farther away. Deaths from cancer of the stomach, mouth, pancreas, and prostate were also higher. The plant was a chemical company in Baltimore that made pesticides.[8]

In Virginia in 1976 sharp concern was expressed over the contaminated dust emitted from a chemical plant which produced Kepone, a highly toxic pesticide, although admittedly little was known about the qualities of the chemical. It was suspected that Kepone may cause liver cancer, sterility, and neurological disorders.[9] In Grants, New Mexico, in 1976 there was concern by the EPA over wind-carried radioactive dust emanating from 20 uranium mines and 3 processing mills in the area, which conceivably could cause lung cancer after a latent period of 10 to 15 years.[10] Studies made in the Salt Lake Basin, the Rocky Mountain Areas, Chicago, and New York have reported excessive acute respiratory diseases in communities where the air is heavily polluted with sulfur dioxide and suspended sulfates.[11]

With respect to the effect on materials alone, the American Industrial Hygiene Association has noted that the economic losses resulting from air pollution include corrosion of metals, damage to fabrics, soiled clothing, leather decay, the dissolving of paints, damage to marble and to limestone building materials, and shortening the life of rubber tires. Costs have been estimated to range from $2 billion to $12 billion annually. Despite the broad margin for error, there is clear indication of the great economic property losses that result from air pollution.[12]

The U.S. Chamber of Commerce has recommended that the business community should be actively involved in antipollution programs: they should determine the causes and sources of pollution and the seriousness of the problem, examine state and local ordinances that regulate pollution, function cooperatively with other interested community organizations (such as conservation groups, the PTA, the Y's, civic and fraternal groups, youth organizations, the bar association, labor unions, the news media, and government at all levels), evaluate the capabilities of government agencies in dealing with the problem, examine the costs of pollution abatement, and then define a course of action.[13]

The correction of air pollution obviously lies in many directions, since the causes and contributors are so varied. Changes may be called for in agricultural processes. Individuals may be called upon to become more conscious of the consequences of their acts. Where industrial operations are determined to be the cause, altered processes, substituted materials or chemicals, and revised methods for handling the pollutants, including the use of filters or wet scrubbing, may be required. Of basic importance is that every industrial plant with a potential air pollution problem should make a realistic appraisal of its emissions, should designate staff responsibility for any needed corrective measures, and should

THE ASPECTS OF INDUSTRIAL HEALTH PROGRAMS   175

diligently investigate all complaints. It should then evaluate its control program on a continuing basis.

Closely allied to the subject of air pollution is creation of *noxious odors* by industrial processes such as fertilizing, tanning, brewing, ore smelting, petroleum refining, food or rubber processing, and the manufacture of chemicals, textiles, paint and varnish, plastics, soap and detergents, paper, and pharmaceuticals. Noxious odors are annoying and cause public resentment, but seldom present a health hazard. They can, however, cause nausea, headache, and loss of sleep and appetite.

Abatement methods for odor control depend, naturally, upon the industrial processes and materials involved. They include complete combustion for odorous gases, ventilation to dilute the gas with air, absorption to assimilate the gas with liquids, masking by substituting another odor, counteraction with another odor, dilution with odor-free air, chemical counteraction, the removal of particulates that create odor, and elimination of the source by substituted processes or materials, if possible. Through one or a combination of these approaches, most noxious odors can be eliminated.

Noise pollution is another community concern whose source lies in industrial processes. It can result in hearing impairment, weight loss, sleeplessness, fatigue, and indigestion. It can contribute to the occurrence of accidents. The means for abatement have been discussed earlier in this chapter with respect to occupational health hazards.

A final community concern involving places of employment is the *conservation of fuel and energy*. The National Safety Council has cautioned that employers, in trying to conserve energy, are forced to balance that need against the indirect creation of safety and health hazards. For example, reduced illumination can lead to accidents and faulty workmanship; reduced temperatures can cause illnesses and accidents; reduced voltage levels can result in faulty equipment operation with resultant accidents; reduced ventilation can cause explosions and fires and result in harmful concentrations of chemicals, dusts, and vapors; and reduced power can result in excessive levels of air, liquid, and solid pollution.

That places of employment have a responsibility to their communities to control health hazards to the general public is well recognized. Any failure to do so can result in an adverse public reaction, damaging publicity, deteriorated public relations, a decline in employee morale, and reaction in the marketplaces of products. Beyond that is the fact that the employees live in the surrounding community and they and their families can suffer from any health hazards created by the place of employment, with resulting costs of consequences to the business management. As one of the leading industrial health organizations in the nation has said: "We know that our present American system of free enterprise will be retained only if industry continues to merit the confidence and respect of its neighbors."[14] In the fulfillment of such fundamental responsibilities the industrial health program can play a significant role.

REHABILITATION

A final aspect of an industrial health program is the rehabilitation of disabled employees so that they can return to gainful employment. Closely related to this is the employment of handicapped persons.

Reemployment of the disabled worker following rehabilitation succeeds only when it is official policy and when management firmly supports that policy. The industrial health program should urge the formulation of that policy, which should guide the processes of rehabilitation following injury or chronic illness. Upon the rehabilitated worker's return to work, the program should be fully concerned with the possiblity of transfer to another job and should provide training for this, as well as make provision for any additional safety precautions that might be necessary. The rehabilitated employee needs all possible encouragement, and the company's industrial health program must be geared to that need.

Similarly, the industrial health program should guide management policy on employment of handicapped persons and should provide for their proper job placement. necessary training, and special necessary facilities, such as ramps and other safety precautions.

The criteria frequently applied for the employment of handicapped persons are these: the physical handicap must not be progressive, must not interfere with the work performance, and must not present a safety hazard to other employees.

Such an endeavor is a rewarding responsibility for an industrial health program.

SOURCES

Material from the following organizations was used in the preparation of this chapter.
    National Health Council
    National Safety Council
    Manufacturing Chemists Association

REFERENCES

1. Nicholas A. Ashford, *Crisis in the Workplace: Occupational Disease and Injury* (Cambridge, Mass.: M.I.T. Press, 1976).
2. The Conference Board, "Top Executives View Health Care Issues," New York, 1972.
3. Robert Hilker, M.D., *Journal of Occupational Medicine*, 14(5)–371, 1972.
4. U.S. Environmental Protection Agency, "Health Effects of Environmental Pollution," 1973.
5. Ralph Nader and Mark Green, *Corporate Power in America* (New York: Grossman Publishers, 1973).
6. "Drinking Water," *National Safety News*, June 1975.
7. National Cancer Institute, 1974, as printed in *The New York Times*, June 17, 1975.
8. Genevieve Matanoski, "Pilot Study of Cancer Mortality in the Three-Year Period 1971-1973 in the Census Tract Near a Chemical Company," Johns Hopkins School of Hygiene and Public Health, Baltimore, 1974.
9. American Public Health Association, "The Nation's Health," February 1976.
10. *The New York Times*, May 22, 1976.
11. Jean G. French et al., "The Effect of Sulfur Dioxide and Suspended Sulfates on Acute Respiratory Diseases," *Archives of Environmental Health*, September 1973.
12. American Industrial Hygiene Association, "Air Pollution Manual," Akron, Ohio, 1968.
13. Chamber of Commerce of the United States, "Improving Environmental Quality," Washington, D.C., 1970.
14. American Industrial Hygiene Association, op. cit.

14

The Role of Employers

Employers have often played a quite significant role in the prevention of occupational injuries and diseases and in the preservation of the health of their employees. Through the years, they have shown a growing concern for the health of their employees and their families as it can be affected by nonoccupational hazards of various types, including deleterious lifestyles. They have increasingly engaged in health education and counseling, made health insurance programs available, provided or financed health screening or physical examinations, provided care for nonoccupational illnesses and injuries, and established control programs to cope with the abuse of alcohol and drugs. They have at times played a significant role in the encouragement and support of health-related research, either directly or through foundations they have established. On a broader scale they have done much to improve the environment of the community surrounding their operations and to assure the safety of the products made available to the public. They have become active participants in the process of rehabilitating disabled employees and increasingly offer employment to persons with physical handicaps.

This is by no means intended to suggest that such activities are universal among employers. They are not. But many leaders in business and in public health are well aware of the broad interrelatedness of the health of employees, their families, and the community as a whole, and of the importance of the prevention of injuries and illnesses. Dr. George James, an important figure in public health, brought the matter into sharp focus when he said that "industrial plants represent a natural point of first contact with the health care system for the employed population" and that "a health center at the workplace would have a high potential for being used as the multipurpose primary care center for emphasizing health promotion and preventive medicine."[1]

The point receives further emphasis in a statement by Dr. Robert O'Connor, medical director of a large steel company and a director of the Academy of Occupational Medicine, who asserted that "the only concerted, planned, large-scale, personal, adult health maintenance programs in this country today are in the occupational health programs of American industry. The workplace is probably the ideal locale to practice real preventive medicine on a wide and effective scale."[2]

A few statements by corporate executives will serve to illustrate the breadth of the health awareness of some employers. Almost two decades ago Dr. James H. Sterner, as medical director of the Eastman Kodak Company, in addressing the 1959 National Health Forum, said:

> There is increasing recognition of the importance of health to our economic well-being. But, at the same time, new threats to health, physical and mental, are to be found in new products, new forms of energy, and new patterns of production and distribution. To apply our increasing knowledge of preventive medicine to improve the health of people who work . . . are the tasks of occupational health." Elsewhere, he noted: "We now recognize that even if we were to achieve the highest possible degree of health protection in the plant setting, we would still suffer serious losses because the health of the worker is affected by what happens . . . in the home or in the community.

A few years later Kenneth H. Klipstein, president of the American Cyanamid Company, said:

> Good employee health is a universal industrial policy. In assuming responsibility for occupational health, industry only serves its own enlightened self-interest. But benefits to society nonetheless radiate out: (a) millions of people are given physical examinations who otherwise would probably never receive one; (b) the body of accumulated knowledge of environmental and occupational health grows, and is available for all to use. . . . Industry has assumed responsibilities in the field of toxicology. . . . It recognizes its responsibility to aid education. . . . For the last 20 years the Nutrition Foundation, which is supported by 51 important food processors and their suppliers, has

been conducting scientific research and educational projects in food and nutrition. . . . Several large insurance companies are using their accumulated knowledge to promote better health. . . . Another major corporation which has been contributing its professional skills on behalf of better health is Eastman Kodak. . . . My own company has for the past 10 years made a contribution to the cause of health by sponsoring the Lederle Medical Symposia.[3]

N. H. Collisson of the Olin Mathieson Chemical Corporation said: "There has been much improvement in the human relations aspect of industry. The concept of industrial health . . . has moved slowly from 'post facto' medical care to the beginning of a concept of preventive medicine. . . . A good medical program persuades the worker that management is sincere in its concern for him. It promotes mutual respect. . . . An industry has a bona fide interest in the health of its employees . . . their good health is good business."[4]

That unanimity does not prevail is evident from the findings of a survey of senior business executives published by The Conference Board in 1972. The survey found that a sizable majority of executives believed that social issues other than health care have equal or higher claim on their attention. While nine out of ten advocated some participation by business in community health matters, only one in ten foresaw any increase in such attention in the coming years. These executives did express an awareness of and concerns about such matters as the importance of the prevention and treatment of disease and disability, the status of the health care delivery system, the escalating costs of health care, the shortages and maldistribution of health manpower, and the inefficiencies that occur in the utilization of health services. Several executives urged the creation of joint occupational health facilities by groups of employers, whereas others opposed the creation or the expansion of industrial health services. One in five, however, including several executives from the smallest companies (fewer than 500 employees), urged the creation or expansion of in-house health facilities.

Negative opinions concerning industrial health programs, the survey showed, centered on the costliness of medical facilities, the lack of expertise in health affairs on the part of business management, the inefficiency that could result unless the employee population was sufficiently large and concentrated, and the employee dissatisfaction that could result if anything less than complete health care was provided for. On the other hand, a number of executives stated the belief that the company could control the costs of health care it provided directly. The survey quoted an unidentified president of a rubber company as saying: "the possibility of business doing in-house care is probably the most effective brake in the escalating costs of outside health services."

The survey revealed considerable sentiment for businesses actively involving themselves in the functioning of hospitals, in the operations of voluntary health agencies, in the development of community health planning, and in the establishment of health-maintenance organizations (HMOs).

These views on the part of business management as to the role of employers in the various aspects of industrial health are revealing in many respects. Most certainly there are forward-looking concepts, though these might not always predominate. Most certainly the question of what the role of the employer should be in industrial health is one that has not yet been resolved by a preponderance of opinion.

OCCUPATIONAL SAFETY AND HEALTH

Occupational safety and health is a complex, many-faceted subject, differing among places of employment and according to the nature of the operations. Thus the findings of the Institute for Social Research at the University of Michigan, reported to the 1959 National Health Forum by Dr. Robert L. Kahn, are of interest. The institute, through interviews with the senior executives of 262 places of employment, found that 86 percent wanted more information that could be useful in considering the establishment of an industrial health program or in expanding an existing program. Interestingly, it was found that such information could best be conveyed by personal conversations with physicians. Conversations with government officials or nonmedical insurance people were found to be ineffective, as was the printed word. Significantly, the study found that top management that had had some experience with industrial health programs valued such programs highly. By contrast, understanding of such programs was found to be poor on the part of managements that did not have such programs, even though these managements generally accepted a heavy responsibility for the well-being of their employees. These findings would appear to clearly indicate a significant communications gap.

The importance of occupational health was recognized by Dr. William Jend, Jr. of the Michigan Bell Telephone Company in 1973 when he aptly commented:

> The physical and mental health of our employees is a major consideration in the ability of any business to provide good service or a good product. . . . The prevention of unnecessary exposure to possible or potential hazards, as well as the treatment and rehabilitation of those who become ill or injured as a result of exposure, becomes more a

question of providing good preventive and therapeutic care . . . than an exercise where the primary emphasis is on placing the blame for what has occurred. The emotional environment of the workplace is also of increasing concern and the occupational physician, nurse, or other professional can do much to prevent emotional problems and provide counseling and first-aid when they do arise.[5]

The following year a study by The Conference Board noted that three of the options open to industry in the consideration of an industrial health program are (1) greater protection for employees through industrial hygiene and medical care programs, (2) more selective recruitment of personnel in relation to the job to be performed, and (3) more extensive in-house services to prevent and care for nonoccupational illness and injury, including periodic health examinations, treatment, education, counseling, and primary care.

Certainly an essential element in any industrial health program is the selective recruitment of personnel, where potential employees are judged not alone on job capability, but also on their general health in relation to the occupation. In most instances reliance for the health evaluation is placed in preemployment health-screening examinations. Some employers, on the other hand, rely entirely on well-conceived health questionnaires. Some use both approaches. The findings of such preemployment examinations can then, and should, furnish a baseline for a continuing health record of the employee and the possible detection of any subsequent occupational health problems.

Placement health examinations are used by most large companies to varying degrees in order to eliminate job applicants who have disabilities or diseases that would prevent them from performing their required duties or would endanger the health of other employees. Such examinations also assist in identifying health conditions that need prompt attention, and furnish a baseline record for protection against future workmen's compensation claims based on preexisting conditions. Of the companies surveyed by The Conference Board and reported in 1974, 71 percent provided the examinations for some employees and 58 percent provided them for all new employees. The content of such examinations differed by such things as the type of work, age, and job status. Of the transportation and utilities companies, 92 percent provided such examinations. For manufacturing the proportion was 80 percent; for financial companies 60 percent; and for wholesale and retail trade 57 percent. The proportion providing such examinations grew smaller as the size of the establishment diminished (94 percent of the companies with over 10,000 employees gave them). Of the companies surveyed, 21 percent had made such examinations more thorough in recent years.

Companies that commonly have a large turnover within the first months of employment, however, find preemployment examinations a poor investment. Accordingly, the Boeing Company ceased the practice of preemployment examinations 25 years ago, placing reliance instead in health questionnaires. The decision was coupled with the inception of an increased in-plant environmental control program and did not limit a liberal program for hiring handicapped persons. Preemployment examinations are provided for a few jobs involving special hazards, such as firemen, guards, test pilots, and drivers; after-employment preassignment examinations are used if the worker is to be transferred to a job involving such hazards as toxic fuel, chemicals, unusual atmospheric pressures, or the use of beryllium. The questionnaire used is quite explicit, and strict penalties are applied for falsifications. Boeing made this decision after observing the high rate of turnover among its workforce (usually exceeding 50 percent each year of those hired, and most of which occurred in the first six months after employment for reasons unrelated to health) coupled with the cost of the preemployment examinations, which in 1952 averaged $13.50 per examination (they would have cost $5,640,000 for the 470,000 persons hired between 1952 and 1969). This cost, according to Dr. Sherman Williamson, then of Boeing, "would not have discovered any more physical conditions that would have affected their hiring than did the questionnaire." He also makes the point that between 1952 and 1969 only five or six workmen's compensation claims might have been avoided had there been preemployment examinations. He further maintains that more and more employers are beginning to have doubts about such examinations and concludes, "I feel that pre-hire examination practices need to be critically reviewed and evaluated for what they accomplish."[6] In 1962 Dr. Hawkes of Boeing said that "physical examinations can be important [as an] adjunct to environmental control measures [but that claims for workmen's compensation are not significantly affected by the use of a physical examination as opposed to that of a health questionnaire].... Seldom do human physical factors enter into a shop safety record. Rather it is the attitude engendered by shop management and reflected in day-to-day concern for safety.... Environmental control and training remain the major effective devices for limiting claims.... The money-saving advantages of job placement and claims prevention through examination have been overemphasized ... examinations contribute little to reduction of absenteeism.... Any claim that psychologic testing provides a substitute [for personnel observation] is pure pretense.... In occupational health and safety [the examination] plays a distinctly subordinate role to other preventive measures, particularly to environmental control measures, safety training, and safety certification."[7]

The Aluminum Company of America is reported to have taken an action essentially similar to that of Boeing and for essentially the same reasons.

The experience of the New York Telephone Company in this respect is of interest because of contrasting evidence. The general medical director of the company, Dr. Collings, has said that the New York Telephone Company similarly experienced a very high personnel turnover rate. In 1968 medical personnel were examining 20,000 young women applicants a year, 60 percent of whom had left the company within the first year of employment. The practice seriously taxed the capacity of the medical department and cost the company over a half million dollars a year. Observation of such facts caused a reevaluation by the company. There was concern that discontinuance of the examinations could increase the incidence of communicable diseases within the employee population, as well as increase the number of health defectives in the permanent population. The decision was made, however, to discontinue the examinations and to substitute instead a brief questionnaire coupled with personal observation as respects young women applicants. For a small number of special cases examinations were provided where these were felt to be warranted.

After one year of employment a "baseline" examination was then made for those who continued to be employed. These examinations indicated that 76 percent of those employed for one year had no health problems, 1 percent had limited problems as respects certain types of jobs, 9 percent had a health problem that could be corrected while work continued, 7 percent had health problems that should be corrected before they resumed work, and 7 percent had health problems that contraindicated employment. When supervisors were questioned about the health problems of these employees, the 14 percent in the two latter categories (who would not have been employed if recourse had been had to preemployment examinations) presented a problem during the first year of employment four times as often as those in the other categories. Their number of absences during the first year was 14 percent higher than for the other employees but their days of absence that year averaged 12.1 compared with an average of 6.4 for the other employees. Dr. Collings, in 1971, concluded that (1) employee turnover during the first year did not favor the loss of or the retention of individuals with health problems; (2) preemployment examinations identify about 14 percent of young women as having significant health defects; (3) if hired, these "defectives" cause four times as many problems for supervision and two times as much absence in the first year as healthy persons; and (4) the excess absence costs due to "defectives" in the first year or year and a half of employment virtually equal the cost of the examinations that would have prevented these and other costs.[8] The company subsequently reinstituted the preemployment examinations.

Such considerations as the foregoing have led the National Institute for Occupational Safety and Health (NIOSH) to observe that, although the merits of the standard preplacement examination have been argued, the virtue of the preplacement questionnaire has gone almost unchallenged. It concludes, then, that "all applicants do not need an in-depth preplacement examination and, while this would be most desirable, such an approach would be costly, time-consuming, and, given the current dearth of physicians in America, probably impossible to carry out."[9]

Once hired, the employee may be subject to a variety of safety and health hazards. This places upon the employer the responsibility of preventing occupational accidents and illnesses to the maximum extent possible, through the periodic monitoring of working conditions, including the construction, design, and layout of the plant; consideration of ventilation, lighting, noise, and temperatures; attention to machine design, construction, and operation; provision of safety devices, safety education, job training, and job supervision; concern for the presence or use of hazardous materials, substances, dusts, fumes, or gases, and the provision of safety measures against their effects; and the provision of periodic health screening of employees to determine whether their health is being harmed. In the execution of this responsibility, the employer should be guided by the various standards made available by many types of organizations. Such preventive measures failing, adequate first aid and treatment should be readily available.

That much is increasingly being done out of recognition of this responsibility goes without saying, although complete documentation does not exist. On the other hand, it is evident from some available studies, as well as the findings of the OSHA investigations noted in Chapter 16, that full recognition and acceptance of such responsibilities is not always forthcoming.

A few years ago The Conference Board, in a survey of 858 companies employing 500 or more persons, found that 47 percent of those companies believed that some of their employees were in occupations involving potential hazards to their health. By type of employer operation the proportion of employees possibly endangered was judged to be considerably greater in manufacturing, transportation, and utilities than in other types of operation. The potential hazards most frequently identified were noise, followed by dust, gases, vapors and fumes, radiation, and toxic substances.

An important aspect of an industrial health program directly related to such hazards is that of providing periodical monitoring or work-related examinations to determine whether the worker continues to be physically equipped to perform the assigned job, and whether any hazards inherent in the job or the workplace are resulting in the impairment of the health of the worker. The findings of such examinations should be added to the ongoing health record of the employee, the baseline for which, as has been noted, would be the preemployment examination.

The findings of the Conference Board survey are significant in this respect. The survey found that only 53 percent of the respondents provide such work-related periodic health examinations to determine the compatibility of the employee for the job, or to determine any harmful effects inherent in the work environment. The proportion of the employees tested for this latter purpose differed, as might be expected, in relation to the potential hazard present. It also differed in direct relation to the size of the place of employment. Some employers, such as Du Pont, combine a statistical analysis of absenteeism with the results of these physical examinations. The purpose is to obtain further clues to the presence of potential hazards or to abnormal reactions to the use of new materials or new methods of operation.

Basic to the functioning of an industrial health program is, of course, the adequate staffing of the program with trained professionals. These, unfortunately, are in short supply. The extent to which professionals are engaged to carry out the responsibilities of an industrial health program was looked into in the Conference Board survey, which showed the following with respect to the employment of physicians and nurses by places of employment in 1972:

	None	Nurses Only	Physicians
Manufacturing	29%	33%	38%
Transportation and utilities	66	13	21
Wholesale and retail trade	81	11	8
Financial	38	25	37
Others	66	10	24

For companies employing over 50,000 employees, 100 percent had physicians on staff. The smaller the number of employees, the smaller the percentages. Only 16 percent of companies employing fewer than 1,000 persons employed physicians and 17 percent employed nurses only.

For that same year the Bureau of Labor Statistics reported that 69 percent of the 8.4 million employees in manufacturing establishments were provided with nursing services, compared to 15 percent of the 1.2 million employed in the service industries, 7.9 percent of the 961,700 employed in wholesale and retail trade, 5.8 percent of the 698,600 in transportation and utilities, 5.6 percent for the 681,600 in finance, insurance, and real estate, and 1.5 percent of the 1.2 million employed in contract construction.[10] These figures vary with the findings of The Conference Board, owing to the influence of the size of employers in each of the fields shown.

While these findings show a considerable availability of industrial health professionals in places of employment, they also make it very clear that the advancement of industrial health has a long way to go. When it is considered that the Conference Board findings were limited to employers with more than 500 employees, it could be accepted as fact that the availability of industrial health professionals over the entire range of places of employment would be much less than what has been shown. The Bureau of Labor Statistics data present some indication that this is so.

In the Conference Board survey, furthermore, only 14 percent of the companies surveyed employed an industrial hygiene director. As might be expected, the proportion was considerably higher where the potential hazards were greater (57 percent). The proportion was also significantly higher in larger places of employment (75 percent of companies having 10,000 or more employees had an industrial hygiene director). Of further significance in this survey was the fact that of the companies surveyed, only 29 percent engaged in research for the purpose of gaining a better understanding of their potential occupational hazards, which in turn could lead to improved methods for controlling such hazards.

The Bureau of Labor Statistics also provided an indication of the proportion of employed people in 1972 who had available to them the services of industrial hygienists. The Bureau at that time reported that only 18.2 percent of employees in the private nonfarm sector of the economy had the services of industrial hygienists available to them. In manufacturing the proportion was 35.7 percent, in transportation and utilities it was 23.4 percent, in finance, insurance, and real estate it was 7.8 percent, in contract construction 6.9 percent, in services 6.8 percent, and in wholesale and retail trade the proportion of employees provided industrial hygienist services was 6.5 percent. Such services were only provided by 7.7 percent of transportation and public utilities companies, 7.2 percent of manufacturing companies, 2.7 percent of wholesale and retail trade organizations, 1.8 percent of finance, insurance, and real estate establishments, 1.6 percent of contract construction companies, and 1.3 percent of the service industries.[11] Again, differences of the findings are principally due to the sizes of the places of employment.

Directly related is information having to do with the safety of employees. In 1974, NIOSH surveyed 274 employer organizations and found that 58.6 percent provided safety and health training for their employees. The greatest proportion was found in the retail and wholesale trades and in the service industries: 66.7 percent. The lowest was in mining: 18.2 percent. That survey also found that only half of the places of employment surveyed had established safety committees. The proportion was highest in manufacturing and mining: 76 and 75 percent respectively. In transportation, public utilities, and communications the proportion was 52 percent; in government it was 47 percent, in construction it was 39 percent, and in retail and wholesale trades and the services industries the proportion having established safety committees was only 27 percent.

Here it should be noted that each year the National Safety Council publishes a listing of those places of employment that have a no-injury record for the preceding year. The listing shows the number of continuous man-hours worked without a disabling injury.

What has been shown here presents some indication of what places of employment are or are not doing with respect to occupational injuries and illnesses and their control or prevention. One quite hopeful sign for the future was found in the Conference Board survey, where it was shown that as a result of the enactment of OSHA in 1970, 71 percent of the companies surveyed had changed their methods of monitoring the work environment and 27 percent had made changes in their research activities. The staffing or organization of the industrial health program was changed in 48 percent of the companies, 40 percent had changed the content or the emphasis of their program, and 30 percent had added to the staff of their program. Other changes included an increased awareness on the part of management, improved record keeping, more intense employee education, and increased expenditures for occupational safety and health.

NONOCCUPATIONAL HEALTH AND SAFETY

It has been shown that nonoccupational illnesses and injuries have a very considerable impact on the operations of places of employment. The differences of opinion surrounding the role of the employer and what it should or should not be have also been discussed in the preceding chapter. Nonetheless, the Manufacturing Chemists Association commented more than 15 years ago that "more and more companies, becoming aware of the cost to them of off-the-job injuries, are aggressively promoting off-the-job safety for their employees. They realize that an off-the-job safety program is of economic importance to business and industrial firms.... It is primarily a top management or plant manager's responsibility to direct that off-the-job safety be considered an integral part of the on-the-job safety program.... Although admitting the benefits, some firms [are of the opinion that systematic promotion] might be interpreted as encroachment of an individual's 'freedom of action.' However, this has not materialized for the companies that are promoting off-the-job safety."[12]

Dr. G. H. Collings, Jr. of the New York Telephone Company takes the view that total health care is a "composite and a continuum" and that industry is often better able to deliver primary care to its employees than are other services in the community. He notes that industry has a long-time medical record for many of its employees and is in a better position to achieve a more comprehensive view and understanding of each patient, and that where industrial health facilities already exist it requires minimal expansion to provide primary care.[13]

Dr. Miles O. Colwell of the Aluminum Corporation of America has made the point that with such conditions as hypertension, obesity, or mild diabetes, if good in-company facilities are available, and if the employee is given free choice of the source of treatment, he or she could receive as good, or better, treatment at the in-company facility than if an outside physician were chosen. Furthermore, Dr. Colwell contends, health education, counseling, and other preventive services can be carried out more conveniently and less expensively in a company facility.[14]

Spokesmen for business have also been quoted by The Conference Board in 1974 as saying:

> We [General Dynamics] are now considering more in the way of treatment of non-work-related diseases. . . . Justification for the expansion into this area has been based primarily on the time savings that accrue for both the individual employee and the corporation via a reduction in the amount of time lost from the workplace.
>
> As industry must pay for more primary health care of its employees, the economics of expanding its in-house medical services becomes increasingly attractive.
>
> Business should reexamine the extent to which their in-house medical programs are providing nonoccupational disability medical care to determine whether such services should be expanded or curtailed, with a decision being made by taking into account a cost-benefit analysis of the alternatives available to him.

Business management not infrequently, however, reacts in opposition to such views, or at least is skeptical of the merits. The grounds for negative reactions may include the feeling that to adopt such views would be an infringement on the employee's freedom of choice in the selection of a physician and that to engage in such concerns would water down current industrial health efforts. They also rest in an anticipation of the resultant administrative burdens, in doubts about corporate management's competency to enter the health service field, and in concerns over the effect on employee relations (including the mistrust of an employer-established program that can develop among the employees, as well as the consequences in instances where an employee feels that inadequate or improper treatment has been provided). Despite this, The Conference Board survey found that a small but increasing number of employers were questioning the traditional role of the employer that limits health care services to injuries and illnesses emanating from the workplace, or to emergency or palliative care for nonoccupational illnesses and accidents. This has led some employers, the study reported, to provide new or greatly expanded services, as well as to establish special health counseling programs.

It is significant that of the 858 companies surveyed by The Conference Board, 83 percent provide employees with some nonoccupational health services. Included are such companies as U.S. Steel, Mobil Oil, Gates Rubber Company, IBM, Chase Manhattan Bank, New York Telephone Company, Metropolitan Life Insurance Company, Westinghouse Corporation, General Dynamics, Aluminum Company of America, Illinois Bell Telephone Company, Exxon, and Du Pont. The nonoccupational health services, it was found, most usually provide diagnostic evaluation of nonoccupational symptoms. Most provide treatment for the more common acute conditions for a short term, although 4 percent of the companies would not provide such services. Treatment for chronic disorders is less frequent, although 7 percent of the companies, through their industrial health program, treat hypertension, 13 percent treat peptic ulcers, and 14 percent treat chronic allergies. In many more cases the industrial health program provides treatment for chronic conditions in association with outside physicians.

Today some employers are convinced that nonoccupational health services as part of an industrial health program result in better work attendance, improved job performance, lower insurance costs, greater work-force stability, and improved employee morale. Several, however, see no benefits at all.

One aspect of an employer's role in the control or prevention of nonoccupational illnesses and accidents is the early detection of the onset of disease through the conduct of *health-screening examinations* or *periodic physical examinations,* on either a voluntary or compulsory basis. The provision of such examinations can, of course, be related to those examinations mentioned previously, the purpose of which is to determine the effects of potential hazards in the workplace.

Dr. Jend of the Michigan Bell Telephone Company, in detailing the industrial health program concept of his company, has described one facet as being: "A program of regular health evaluations for permanent employees, the frequency and technique used depending on age, work location, health history, and other pertinent factors [including] the use of health questionnaires, counseling, multiphasic screening, complete medical examinations, and other services as appropriate."[15]

The Conference Board study revealed that 57 percent of the employers surveyed provided general periodic physical examinations. This proportion was considerably higher than was found in a similar survey made ten years earlier and the quality of the examinations provided was found to have improved. The proportion of employers offering such examinations was highest among financial and insurance companies and was highest (90 percent) among companies with more than 50,000 employees. It was lowest (42 percent) among companies employing from 500 to 1,000 persons. For the 57 percent of employers providing such examinations, eligibility in the majority is limited to certain employees, although 24 percent of these companies provide examinations for all employees. Exclusions, where they occur, are usually based on wage or job level. Others are based on age (22 percent), health history (5 percent), or length of employment or type of work performed (18 percent). In two-thirds of these companies all eligible employees are examined with equal frequency. Where differences occur they are based on age (73 percent), health status (31 percent), job level (24 percent), or other considerations (22 percent). The survey found that examinations have been made more comprehensive in recent years. In two-thirds of the companies, examinations were described as voluntary. The importance of adequate followup was recognized. The study noted the insufficient supply and poor distribution of primary-care physicians and facilities, and the fact that neither the public nor the medical profession is sufficiently oriented to the potentialities of prevention and early care.

The content of such examinations can be a subject on which there are differences of opinion. Dr. Norbert J. Roberts of Exxon has discussed such aspects as proctosigmoidoscopy for employees over age 40, breast examinations for females, pelvic examinations and cervical cytologic studies for employees age 30 or older, prostate examinations for males, tonometry for employees over 40 and earlier if there is a family history of glaucoma or gall bladder, radiography, flat plate of the abdomen, electrocardiograph, chest X-rays, and caliper measurements of skinfolds. He notes that in some populations such tests as stool specimens, urinalysis, cytologic studies of urine or sputum, hemoglobin measuring, the measurement of the erythrocyte sedimentation rate, blood glucose determination, cholesterol measurement, uric acid determination, serum calcium determination, cervical smears, pulmonary function tests, mammography for women over 40, medical thermography, examination of the gastrointestinal tract, or intravenous pyelograms can be justified or even advisable. He says that at Exxon such examinations are suggested annually for employees age 40 or over, biennially for those between ages 30 and 39, and every third year for those under 30, unless clinical judgment dictates otherwise in individual cases. The examinations are voluntary and the results are confidential.[16]

Dr. F. W. Holcomb of IBM has described a voluntary multiphasic screening program inaugurated by his company in 1968 for all its employees in the United States, at the plant, at laboratory locations, and in the field. The objective was to stimulate interest in health maintenance, to bring about the early detection of subclinical problems that could be controlled before serious illness developed, and to accumulate health data on the workforce. At first the program was offered to all employees over age 40, but subsequently it was extended to certain other age groups. Between 1968 and 1973, 46,046 examinations were conducted on 42,688 individuals. The examinations included blood pressure, height, weight, vision, hearing, blood serum, urinalysis, chest X-ray, electrocardiogram, ears and throat, and for female employees breast examinations and Pap smears. The employee is advised of the results and a copy is sent to his or her personal physician if consent is given. The results are confidential. Participation in the program has averaged 71 percent of the eligible employees (74 percent for plant and laboratory employees and 62 percent for field employees). Of interest is the fact that only 27 percent of the employees examined were found to be free of any demonstrable medical problems. Among the problems identified of which the employee had not been previously aware were diabetes, hypertension, ischemic heart disease, and other cardiac conditions.[17]

One example of an employer making available physical examinations through the use of means other than the company's own industrial health program has been noted by Robert J. Clarke of the Seamen's Bank for Savings in New York. In 1941 the bank instituted such a program in collaboration with the Grace Clinic, a private group of doctors in Brooklyn, for voluntary examination of the bank's employees and their families, offered at no cost to them. The examinations can be required, however, if job deterioration is observed. The results of the examination are confidential.[18]

Sources of examinations other than on an in-plant basis that are available to employers include such clinics as the Life Extension Institute in New York, the Executive Health Examiners in New York, the Mayo

Clinic at Rochester, Minnesota, the Leahey Clinic in Boston, the Geisinger Medical Clinic at Danville, Pennsylvania, the Beverly Hills Medical Clinic in Beverly Hills, California, the Strang Clinic in New York, and the American Health Foundation in New York. It is of interest, with respect to the American Health Foundation, that in 1976, when it conducted a health-screening program for employees of the Federal Reserve Bank, 99 percent of the employees participated. Additionally, it should be noted, those prepaid group practice plans, or HMOs, available to employees provide either multiphasic health screening or periodic health examinations without additional cost.

That such examinations produce savings in mortality became evident over 50 years ago when the Metropolitan Life Insurance Company offered physical examinations to its policyholders at the Life Extension Institute in New York and found that between 1914 and 1921, as a result of 95,000 examinations, there was a 28 percent reduction in mortality.

Another aspect of an employer's role is the provision of *treatment for nonoccupational illnesses and injuries.* It was shown earlier that this is perhaps the most controversial area in the consideration of an industrial health program. In some instances the care provided is, in a sense, emergency care or first aid for acute conditions and is provided on a temporary basis, followed, if the needs so dictate, with recommendations for further treatment by the employee's personal physician. In other instances the treatment can be more thorough and can be given over a longer period of time, perhaps in conjunction with the personal physician. Where an HMO is involved, whether established by the employer or opted by the employee, referral of the employee to the HMO for further treatment can be the avenue taken.

Dr. William Jend of the Michigan Bell Telephone Company has written that "treatment in its own facility by industry of not only occupational disease and injury but also diseases for which an employee either may not seek treatment or may not have good medical resources available to provide treatment can be an important part of the total health care delivery system. Hypertension and diabetes are examples of diseases that lend themselves to this type of treatment. . . . Pre-hospitalization testing . . . and post-hospitalization follow-up, care, and rehabilitation either at home or in the workplace is a logical extension of this idea. . . . The physician or physician-surrogate in industry may be the ideal primary physician for this [employee] constituency."[19]

Yet another aspect of an industrial health program and its effect on nonoccupational diseases and accidents that receives increasing attention today is *health education.* This might include such activities as the distribution of literature, the showing of films, the display of posters, and the conduct of group discussions. Most effective, however, is personal counseling in connection with the findings of health examinations.

The Conference Board survey found that 34 percent of the companies surveyed engaged in health education activities in 1974, with greater effort among the larger employers (58 percent for those with over 5,000 employees), and a higher proportion among transportation and utility companies (63 percent), manufacturing companies (53 percent), and financial companies (42 percent). Among wholesale and retail trade companies the proportion was only 10 percent. Among companies with established health education programs it was found that 56 percent provided health counseling in connection with periodic health examinations, 72 percent held first-aid classes, 57 percent had film showings, and 85 percent distributed printed matter in the form of pamphlets, posters, and articles in employee newspapers and journals. A number of companies held seminar discussions on specific aspects of health education. Employee interest in such discussions was found to vary, usually depending on the subject matter. The survey took cognizance of the fact that places of employment present a unique opportunity for health education and noted the conviction on the part of some employers that highly personalized counseling is the most effective tool in health education.

One study, which attempted to measure the effectiveness of health education in places of employment, found that 65 percent of the employees had read the pamphlets that were distributed, 47 percent remembered health posters they had seen, and 65 percent of women attended films on self-examination of the breast. Of the employees in the study, 22 percent sought health counseling from the industrial health program, 47 percent said they would consider seeking such counseling, and 31 percent said they would not. The study concluded that improvements in the methods of health education being employed are needed, and that greater effectiveness would result from placing more stress on person-to-person techniques.[20]

Dr. Jerry Cassuto of the Western Electric Company has commented that most health education programs in places of employment are traditional stereotypes (posters, pamphlets, films), many of which are excellent, but that in terms of actually reaching employees and their families they are "often sadly lacking." He notes that the Industrial Medical Association in 1967 surveyed 400 directors of industrial medical programs (of whom only 200 replied) and found that two-thirds had a specific health education program. Of these, 89 percent utilized individual counseling supported by the more customary means of health education and 67 percent directed the program at the entire employed population (only 2.6 percent limited the program to selected groups of employees and only a very few limited the program to management personnel only). Only one company employed a health educator specialist. Less than 20 percent of the programs used official or voluntary agencies from the outside. The need for more help was recognized by 69 percent of the respondents. Dr. Cassuto notes the value of individual or group counseling but comments that "the percentage of employees currently being reached is relatively small." He laments the few attempts to inaugurate new concepts, the apparent reluctance to utilize the professional health educator in industry, the fact that outside agencies are also largely ignored, and that little attempt is being made to objectively analyze the efficacy of existing programs. He cites as imaginative approaches to health education those taken by such companies as the New York Telephone Company and the Illinois Bell Telephone Company.[21]

A continuum is evident from these few findings and comments: There is a growing interest in the health education of employees, there is dissatisfaction with or skepticism regarding the value of the means of health education most usually used, and there is unanimity that person-to-person health counseling is the most effective means of health education.

A unique approach to health education was developed by the Metropolitan Life Insurance Company for its 14,000 home office employees in 1974. The program focused on high blood pressure and was combined with a blood pressure screening program and the provision of care. Participation in the program was voluntary. The purpose was to enhance employee awareness of a major health problem, including such associated aspects as family history and lifestyles. All aspects of the program were confidential. Of the employee population, 39 percent participated in the program, and of these 60 percent had their blood pressure taken. The program was considered worthwhile by 94 percent of the participants. The testing revealed 626 cases of hypertension (20 percent of those tested), of whom 248 were unaware of the condition. Of the remainder who had known of the condition, only 51 percent claimed to be under treatment. Those found to be hypertensive were then provided an EKG, chest X-ray, and urinalysis.

As a consequence of the program, a clinic was established within the company in January 1975 starting with 415 employees and now serving over 600. A regimen is prescribed, including drugs, diet, and abstinence from smoking. Of the original participants in the clinic, 331 continue, 11 developed normal blood pressure, 4 use their own doctor, 40 left the company, 1 died, and 28 discontinued attendance. The company concludes that "an occupationally based clinic which supplements the care of private physicians, offers diagnostic services and blood pressure monitoring, counseling, and health education services is an important employee benefit as well as a benefit to the company."[22]

Another unique approach for making health education available to several employers in a community has been developed by the Morristown (N.J.) Memorial Hospital. Through its Department of Community Health Education the hospital has made an effort to reach workers in the community. In so doing it had the cooperation of Bell Laboratories, Inc. Five companies in the area have participated in the program, with five other companies participating in certain aspects of the program. The program focused on stress, cardiac risk factors and emergencies, cancer, breast examination, and dentistry. The objectives were to present medically accurate health information and to help employees to recognize symptoms of disease and to seek medical care. More than 600 of 2,500 eligible employees participated in the program, 80 percent of whom were males between ages 30 and 54. The program was rated good or excellent by 96 percent of the participants.[23]

Here, one study of the relationship of a multiphasic screening experiment to health education on a multi-employer basis is of interest. In 1951 a multitest clinic was made available in Richmond, Virginia. Eight employers made the screening program available to their employees, the eight ranging in size from a freight-equipment repair shop with 35 employees to a department store with 2,000 employees. Four did not have a nurse on their staff, using a physician for emergencies only, in addition to having first-aid kits available. One had a nurse and a dispensary, and provided physical checkups at intervals. A few gave preemployment examinations, a few provided home visitation to ill employees, and one had its own hospital. The established programs had at that time been in existence for anywhere from 7 to 55 years.

Management attitudes displayed a concern over nonoccupational ailments, particularly respiratory conditions, over the difficulty in inducing employees to report to the medical department for nonoccupational illnesses, over the prevention of work accidents resulting from carelessness, over absenteeism, over the problem of sanitation in eating facilities, and over the cost of an industrial health program. One employer had difficulty in encouraging warehousemen to deal promptly with strains due to lifting, these resulting in a high rate of labor turnover.

The attitude of employees to the testing program was generally enthusiastic, although two-thirds were afraid of the findings (a situation that was relieved when they found that the records were confidential). Employers were concerned, however, that employees would look on the examinations as an invasion of privacy. In the main, however, the employers felt that such a testing program helped the cause of health education and that it would relieve the problem of their own shortage of funds. One employer invited the employees of neighboring firms to participate. One invited the people in the neighborhood to participate but most did not. One was not in favor of the testing because of its own medical program.[24]

These unique approaches to health education are worthy of a broader consideration and applicability. Particularly is this so in the instances of multi-employer participation. Much more could be accomplished, particularly on behalf of the employees of smaller employers, if such approaches and experiments were more generally available in communities throughout the nation.

A final aspect of the role of employers in helping to protect the health of their employees, and one of particular interest in recent years, is the positive approach being taken by employers of all types in the early identification, treatment, and *control of alcoholism* among their employees. Since the incidence of alcoholism among employed people ranges from 4 to 10 percent of all employees, since alcoholic employees most usually have been employed by the same employer for some 12 years and are therefore skilled at what they are doing and valuable to the employer, and since illnesses, absenteeism, accidents, and premature deaths are considerably higher among alcoholic employees, this is a disease of consequence to employers. Where well-conceived programs have been developed in places of employment for the control of this problem the recovery rates have ranged from 60 to 80 percent of the alcoholic employees. Today over 600 employers have instituted such programs and they are proving to be not only successful but distinctly cost-effective for both individuals and places of employment. Of further interest is the increasing number of instances where such programs are being extended in concept to include a concern for any troubled employee, be the cause mental or emotional disturbances, debt accumulation, marital difficulties, or drug abuse. The Illinois Bell Telephone Company is one of the employers that have taken this broadened approach.[25]

The approaches to the prevention or control of nonoccupational hazards to employees, then, are several. More and more, employers are becoming interested in attempting to come to grips with such problems, and increasingly are such attempts being found to be worthwhile, from any standpoint.

THE COMMUNITY AT LARGE

The health interests of employers as respects the community at large are varied, at times exceedingly complex, and of increasing concern to the public. As has been shown, such interests include the safety of persons on the premises of the employer or being serviced by employees, the safety of the public using the products offered for sale, the effects on the community health of the operations and processes of the employer, and the relationship of the employer to the health services available to the community.

While an ever increasing active interest by employers over the safety and health effects of their products is readily evident, not too much is available in the way of an overall documentation of such efforts. The same is true as respects the interests of employers in safeguarding the public health through the establishment of controls on such matters as air pollution, water pollution, waste disposal, excessive noise, or noxious odors. Included in such controls are environmental monitoring and a concern with ergonomics, as well as the support of soundly conceived health legislation. While such efforts have not been documented it is quite clear from what has been shown in the previous chapter that instances of failure or neglect are quite frequently highly publicized. As has also been shown, the subject of environmental health hazards not infrequently lacks a solid scientific base and therefore can be the subject of considerable controversy. It has also been shown that corrective measures often involve swapoffs in relation to the good that flows from such processes (for example, pollution control versus the economics of a community). Unfortunately, here again, generalities often prevail in the absence of more concrete knowledge.

There are, however, some data on the expenditures by industry in an effort to curb environmental health hazards and these are shown later. The sheer magnitude of such expenditures, in themselves, provides an indication of the breadth of the effort being made by industry today to curtail harmful environmental influences and to better the health of the community. Unquestionably, the effects of OSHA and the concerns by federal, state, and local government agencies in the areas of environmental protection and product safety, shown subsequently, can only serve to increase the health-protective efforts by industry in the years immediately ahead.

An illustration of what can be done by employers is furnished by the activities of the Du Pont Company as described by Edwin A. Gee, a senior officer. Commenting that "there is a need for more testing of many [chemical] compounds," Mr. Gee notes as a hopeful sign the recent formation of the Chemical Industry Institute of Toxicology, which will test chemicals of interest to its member companies. Mr. Gee recognizes that many substances are potentially dangerous but that many are indispensable (for example, ammonia for fertilizers and medicine; chlorine for water sanitation; or sulfuric acid for steel production). The answer, he maintains, lies in handling such substances with respect and in using proper and necessary safeguards. At Du Pont, the company established its first industrial safety research facilities over 40 years ago: the Haskell Laboratory of Toxicology and Industrial Medicine. The laboratory identifies not only acutely toxic materials but also those involving chronic toxicity with long-term effects. (Similar laboratories have been established by other companies, such as Dow Chemical and Eastman Kodak, as well as joint employer efforts through such organizations as the American Petroleum Institute and the Manufacturing Chemists Association.) Once it is determined that a substance is a carcinogen, a program is instituted to safeguard workers through a revision of work practices, the provision of additional protective equipment, process changes, or engineering controls.

Next a program for monitoring and measuring exposure levels is instituted, along with an examination of employee medical histories. Employees, customers, and appropriate government agencies are then notified of the actions taken. If safe operation cannot be assured, production is shut down and sales halted until adequate safeguards are developed. An additional aspect of the program is one called "defensive shipping" and is concerned with all aspects of the packaging and transportation of materials.[26]

A final relationship of employers to the health of the community is one concerned with the availability, distribution, and cost of health care facilities and services in the community. Public health officials, labor leaders, health insurers, and physicians have not infrequently expressed disappointment and criticism of the present role of employers with respect to the health care system. They sense a faulty grasp of the subject by employers, even in relation to the economic stakes employers have in improving the levels of health in a community. The Conference Board study quoted an unidentified senior executive of a major steel company as having said that "industry does not have a good track record on nonlegislative activity but there is a growing awareness on the part of business that they should become more active in this field. It is important to remember that hospital planning councils were started by business many years ago to attempt to prevent duplication of medical care facilities." Certainly the community health efforts of a company such as Eastman Kodak in developing home health services in Rochester, New York, provide an example of what an employer can do in this respect. The Conference Board has also quoted an unidentified board chairman of a leading manufacturing company as saying: "Industry's role should be broadened beyond being simply the source of funds and a source of board members for hospitals and other institutions."

The Conference Board survey has also furnished some useful information on the activities of employers in this respect. The survey showed that 59 percent of the companies surveyed play some role in community health affairs, with manufacturers and public utilities being the more active contributors. The survey showed that 43 percent of the companies were involved in the creation of needed health facilities or services, most usually in the form of financial contributions. In some instances support has been given to the development of health services in low-income areas. In 18 percent of the companies assistance has been given to the creation of prepaid group practice, or HMO, services. Notable here is the HMO developed by the Gates Rubber Company in Denver for its employees, the HMO developed by the Connecticut General Life Insurance Company in Columbia, Maryland, and the HMO developed in the Twin Cities with the support of such companies as Honeywell, Pillsbury, Hoerner-Waldorf, Cagill, General Mills, 3M, Control Data Corporation, the Dayton-Hudson Foundation, the Northern States Power Company, the Northwestern National and First National banks, and several insurance companies. There are, however, some rather strong differences of opinion among employers as to the efficacy of HMOs as a source of health services, and certainly the high degree of interest in HMOs that pervaded the health scene in the early 1970s has cooled considerably in more recent years.

The Conference Board survey further showed that 17 percent of the companies support the training of health manpower, most usually by financial contributions, and that 30 percent lend their efforts to increasing the efficiency, coordination, and cost controls of hospitals and other health facilities and services. Today the active interest on the part of employers in community health planning is exemplified by the efforts of the Employers Health Council in Tulsa, Motorola in Phoenix, the New York Telephone Company, the Mesta Machine Company in Pittsburgh, the Goodyear Tire and Rubber Company in Akron, Sandia Corporation (a subsidiary of Western Electric) in Albuquerque; by joint employer efforts in Rochester, New York, including Eastman Kodak, Xerox, and Sybron Corporation; and by joint efforts in Syracuse on the part of such companies as General Electric, Crouse-Hind, Bristol Laboratories, Agway Corporation, and the Niagara Mohawk Power Corporation.

Other community contributions by employers were shown by the survey to include the influencing of public policy and the making of their own industrial health facilities available to the community as a neighborhood health center.

Important contributions, furthermore, are made by many employers in the field of health research. This can come about through the creation of such foundations as Rockefeller, Ford, Johnson, Milbank, and the Haskell Laboratory for Toxicology and Industrial Medicine supported by Du Pont. In other instances organizations are jointly created to conduct research activities or to establish health and safety standards. Included here are the Manufacturing Chemists Association, the National Safety

Council, and the American National Standards Institute. The contributions of insurance companies to health research also cannot be overlooked, whether done individually, such as the contributions made by the Metropolitan Life Insurance Company, or through joint effort and contribution. Certain of these have previously been noted.

Despite such evidences of active interest in the health of the community, the Conference Board study commented that relatively few companies appear to be reevaluating the traditional relationship between company medical departments and the wider health care system and that the approach being taken by employers has been less reasoned or energetic than their cost interest or social responsibilities would seem to dictate.

REHABILITATION

Today a great many employers give considerable attention to the rehabilitation and reemployment of disabled employees. In part, the interest of employers is spurred by the nature of workmen's compensation insurance, and their interests are aided and encouraged by the insurance companies. Assistance can also be provided by various government agencies, particularly those concerned with vocational rehabilitation.

Increasingly, and prodded by the employment of the handicapped legislation, employers are hiring handicapped persons. That employment or reemployment of handicapped persons is not only humane but feasible is evident from the experience of the Du Pont company, which since 1947 has employed 1,452 physically handicapped people among its 100,000 employees. It has found that most of the handicapped achieve average or better-than-average job performance, safety records, and attendance records, and that their turnover rate is low. Of the employed handicapped people at Du Pont, 562 are craftsmen; 334 are classified as professional, technical, or managerial; 233 are classified as operators; 224 have office or clerical positions; 83 are service workers; and 16 are laborers. Included in the handicapped workers are those who are blind, those who are deaf, those who have had amputations, paraplegics, and persons with heart disease and epilepsy.

Unfortunately, a complete documentation of the activities and experiences of employers in the rehabilitation of disabled employees and in the employment of handicapped persons is not available.

THE FUTURE

All indications are that employers will increasingly develop or expand their industrial health programs as respects both occupational safety and health (owing, in part, to the effects of OSHA) and the prevention or early treatment of nonoccupational diseases and accidents. Concerns over the effects of their products or their operations on the health of the community will also result in added precautions and actions, again owing in part to the effects of OSHA and other government activities.

SOURCES

Material from the following organizations was used in the preparation of this chapter.
  The Conference Board
  National Health Council

REFERENCES

1. George James, M.D., "The Teaching of Prevention in Medical and Paramedical Education," American Public Health Association, Washington, D.C., 1970.
2. Robert B. O'Connor, M.D., "The Role of Industry in the Health of the Nation," *Journal of Occupational Medicine*, October 1968.
3. Kenneth H. Klipstein, "Your Health and Industry," *National Health*, March 18, 1963.
4. N. H. Collisson, "Management and an Occupational Health Program," *Archives of Environmental Health*, February 1961.
5. William Jend, Jr., M.D., "Where Do We Want to Be in Occupational Medicine?" *Journal of Occupational Medicine*, July 1973.
6. Sherman M. Williamson, M.D., "Eighteen Years' Experience Without Pre-Employment Examinations," *Journal of Occupational Medicine*, October 1971.
7. Thrift G. Hawkes, M.D., "The Physical Examination in Industry: A Critique," *Archives of Environmental Health*, October 1962.
8. G. H. Collings, Jr., M.D., "The Pre-Employment Examination: Worth the Cost?" *Journal of Occupational Medicine*, September 1971.
9. Steven R. Cohen, M.D., "Another Look at the In-Plant Occupational Health Program," *Journal of Occupational Medicine*, November 1973.
10. U.S. Department of Labor, Bulletin No. 1830, 1972.
11. Ibid.
12. Manufacturing Chemists Association, "Off-the-Job Safety," Washington, D.C., 1961.
13. G. H. Collings, Jr., M.D., paper presented before the American College of Preventive Medicine, November 13, 1972.

14. Miles O. Colwell, M.D., paper presented before the Industrial Health Foundation, October 13, 1970.
15. Jend, op. cit.
16. Norbert J. Roberts, M.D., "Periodic Health Evaluation: Some Developments and Practices," *Journal of Occupational Medicine*, November 1973.
17. F. W. Holcomb, Jr., M.D., "IBM's Health Screening Program and Medical Data System, *Journal of Occupational Medicine*, November 1973.
18. Clarke and Ewing, "New Approach to Employee Health Programs," *Harvard Business Review*, July 1950.
19. Jend, op. cit.
20. Lorraine V. Klerman, "Health Education in Industry," *Industrial Medicine and Surgery*, July 1965.
21. Jerry Cassuto, M.D., "New Health Education Programs," *Journal of Occupational Medicine*, December 1967.
22. Fernbach, et al., "High Blood Pressure Education: An Employee Health Service Challenge," paper presented before the American Public Health Association, November 18, 1975.
23. "Industrial Health Education: A Program That Works," paper presented before the American Public Health Association, November 17, 1975.
24. Walter E. Book, *An Analysis of the Multi-Test Clinic of Richmond, Virginia*, Health Information Foundation, 1951.
25. J. F. Follmann, Jr., *Alcoholics and Business* (New York: AMACOM, 1976).
26. Edwin A. Gee, "Report on Safety," *Du Pont Context*, January 1975.

15

The Role of Unions

A decade ago Theodore Goldberg of the UAW said: "Organized labor in the United States has had a long and consistent record of concern for the welfare of those who suffer a disabling illness or injury, whether on or off the job."[1] Almost 20 years ago Jerome Pollack, then with the UAW, commented: "There are prospects for controlling disability . . . by employing a high standard of medical care, rehabilitation, and prevention. . . . This is an effort in which labor can join wholeheartedly. . . . Of all branches of disability control, prevention holds the greatest promise but it requires the greatest further development."[2]

At that same time Mr. Pollack, in addressing the 1959 National Health Forum, made the point that: "The distinction between occupational and nonoccupational health, which was always [a] legal . . . rather than a scientific distinction, is narrowing, and society is becoming increasingly determined to care for nonoccupational conditions." Addressing that same forum, Dr. William A. Sawyer, then with the International Association of Machinists, said: "Generally, organized labor does not consider its role in occupational health to be particularly significant. Since occupational health programs are the outgrowth of workmen's compensation law requirements, and must be provided by the employer, union labor has had little part in them . . . labor regards itself as the recipient rather than the participant. . . . [However] with labor union's concern about the inadequacies of in-plant occupational health programs . . . there is gradually emerging a belief that the overall health of workers should not be arbitrarily and inefficiently divided as it is now. . . . As a result, more than seven million trade union members have collective bargaining agreements which contain safety and health clauses . . . such programs today are concerned with the total health of the worker."[3]

Melvin Glasser of the UAW wrote: "An alternative must be found to the present company doctor system. The workplace should be used more extensively as a focal point for providing health services, not only for job-related diseases but for preventive health services, health education, and indeed all health problems."[4]

Such statements by spokesmen for important labor unions give evidence of the deep and long-standing interest on the part of organized labor in industrial health and preventive medicine. Corollary roles have been played by the unions through the years in wage and hour improvements, bettered working conditions, and in bargaining for health and disability insurance programs. The unions have also supported such forms of legislation as measures dealing with child labor, those establishing systems of workmen's compensation insurance, those providing for unemployment compensation, amendments to the social security system (OASI) to provide benefits in instances of total and presumably permanent disability, state legislation for temporary disability benefits, and minimum wage laws.

In several instances, furthermore, unions have established their own health services. Some, such as the programs developed by the garment workers in New York and Philadelphia, the United Mine Workers, and the St. Louis Labor Health Institute, provide a broad range of services. In other instances local unions have often provided specific types of services, which differ from union to union. In 1954 the American Labor Health Association was formed by the administrators of union direct service programs that had as their purpose the early diagnosis of disease, effective treatment, rehabilitation, care for the chronically ill, and the prevention of disease. The ALHA, now part of the Group Health Association of America, recognized that that which endangers the health of the nation must also be a problem for workers and that the key to potential long-term economies is research into the prevention of diseases and accidents. Closely related to the development of union health programs has been the long-standing support of organized labor in the development of prepaid group practice plans, or HMOs. In some instances, as in Detroit and Cleveland, organized labor has been instrumental in the development of such plans.

That there is a general concern among workers over occupational health hazards was made evident from a 1969 survey by the U.S. Department of Labor, which reported that 71 percent of workers considered occupational safety and health a more important matter of concern than wages.

There can be a very considerable difference, however, between such expressions of interest and concern and the playing of an active role in occupational and nonoccupational health and safety by the unions. There are, in fact, too many instances where the stance of the unions has served to hinder the development or the functioning of individual health programs. Not infrequently friction or suspicion develops between the employer and employees concerning the functioning of such programs, particularly when periodic physical examinations are involved. Employees become concerned that through a violation of confidentiality, advancement, promotions, pay increases, or the job itself can be placed in jeopardy. Such attitudes serve as a strong inhibition to the further development of industrial health programs. This was made clear by spokesmen for organized labor noted in the Conference Board study and, as was noted in the preceding chapter, is a reaction of which employers are aware.[5] The problem is not insurmountable, however. Where a joint employer-employee effort is made, with guaranteed confidentiality, the inhibition can be overcome. Experience has shown, for example, that when an alcohol control program is approached from the very beginning as a joint employer-union effort, with full and equal union participation, a workable program emerges. When these conditions do not exist, the program frequently fails.[6]

In 1972 Sheldon W. Samuels of the AFL-CIO interpreted the views of most union officials toward industrial health programs as being (1) that company doctors are no less a part of management than other company officials, (2) that the company doctor often cannot protect the confidentiality of patients' records, (3) that company medical programs must justify themselves by dollar benefits, (4) that in-plant medical care is poor in quality and quantity, and (5) that there is growing difficulty in accepting plant programs as resources unrelatable to and separate from the total medical care delivery system.[7] This is an indictment of significance and it must be heeded.

Despite such views, it must be noted that organized labor has been increasingly playing an active role in the reduction of occupational hazards to health and safety and has been sending its representatives to attend university courses in occupational health and safety. In some instances these courses are union-sponsored.

Among the unions that have been in the vanguard of active interest in job safety and health are the United Steel Workers of America, the United Automobile Workers, the United Paperworkers International Union, the International Association of Machinists, the United Rubber Workers, the Boilermakers Union, the United Mine Workers, and the Oil, Chemical, and Atomic Workers International Union. This latter union, through collective bargaining, requires that a survey of plant health and safety conditions be made by independent consultants at the expense of the employer, that physical examinations and medical tests of the workers be provided, and that the employer annually furnish the union with the statistics on the incidence of disease and death among the workforce.

Another example of union interest is the contract negotiated by the United Rubber Workers Union with the Goodyear, Firestone, Uniroyal, and General Rubber companies that led to a study of the health hazards of 18,846 workers in the rubber industry. The research project is jointly financed by company management and the union.

Still another example of union concern is the introduction by the UAW in its contract negotiations in July 1976 of the subject of controlling excessive noise in the workplace. The union had found that 20 percent of a random sample of workers had suffered severe hearing loss and that 60 percent had a measurable hearing impairment. It further suspected that noise could be a co-causative factor in diseases associated with toxic materials and that it reduced visual acuity and caused accidents as well as shoddy workmanship. The United Steel Workers of America, the United Rubber Workers, and the Boilermakers Union have also expressed concern over occupational noise.[8]

Yet other examples of the role of unions are the concerns of the International Association of Machinists over the control of dermatitis and the actions of the unions in the roofing, paint, textiles, printing press, typography, oil, chemical, paper, steel, automobile, and atomic energy fields in setting up scientific studies and surveillance systems to detect cancers and other diseases attributed to occupational hazards. One such effort is proceeding at Mount Sinai Hospital in New York where an epidemiological approach is being taken to determine the effects of PCBs.[9] Contract negotiations have also been developed by the United Paperworkers International Union to make available to the union the company medical records in order to determine the extent of the dangers of working with asbestos.

There are those, however, who do not feel that an industrial health program belongs in the area of labor negotiations or is an appropriate subject for collective bargaining. One such person is N. H. Collisson of the

Olin Mathieson Company. While taking cognizance of the fact that a "union can be, and very often is, an important contributor to the success of an industrial [health] program," Mr. Collisson says that such a program "has no place in collective bargaining discussions . . . [to do so] invades the personal and confidential area of relationship between worker and management . . . [which is] an individual relationship . . . foreign to the whole collective bargaining concept."[10]

The rights, and the responsibilities, of employees have, however, been delineated and strengthened by the conditions of OSHA, and this development will be discussed in the subsequent chapter. As was noted by the Conference Board survey published in 1974, the enactment of OSHA "appears to have stimulated a long-simmering controversy of the relationship of labor and management to both the occupational and nonoccupational health functions."[11] The study noted that prior to the enactment of OSHA, certain differences between management and labor over industrial health programs were bargaining issues.

In passing, it is of interest that in 1975 the Swedish unions presented to the ILO their program for a national occupational health policy. The proposal concentrated on chemical technologies, the threshold-limit values of hygiene, employee cooperation in the establishment of standards, the use of data processing, the safety and ergonomic aspects of design and production, and the sociopsychological aspects of the workplace environment. Recently enacted laws in Sweden include occupational safety and health, employment security, jobs for handicapped workers, worker representation on corporate boards, a work environment fund for research and education, and a revision of collective bargaining rights. Industrial health centers will be operated by companies individually or on a cooperative basis for small employers.[12]

It also should be noted that the International Labour Organisation has, since 1919, been an active force in the prevention of occupational health risks. The ILO has, for example, made recommendations in such areas as the safeguarding of machinery, protection against ionizing radiation, the prohibition of the use of white phosphorus in the match industry, ensuring the disinfection of wool contaminated with anthrax spores, protection against lead poisoning, the labeling of dangerous materials, setting tolerable limits of toxic substances, preventing air pollution in workplaces, safeguarding against benzine poisoning and dust hazards, installing safety appliances, preventing industrial accidents, eliminating the hazards of coal mining and the building industry, and establishing hygiene and occupational health services.[13]

Going beyond the workplace to environmental hazards in the community at large, the role of the unions does not appear to have clarified, and in most instances perhaps not unfairly might be described as "hands off." Not infrequently, where steps are contemplated to eradicate actual or suspected health hazards from the environment, workers display a concern over job layoffs, plant shutdowns, or complete job loss and stand in opposition to such actions.

A study made for the Ford Foundation by Nicholas Ashford at MIT has recognized the concern of workers over the loss of jobs that can result when environmental controls bring about layoffs in the workplace, shutdowns while corrective steps are taken, or the relocation of the plant to another area. Dr. Ashford takes the position that such concerns are ill-founded and that, in fact, more jobs may be created. He points to the fact that when it was discovered that vinyl chloride caused liver cancer it was contended that banning the use of the chemical would cost 500,000 jobs. "In fact," he says, "Goodrich designed a whole new technology ... and throughout the country less than a hundred people lost their jobs. And the price of polyvinyl chloride rose by just 3 percent. ... Environmental compliance is not the kind of financial ogre that it is alleged to be."[14]

Stellman and Daum have similarly noted that workers have a fear of plant shutdowns or the loss of their jobs as a result of the ecology movement or environmental health and safety improvements, but that "those plants which were shutting down were marginal [unproductive] operations" and that "according to the latest federal data, unemployment due to pollution control numbered fewer than 1,500 people in all.[15]

In summary, it is perhaps fair to say that the labor unions, while vastly interested in occupational health and safety, and while they have supported many types of legislation that have benefited their members as well as other workers, could do much more to improve both the occupational health and safety of workers and the nonoccupational aspects of such health and safety. Whether collective bargaining is the best approach to reach this goal is perhaps questionable. Most certainly an industrial health program that is cooperatively conceived by management and labor can be the most productive form of such a program. The labor unions are also called upon to devote much more attention to the improvement of the health of workers through health education than is presently the case. Their members, it is quite conceivable, will place more credence in such efforts if they emanate from the union rather than from the employer. At the same time the unions could devote more effort to coping with troubled employees—those with mental or emotional problems, alcoholics, or drug addicts. Here a cooperative effort with management would appear to be a matter of primary necessity.

SOURCES

Material from the following organizations was used in the preparation of this chapter.
American Labor Health Association
Sidney Hillman Health Center of New York
U.S. Department of Health, Education, and Welfare

REFERENCES

1. Theodore Goldberg, "Labor's Interest in a Proper Program for Long-Term Disability Protection," paper delivered before the American Management Associations, October 3, 1966.
2. Jerome Pollack, "Labor's Role in the More Effective Control of Disability," *Compensation Medicine*, Vol. 11, No. 2, 1959.
3. William A. Sawyer, M.D., "Role of Organized Labor in Occupational Health," paper presented before the 1959 National Health Forum, March 19, 1959.
4. Melvin A. Glasser, "Worker's Health," *American Journal of Public Health*, June 1976.
5. The Conference Board, "Industry Roles in Health Care," New York, 1974.
6. J. F. Follmann, Jr., *Alcoholics and Business* (New York: AMACOM, 1976).
7. Sheldon W. Samuels, paper presented before the New York Academy of Medicine, May 31, 1972.
8. *The New York Times*, May 2, 1976.
9. *The New York Times*, March 27, 1973.
10. N. H. Collisson, "Management and an Occupational Health Program," *Archives of Environmental Health*, February 1961.
11. The Conference Board, op. cit.
12. Birger Vicklund, "The Politics of Developing a National Occupational Health Service in Sweden," *American Journal of Public Health*, June 1976.
13. Luigi Parmeggiani, "The ILO and the Prevention of Occupational Risks," *International Social Security Review*, 1970.
14. Nicholas A. Ashford, *Crisis in the Workplace: Occupational Disease and Injury* (Cambridge, Mass.: MIT Press, 1976).
15. Jeanne M. Stellman and Susan M. Daum, *Work Is Dangerous to Your Health* (New York: Vintage Books, 1973).

16

The Role of Government

Government at all levels plays a very considerable role in a great many areas directly related to the responsibilities of industrial health programs. Particularly is this the case if the broad concept of an industrial health program discussed earlier is to become the trend in the years ahead.

Unquestionably the enactment of workmen's compensation laws by the states, commencing in 1911, provided the first impetus for the establishment of industrial health programs in an attempt to control occupational accidents and diseases. In 1936 enactment of the Walsh-Healey Act established safety standards for plants doing work for the federal government. Meanwhile labor legislation, commencing with the National Labor Relations Act (the Wagner Act) in 1935 and the Labor Management Relations Act (the Taft-Hartley Act) of 1947, established and strengthened the position of organized labor as respects the rights of collective bargaining and laid the groundwork for its more recent demands for occupational health and safety.

In a sense supplementing these statutes, and dealing with worker health and safety, were the Longshoremen's and Harbor Workers' Act, the Maritime Safety Act, the Migrant Health Act, the Metal and Nonmetallic Mine Safety Act, and the Coal Mine Health and Safety Act. The Rehabilitation Act of 1973 was the most recent piece of federal legislation to deal with the rehabilitation of disabled workers and the employment of handicapped persons. The Atomic Energy Act of 1954 has applicability to certain places of employment using nuclear energy. It was with the enactment of the Occupational Safety and Health Act in 1970 (OSHA), however, that the federal government assumed broad and extensive authority having to do with the safety and health of employed people.

Agencies of the federal government concerned with aspects of industrial health include:

Department of Labor
　Occupational Safety and Health Administration
　National Advisory Committee on Occupational Safety and Health
　Occupational Safety and Health Review Commission
　Bureau of Mines
　Bureau of Labor Statistics
　National Commission on State Workmen's Compensation Laws
Department of Health, Education and Welfare
　National Institute of Occupational Safety and Health (NIOSH)
　Bureau of Industrial Hygiene
　Public Health Service
　National Institutes of Health
　National Institute on Alcohol Abuse and Alcoholism
National Center for Productivity and Quality of Working Life

State governments also play, and have traditionally played, a considerable role as respects industrial health through their labor, health, and mental health departments. It is of historic interest that the first industrial health programs in government agencies came into being in 1913 in New York and Ohio. As will be shown, the states can play a significant role in the administration of OSHA, and many of them do.

Local governments have additionally played a traditional role as respects industrial health through the maintenance of health departments, the establishment of building codes, fire and wiring inspection, the maintenance of sewage disposal and sanitation means, pest and rodent control, and the control of various forms of environmental pollution.

The role of government is discussed here in five aspects: (1) OSHA, (2) environmental control, (3) product safety, (4) public health efforts, and (5) other legislative approaches.

In approaching these various roles played by government, it should be understood at the outset that one reason that categorizing the activities of government is so difficult is the considerable degree of overlap not only among the various levels of government but also among the various agencies within any one level of government. In part, at least, this results from the plethora of legislation over a period of years, which, for whatever reasons, fails to take cognizance of responsibilities already on the statute books.

OSHA

In 1970 Congress took a major step in the improvement of safety and health conditions in places of employment with the enactment of the Occupational Safety and Health Act (OSHA) PL 91-596, superseding the Walsh-Healey Act of 1936. It is a potentially far-reaching law, the effects of which will not become fully apparent for some years. A copy of the Act is contained in Appendix A.

The Act states: "The Congress finds that personal injuries and illnesses arising out of work situations impose a substantial burden upon, and hindrance to, interstate commerce in terms of lost production, wage loss, medical expenses, and disability compensation payments." Its purpose is "to assure as far as possible every working man and woman in the Nation safe and healthful working conditions and to preserve our human resources" by:

1. Encouraging employers and employees in their efforts to reduce the number of occupational hazards and stimulating the institution of new industrial health programs.
2. Providing that employers and employees have separate but dependent responsibilities for safety and health.
3. Authorizing the Secretary of Labor to set mandatory standards for safety and health.
4. Building upon advances already made in the field of industrial health.
5. Providing for research and the development of innovative methods and techniques in industrial health.
6. Exploring ways to discover latent occupational diseases.
7. Providing medical criteria on which judgments might be made.
8. Providing training programs for needed occupational health personnel.
9. Providing for the promulgation of occupational safety and health standards.
10. Providing for the enforcement of such standards.
11. Encouraging the states to assume responsibilities in occupational safety and health through the provision of grants.
12. Providing for reporting procedures for occupational injuries and diseases.
13. Encouraging joint labor-management efforts in industrial health.

The Act provides that each employer shall furnish a place of employment "free from recognized hazards that are causing or likely to cause death or serious physical harm to employees" and that each employer shall comply with all standards, rules, regulations, and orders having to do with occupational safety and health.

The Act applies to any workplace in a state, the District of Columbia, the Commonwealth of Puerto Rico, the Virgin Islands, American Samoa, Guam, the Trust Territory of the Pacific Islands, the Johnston Islands, Wake Island, the Outer Shelf Islands, and the Canal Zone. It applies to all federal agencies as well as to places of private employment. The only exceptions are workplaces under the Atomic Energy Act of 1954; mines, railroads, and trucking (which are covered by other legislation); and self-employment. The provisions of OSHA do not supersede or affect workmen's compensation laws nor do they effect the processes of employers' liability.

The Act provides that the occupational safety and health standards established by the Secretary of Labor shall be (1) those national consensus standards promulgated by nationally recognized standard-producing organizations that have substantial agreement and that are designated as acceptable standards by the Secretary of Labor; (2) those standards established by any agency of the federal government or by an act of Congress; (3) those standards promulgated under the Walsh-Healey Act of 1936 and the National Foundation of Arts and Humanities Act; or (4) those standards promulgated by the Secretary of Labor.

The Act further establishes a procedure for standard setting by the Secretary of Labor, which includes consultation with advisory committees, the publication of proposed standards, and consideration of all comments received. It is specifically provided that where toxic materials or harmful physical agents are concerned, the standards shall be those that most adequately ensure the safety of workers. The standards shall, furthermore, prescribe the protective equipment that is to be used, technological procedures for monitoring hazards and conditions, and the type and frequency of any medical examinations deemed necessary, such examinations to be made available by the employer at his own expense. In the promulgation of standards, priorities are established with due regard to urgent needs for standards in certain industries, trades, crafts, occupations, and businesses.

The Secretary of Labor is authorized to inspect and investigate workplaces in order to determine if a hazard exists. In so doing he shall have access to all records concerned with exposures to potentially toxic materials or harmful physical agents. Employees or their representatives are privileged to observe all monitoring procedures and to review all records maintained. They can, additionally, request an inspection of the workplace. The employer, on his part, is required to notify employees of any hazards to which they are exposed.

An enforcement procedure is established for violations of the standards promulgated under the Act, including notification by the Secretary of Labor to the employer, the issuance of a citation for failure to correct harmful conditions, and the means for contesting such actions. Penalties are prescribed for willful violation of the Act in the form of fines and imprisonment, and provision is made for judicial review. In the enforcement of the Act provision is made for preserving the confidentiality of trade secrets.

In the establishment and enforcement of standards, provision is made for reasonable limitations, variations, tolerances, and exemptions necessary to avoid serious impairment of the national defense.

In instances where steps are deemed necessary to counteract "imminent dangers," the Secretary of Labor may petition the U.S. district courts for a restraining order. Should an employee be injured in the absence of such action, he might bring action against the Secretary of Labor.

The Act is clear that it shall not serve to prevent a state from asserting jurisdiction in any instances where OSHA has not established standards. Furthermore, if a state desires to establish a plan by which it would assume responsibility for the enforcement of the provisions of OSHA it may do so, and such plan shall be approved by the Secretary of Labor provided that it appears to be at least as effective as federal enforcement procedures.

OSHA makes provision for four new bodies of government at the federal level:

1. A National Advisory Committee on Occupational Safety and Health appointed by the Secretary of Labor from the ranks of business management, labor, occupational safety and health professionals, and the public to advise, consult with, and make recommendations to the Secretary of Health, Education, and Welfare.

2. An Occupational Safety and Health Commission in the Labor Department to be concerned with the enforcement of standards, to handle employee complaints, and to hold hearings concerning alleged violations of the Act.

3. A National Institute for Occupational Safety and Health (NIOSH) within the Department of HEW to conduct research and develop standards and regulations.

4. A National Commission on State Workmen's Compensation Laws in the Labor Department to study and evaluate such present laws and report its findings.

OSHA also provides for grants in three areas: (1) for the development of educational programs to provide an adequate supply of qualified occupational safety and health personnel; (2) to the states for the identification of their needs, for the development of state plans, and for improvement in the administration and enforcement of state occupational safety and health laws; and (3) for experiments, demonstrations, and statistical gathering related to occupational safety and health. In addition it makes provision for economic assistance to small businesses in the form of loans where such assistance is determined to be necessary to comply with OSHA or state standards and if the business "would suffer substantial economic injury" without such assistance.

The Act further provides for the conduct of necessary research and for the collection and maintenance of related statistics of work injuries and illnesses other than those of a minor nature or not involving medical treatment.

In the enforcement of OSHA it is necessary that standards be established by the Occupational Safety and Health Administration of the U.S. Department of Labor and by the National Institute for Occupational Safety and Health of the U.S. Department of Health, Education, and Welfare. In so doing, these agencies frequently seek the counsel and assistance of such voluntary organizations as the American National Standards Institute, the American Iron and Steel Institute, the Manufacturing Chemists Association, the National Safety Council, and the American Industrial Hygiene Association, as well as representatives from industry, labor, the professions, the states, and universities. Such standards are subject to modification, expansion, or revocation from time to time and are updated on a continuing basis. Each standard must include identification of the agent, permissible levels of exposure, the methods of measurement and monitoring, the medical examinations and treatment required, and the approved method for control of the hazard.

The first standards for compliance with OSHA were promulgated March 29, 1971. By March 1975 the U.S. Department of Labor had established standards for general industry in the following areas:

Abrasive blasting and grinding
Accident record keeping
Air receivers and tools
Aisles and passageways
Belt sanding machines
Boilers
Calenders, mills, and rolls
Chains, cables, and the like
Chipguards
Chlorinated hydrocarbon
Compressed air
Conveyors
Cranes and hoists
Cylinders-welding
Dip tanks

Dockboards	Medical services and first aid
Drains	Noise exposure
Drill presses	Portable electric tools
Drinking water	Power transmission
Drive belts	Pressure vessels
Electrical installations	Protective equipment
Elevators	Pulleys
Emergency flushing	Punch presses
Exits	Radiation
Explosives and blasting	Railings
Eye and face protection	Revolving drums
Fan blades	Saws
Fire doors and protection	Scaffolds
Flammable liquids	Spray finishing
Floor, openings, loading, and protection	Stairs
Forklift trucks	Stationary electrical devices
Gears	Storage
Guards	Tanks
Hand tools	Toeboards
Head protection	Toilets
Hooks	Toxic vapors, gases, mists, and dusts
Housekeeping	Trash
Insulation	Washing facilities
Jointers	Welding
Ladders	Wire ropes
Lighting	Woodworking machinery
Lunchrooms	
Machine guarding	
Machinery	

Under consideration were standards for heat stress, lead, asbestos, beryllium, trichloroethylene, toluene, sulfur dioxide, ammonia, various hexavalent chromium compounds, as well as more complete standards for 400 toxic substances currently regulated. In May 1976 a proposal to modify the existing standard for temporary labor camps was withdrawn because there was "no adequate basis for a final standard." The OSHA standards for ammonia, arsenic, beryllium, coke oven emissions, cotton dust, lead, noise, sulfur dioxide, toluene, trichloroethylene, and ketones were, however, held up in mid-1976 owing to the need for compiling economic impact statements.

Under OSHA the Bureau of Labor Statistics (BLS) is charged with the responsibility for collecting, compiling, and analyzing data concerning occupational injuries and illnesses. Accordingly, BLS conducts surveys and collects other related information. Employers are required to maintain records of the incidence and nature of injuries and illnesses, as well as those concerned with monitoring and exposure to potentially toxic materials or harmful agents.

Under the Act, NIOSH performs functions in several areas. It conducts research in an endeavor to establish criteria on which standards for toxic materials and harmful physical agents can be based. In so doing it establishes limits on maximum permissible exposure for both workers and the public and provides for monitoring, protective equipment, caution signs and labels, waste disposal, storage, and record keeping. NIOSH also provides technical assistance to employers, employees, or government organizations on request. Such assistance concerns hazard evaluation, the availability of technical information, the means for accident prevention and industrial hygiene, and the establishment of medical services. NIOSH has a further responsibility of providing training to assure an adequate supply of qualified occupational health personnel in industry. In 1972 it was estimated that there was a shortage of from 5,000 to 10,000 safety professionals, 1,000 industrial hygienists, 3,000 occupational health physicians, 10,000 occupational nurses, and 8,000 occupational health scientists. The intent of NIOSH is to overcome such shortages by providing grants and contracts for the development of the necessary educational programs.

By July 1, 1975, recommended standards for compliance with OSHA had been developed by NIOSH as follows (the date of promulgation is also shown):

Ammonia	July 15, 1974
Arsenic	January 21, 1974
Revised	June 23, 1975
Asbestos	January 21, 1972
Benzene	July 24, 1974
Beryllium	June 30, 1972
Carbon monoxide	August 3, 1972
Chloroform	September 11, 1974
Chromic acid	July 17, 1973
Coke oven emissions	February 28, 1973
Cotton dust	September 26, 1974
Fluorides	June 30, 1975
Hot environments	June 30, 1972
Identification system for occupationally hazardous materials	December 20, 1974
Inorganic lead	January 5, 1973
Inorganic mercury	August 13, 1973

THE ROLE OF GOVERNMENT 219

Noise	August 10, 1972
Silica	November 11, 1974
Sulfur dioxide	February 11, 1974
Sulfuric acid	June 6, 1974
Toluene	July 23, 1973
Toluene diisocyanate	July 13, 1973
Trichloroethylene	July 23, 1973
Ultraviolet radiation	December 20, 1972
Vinyl chloride	March 11, 1974
Xylene	May 23, 1975

NIOSH has also proposed guidelines for occupational health and safety program evaluation for hospitals, including preplacement physical examinations, periodic health appraisal examinations, health and safety education, immunizations, care for occupational illness and injury, health counseling, environmental control and surveillance, health and safety records systems, and coordinated planning. It has also issued alerts on PCBs, HMPA (hexamethylphosphoric triamide), and DDM (diaminodiphenyl methane) as respects their potential danger.

In fiscal years 1975 and 1976, OSHA inspections were made in over 152,000 places of employment. Of these, 45.4 percent were in manufacturing establishments and 28.9 percent were in construction. Only 0.3 percent were in finance, insurance, and real estate, 0.4 percent in agriculture, and 0.7 percent in mining. Of those inspected, 79 percent were found to be not in compliance with OSHA regulations. The highest rate of noncompliance was 88 percent for the retail trades, followed by 84 percent for manufacturing, 82 percent for the service industries, 81 percent for wholesale trade, 77 percent for transportation, 75 percent for mining, 74 percent for agriculture, 72 percent for construction, 63 percent for finance, insurance, and real estate, and 58 percent for the maritime industry. In 1975 and 1976 OSHA issued over 118,000 citations, of which 7,828 were contested. In 1974 an essentially similar pattern had prevailed, resulting in 30,883 citations, 162,981 alleged violations, a reported 545 willful, repeated, "imminent danger" cases, and 3,579,742 proposed penalties.

The penalties for noncompliance with the requirements of OSHA range from $1,000 for each serious violation or an optional amount up to $1,000 for nonserious violations, up to $10,000 each for repeated violations. Criminal penalties can attach to willful violations that result in death, in the form of fines not to exceed $10,000, imprisonment not to exceed six months, or both. An example of such fines levied is shown in cases of excessive noise where, between May 1971 and May 1976, 8,000 citations resulted in fines ranging up to $1,000 but averaging $33. Another example is an auto battery plant in California that was fined $4,500 by the State Division of Industrial Safety.[1] In instances of legal dispute, OSHA has most generally won. Where the courts have not upheld OSHA, it has most usually been on procedural grounds.

Expenditures for OSHA were $70.1 million in 1974 and by 1976 were estimated to be $116 million. The budget appropriations in 1974 were $26 million for NIOSH and $69 million for OSHA.

As has been noted, the enforcement of OSHA can be delegated to the states. The state must demonstrate that it has a regulatory body and program compatible with the federal requirements and that it has appropriate technical capability to operate the program. The state in turn may delegate authority to its political subdivisions. Through OSHA, 50 percent of the necessary funds are provided. OSHA also mandates occupational safety and health programs for all public employees in a state, to the extent permitted under state law. Here it should be noted that prior to the enactment of OSHA many states regulated at least certain occupational safety and health hazards, including such activities as licensure, inspection, standard setting, and monitoring. By 1975, OSHA had approved 26 state plans.

From what has been shown, it is obvious that both the administration of and compliance with OSHA are complex matters covering a wide range of concerns and presenting sizable costs to the taxpayer. The effects on industrial health programs, both now and in the future, are clearly far-reaching. The enactment of OSHA was a major development in the field of industrial health and one whose full impact will not be apparent for some time to come.

The administration of OSHA has been the subject of very considerable criticism and controversy in some quarters and there are those who consider the administration of the Act to be ineffectual. It has been pointed out, for example, that the administrative staff has only 500 inspectors and 50 industrial hygienists to examine 4.1 million workplaces. More recently, however, the number of federal inspectors has been reported to be 1,500, plus 1,500 to 1,800 state inspectors. It has been said that enforcement of the act has "been weak, and its effectiveness questionable."[2] Others have said they "don't think OSHA can do the job alone" and that organized labor must play a still greater role if improvements in occupational safety and health are to be made on a broad scale. Thus a recent study said: "There are limitations on how successful the improvement of occupational safety and health can be without participation by the workers in the monitoring or surveillance of conditions in the workplace."[3] It has also been maintained that the "regulation for occupational safety and health is still spotty and tentative." Several other sources have been critical of the fact that OSHA standards lag behind those of the voluntary sector. Thus David Burnham commented in *The New York Times* on December 20, 1976, "Almost everyone finds fault with the Occupational Safety and Health Administration."

Complaints against the role played by most state and local governments have also been voiced in several quarters. These criticisms include such matters as a lack of comprehensiveness in the administration of their responsibility, the omission of various hazards, the lack of sufficient resources and competencies, poor planning, and apathy. It has been said that less than a quarter of the workforce has received strong support from state and local governments. The point has also been made that small plants cannot afford private consultation and that state and local departments can provide consultation to only a small proportion of industries. Thus Dr. John Hall of Georgia State University has said that "although the government wants immediate compliance it does not set up adequate provisions for training employers." He said the state of Georgia was concerned over this and was trying to establish an employer training program for a nominal fee.[4] Of interest in this respect is a seminar conducted by the American Management Associations in May 1972 to assist small companies in compliance with OSHA.

In 1976 the American Enterprise Institute for Public Policy Research suggested that OSHA should concentrate on occupational health, since it is unlikely that it can reduce injuries to any significant extent beyond what has already been accomplished. Its standards, it was contended, are unrelated to the major causes of injuries, and compliance incentives are weak.[5]

Three recent examples of litigation under OSHA are of particular interest. In 1976 the Textile Workers Union of America charged the state of North Carolina with the outright refusal to issue citations for violations of the cotton dust standard, with the reversal of a citation for extremely hazardous noise levels in plants because "feasible engineering controls do not exist," with the establishment of an illegal policy permitting respirators as a substitute for engineering controls of dust hazards, and with "gross incompetence and violations" of federal guidelines. The union also accused the U.S. Department of Labor of delaying efforts to tighten federal controls on cotton dust. TWUA and the North Carolina Public Interest Research Group accordingly filed suit in the U.S. District Court in Washington against the Labor Department to compel formal proceedings for the development of an effective cotton dust standard.[6] In 1974 the Labor Department had revoked its earlier rule concerning "no hands on dies" standards for the protection of workers with power presses. The reason given for the revocation was that the standard was "neither technically nor economically feasible." Instead, a fail-safe standard for such presses was adopted. In 1975 the case was taken before the Third U.S. Circuit Court of Appeals, which ruled that the Labor Department had failed to justify its revocaton of the "no hands on dies" rule.[7] The third case was a suit filed on March 3, 1976, against OSHA with respect to the consideration of economic impact in the promulgation of its regulations. It was alleged that OSHA refused to permit any consideration of the impact of inflation in the development of its new health and safety standards.[8]

From these few citations the complexities involved in the enforcement of OSHA become readily apparent. Another type of problem emerges from a statement by Jack Sheehan of the United Steelworkers at the end of 1975 when he contended:

> A negative reaction to OSHA has set in, in Congress. There is now a very potent political and legislative force which intends to weaken or eliminate key policy objectives mandated by the law. They intend to radically amend the Act. . . . Even now some of the states which we prevailed upon to leave enforcement to the federal government are capitalizing upon the discontent. . . . It is our Democratic friends, elected from key working class districts, who are intent upon subjecting OSHA to floor debate and amendment. . . . OSHA is certainly a key target precisely because of its enforcement and regulatory authority.[9]

It is evident, then, that a considerable element of discontent and concern surrounds the administration of OSHA to date.

ENVIRONMENTAL CONTROL

Another role played by government, by no means unrelated to the functioning of an industrial health program, is that of environmental control. Again, all levels of government play a role, although more recently the federal government has assumed the limelight.

The state and local governments have traditionally played a role in environmental control. While the federal government has assumed increasing jurisdiction in this area, it is to be borne in mind that under the Constitution of the United States all powers not specifically vested in the federal government are reserved to the states. The police power of the states in this respect has consistently been upheld by the courts. Thus the state and local governments have played an active role in smoke abatement, the control of air and water pollution, and noise abatement. Federal acts, furthermore, have left to the state and local governments substantial authority and, in some instances, provide for state administration subject to the approval of the administering federal agency. They additionally provide technical and financial assistance to the states.

All such government regulations have a considerable and increasing impact on the processes of industry. Unfortunately, in the execution of these responsibilities interjurisdictional problems develop among government agencies and can serve to confuse all parties concerned. The spotty nature of state controls can also be a matter for concern. For example, a survey of 25 states revealed that only 11 controlled solid-waste disposal and that none controlled the solid-waste disposal of plastics.[10]

Federal legislation having to do with environmental quality and control has taken several forms over the past two decades. A mere listing of certain of those measures indicates the areas of concern:

> Air Pollution Act, 1955
> Water Pollution Control Act, 1956
> Clean Air Act, 1963
> Solid Waste Disposal Act, 1965
> Clean Waters Restoration Act, 1966
> Air Quality Act, 1967
> Clean Air Act, 1970
> Water Quality Act, 1970
> Noise Pollution and Abatement Act, 1970
> National Environmental Policy Act, 1970
> Energy Supply and Environmental Coordination Act, 1974
> Toxic Substance Control Act, 1976

The federal agencies involved in the administration and enforcement of these acts include the Environmental Protection Agency (EPA), Public Health Service, NASA, Council on Environmental Quality, Department of Transportation, Atomic Energy Commission, National Institutes of Health, HUD, and Federal Energy Administration. Examples of the functioning of some of the agencies involved with environmental controls will suffice.

Under the air pollution acts the Public Health Service has authority to control air pollution from new motor vehicles, to abate air pollution, to investigate and seek to prevent new sources of pollution, to conduct research, to make grants to air pollution control agencies, to prescribe regulations, and to secure abatement through the office of the attorney general. The federal government also operates a Continuous Air Monitoring Program to provide data on concentrations of air pollution.

Noise abatement has been a long-standing concern of state and local governments. At the federal level such concerns are spread over several agencies. The EPA is concerned with construction and transportation noise. Agencies created by the Noise Pollution and Abatement Act conduct research in the subject. The Department of Transportation and the Federal Aircraft Noise Abatement Program, as well as the National Aeronautics and Space Administration, are concerned with transportation noises, particularly aircraft. HUD has certain concerns over noise abatement in the inner cities.

Under the National Environmental Policy Act the federal government created, in 1970, the Council on Environmental Quality to advise the president, and the Environmental Protection Agency to execute the provisions of the Act. EPA has issued ambient air quality standards, has established air quality control geographic regions, has ordered reductions in the lead content of gasoline, and has taken action with respect to such matters as the discharge of asbestos fibers in waste, exposure to vinyl chloride, the use of certain pesticides, municipal sewage disposal, industrial fuel burning, fluid wastes, smoke stack emissions, sulfur emissions, construction and transportation noise, and the discharge of carcinogenic chemicals. The EPA issues regulations, is responsible for their enforcement, and can take emergency action when dictated by circumstance.

The matter of the use and discharge of carcinogenic chemicals is an area in which there is considerable interest today. As a consequence, in 1976 the National Cancer Institute of the National Institutes of Health established a clearinghouse on environmental carcinogens designed to eliminate duplication of effort among federal agencies, to focus attention on potentially hazardous substances, and to alert the public to such hazards. This effort is surrounded with conflict, dispute, and differences of opinion.

Very recently, under the Energy Supply and Environmental Coordination Act of 1974, the Federal Energy Administration has required 30 new electric power plants in 17 states to possess a coal-burning capability. Similar actions were under consideration in several other cases.

These few examples clearly serve to illustrate the breadth of the subject of environmental control as well as its far-reaching effects. In the execution of the acts and the responsibilities of the several agencies it is not uncommon for the disputes that arise to be taken before the courts. One example, in 1976, was the decision of a federal court that ordered EPA to list airborne lead as a pollutant with adverse effects on the public health (the order was issued under the provisions of the Clean Air Act). Another example, also in 1976, was the decision of a U.S. Court of Appeals ordering New York City to effect a clean-air plan.

At the same time a concern has developed on the part of some interests that there has been a relaxation in the control of environmental hazards, most usually taking the form of a slowing down of the pace or the granting of more time for compliance with regulations. It has also been observed that in the execution of environmental controls conflict can arise with OSHA. For example: air pollution in a place of employment may be corrected by an exhaust system, but this in turn might cause air pollution in the community; or dangerous materials might be eliminated from the workplace through sewage but to the detriment of the community water system.

By 1976 there was considerable concern over the effect of the state of the economy on environmental controls. A nationwide survey at that time, however, indicated relatively few instances of relaxation in the establishment or execution of such controls.[11] Nonetheless, in some instances, with increasing concern over state and local government financing and the attitudes of taxpayers, budgets for the enforcement of environmental controls have been reduced. The survey showed that of 14 states contacted, 6 had reduced the necessary appropriations. In fact, the governor of one industrial state referred to the costs of environmental controls as being a "luxury." Some states have, however, increased their budgets in this respect. Of interest is the fact that EPA has estimated that federal environmental controls are costing the people $47 a year per capita and that consumer fuel bills could be increased, owing to the enforcement of controls, 10 percent over the next ten years.

PRODUCT SAFETY

The role of government concerning the safety of products offered to the consuming public is also directly related to an industrial health program. Involved are governments at all levels: the local governments in such areas as sanitation surveillance, more recently state governments in their control of consumer products, and the federal government. There are three federal agencies that today attract considerable attention in the field of product safety. A few words about each will serve to illustrate the scope of their authority and certain problems that emanate therefrom.

The Federal Food and Drug Administration (FDA) has a wide range of responsibility under the most recent enactment of Congress on this subject: the Consumer Products Safety Act of 1972. Accordingly the FDA is responsible for the safety and effectiveness of drugs, the safety of the entire supply of food, the safety of medical devices, the safety of cosmetics, the safety of veterinary products, and the assurance of uniform quality of blood products. Today the manufacture and processing of foods and drugs constantly presents fresh problems as new materials are regularly put into use. In early 1976, for example, the FDA prohibition of Red Dye No. 2 as a food additive and colorant attracted much attention, as did its earlier withdrawal of vinyl chloride as a propellant in aerosol products.

That this is a broad responsibility is evident from a highly critical statement by Senator Edward Kennedy in late 1975 when he said: "On the basis of our investigation of this agency [FDA], I am forced to conclude that as presently constituted, it is unable to effectively carry out its mandate to protect the health and safety of the American people." Noting the budget limitations in relation to the broad responsibilities of the FDA, he challenged the scientific capabilities of its employees. To correct the situation he introduced two measures that would establish, respectively, a Drug and Devices Administration and a Food and Cosmetics Administration.[12]

A second agency, concerned with enforcement of the Radiation Control for Health and Safety Act of 1968, is the Bureau of Radiological Health. The Bureau is a part of the FDA. Its authority extends to all electronic products and in so doing the Bureau establishes standards to prevent harmful radiation emissions. Particularly, it is concerned with the safety of X-rays, microwave equipment, small boat radar equipment, dryers, and similar products.

A third federal agency, one that receives considerable publicity, is the National Commission on Product Safety, established in 1968 to study the scope and adequacy of measures whose purposes are to protect the consumer. More recently, in 1972, Congress enacted the Consumer Product Safety Act, and this led to the activation of the Consumer Product Safety Commission in 1973. Under this Act the Commission has authority to establish and enforce safety standards for more than 10,000 products, including their design, construction, contents, performance, packaging, and labeling. Manufacturers and distributors are required to report defects in their products to the Commission. Affected are 200,000 manufacturing plants and 2 million retail establishments. In three years the Commission has adjudged 25 million items to be unsafe and demanded repairs, replacement, or recall. It achieved "corrective action" on 14 million products and banned many others. Not included in the authority of the Commission are foods, drugs, cosmetics, radiological equipment, flammable fabrics, motor vehicles, tobacco, pesticides, firearms, boats, and aircraft; these are regulated by other agencies.

The Commission has been criticized on the grounds that most of the observed problems were reported by manufacturers themselves and were voluntarily corrected. Another criticism is that the Commission works too slowly. It is also said to be seriously understaffed.

The broad range of interests and concerns for the safety of consumer products is evident from what has been shown. It places upon industry, and on industrial health programs, a responsibility of consequence.

PUBLIC HEALTH EFFORTS

The broad aspects of preventive medicine have already been discussed. Many of these efforts are conducted by the public health agencies of local and state governments, as well as the federal government. Since the purpose is to reduce the incidence of unnecessary illnesses and accidents, the relationship to an industrial health program is self-evident.

Included in these public health efforts are concerns over new occupational health hazards, the study and control of occupational diseases, health promotion through the workplace, and the prevention or control of chronic diseases. In carrying out these activities direct service is often available to places of employment.

Thus Morton S. Hilbert, as president of the American Public Health Association, recently commented:

> It is evident that much more is known than is applied in prevention. In order for knowledge to be applied effectively and efficiently, organized and coordinated public effort is required. Most preventive measures can only be implemented or initiated on a group basis. For these reasons, society entrusts public agencies with the responsibility for safeguarding the public interest. The public provides legal, fiscal, social, and moral support to these organizations and they are, therefore, accountable to the public in carrying out their functions. Public agencies have a responsibility to give high priority to the application of the science of prevention. More specifically, the agencies should do the following:
>
> 1. Develop and adopt national and state legislation, where appropriate, which will reflect the need for and feasibility of applying preventive measures and ensure that relevant existing laws are implemented.
> 2. Provide adequate budget at federal, state, and local levels as well as providing for research.
> 3. Monitor the level of existing prevention programs and activities and, if necessary, allocate responsibility for such programs to other agencies.
> 4. Educate the public, providing information, developing prevention attitudes, and modifying behavior so as to make optimal individual use of prevention.
> 5. Acquaint the public with the health system and the services available to them.

6. Establish standards for health professionals that ensure their knowledge and proficiency in the latest prevention practices.

7. Encourage health-related organizations such as social service agencies, and any other organizations influencing legislation, to apply pressure for enhancing prevention in such programs as National Health Insurance, Comprehensive Health Planning, and others.[13]

OTHER LEGISLATIVE APPROACHES

Other legislative approaches taken by government having a relationship to industrial health programs or to occupational health and safety include, briefly:

1. The activities of the Bureau of Mines, which today functions under the Federal Coal Mine Health and Safety Act of 1969 and which, among other things, is concerned with the incidence of black lung among coal miners.

2. The Rehabilitation Act of 1973 under which employment opportunities are provided to qualified physically and mentally handicapped persons. This Act followed a series of state and federal rehabilitation programs dating back to 1920 and having considerable emphasis on vocational rehabilitation.

3. The Atomic Energy Commission, which is charged at once with promoting the expansion and use of atomic energy and with the establishment of exposure standards. In this latter capacity the AEC has set a legal limit for radiation exposure, although there are those who consider that this limit "in essence, guarantees a higher incidence of cancer for exposed workers."[14]

4. The Migrant Health Act of 1962, which is concerned with the health and living conditions of migratory farm workers, including the establishment of health care facilities.[15]

5. The Comprehensive Alcohol Abuse and Alcoholism Prevention, Treatment, and Rehabilitation Act of 1971, which led to the establishment of the National Institute of Alcohol Abuse and Alcoholism in 1971. Since its formation in 1971, NIAAA has played an active role in, among other things, coming to grips with the problem of alcoholism among employed people.[16]

SOURCES

Material from the following organizations was used in the preparation of this chapter.

National Institute of Occupational Safety and Health (NIOSH)

National Safety Council
Pennsylvania Department of Environmental Resources
U.S. Department of Labor

REFERENCES

1. *The New York Times,* June 6, 1976.
2. Leslie I. Boden, "Vinyl Chloride: Can the Worker Be Protected?" *New England Journal of Medicine,* March 18, 1976.
3. Nicholas A. Ashford, *Crisis in the Work Place* (Cambridge, Mass.: MIT Press, 1976).
4. *Journal of Commerce,* December 20, 1971.
5. *American Industrial Hygiene Journal,* June 1976.
6. AFL-CIO, IUD, Vol. 5, No. 1, 1976.
7. Ibid.
8. *American Industrial Hygiene Journal,* June 1976.
9. AFL-CIO, IUD, Vol. 4, No. 4, 1975.
10. Manufacturing Chemists Association. "Solid Waste Management of Plastics" (Washington, D.C.), undated.
11. *The New York Times,* March 15, 1976.
12. *Congressional Record,* November 20, 1975.
13. Morton S. Hilbert, "Prevention: Our Number One Health Problem," unpublished paper, July 1976.
14. Jeanne M. Stellman and Susan M. Daum, *Work Is Dangerous to Your Health* (New York: Vintage Books, 1973).
15. J. F. Follmann, Jr., "Migratory Workers," Health Insurance Association of America (Washington, D.C.), 1974.
16. _____, *Alcoholics and Business* (New York: AMACOM, 1976).

Part V

What Is Being Done?

As was shown earlier, some industrial health programs had their beginnings almost a hundred years ago, and for centuries before that concerns were expressed over certain occupational health hazards. The enactment of workmen's compensation laws by the states commencing in 1911 added a very considerable impetus to the development of such programs in the United States. Much more recently the enactment of OSHA, discussed in the previous chapter, has had, and will continue to have, a marked impact on the establishment and on the scope of efficacious industrial health programs.

The result is that today, as will be shown in Chapter 19 and Appendix B, a great deal is being done in American places of employment to better the health and safety of employees. Most large employers have established some type of industrial health program, and many of these are quite comprehensive in their scope. Since these are large employers, it means that a considerable proportion of the workforce benefit from their operation. Increasingly such programs are extending their concerns to include the consideration of nonoccupational illnesses and injuries. Many also include in these programs at least partial responsibility for the safety of their products and for the potential environmental health hazards that might emanate from the operations of an industry.

Smaller places of employment have a much more difficult task in developing an industrial health program, particularly those employing only a relatively few employees. Unfortunately, a quite high proportion of the total workforce are employed in these smaller establishments. Progress is being made, however, in making it increasingly possible for small employers, through a variety of approaches, to establish at least the essential elements of an industrial health program.

Today it is generally recognized that a soundly conceived industrial health program can do much for the health and welfare of people in the workforce. From the standpoint of the employer, such a program can reduce absenteeism, the incidence of illnesses and injuries, labor turnover, and insurance costs. It can improve employee morale, labor-management relations, and efficiency. It can increase productivity and better customer relations and public relations. It is, then, a matter of importance to both management and employees, and ultimately to society as a whole.

The concept of an industrial health program and the several aspects of such a program have already been explored. To be considered here are the essential elements in setting up an industrial health program, the costs involved in establishing and maintaining a program, and the sources of guidance, consultation, and assistance available to employers. The cost-effectiveness of industrial health programs is discussed later.

In reviewing this material it should be noted that in 1950 the Occupational Health Institute, an affiliate of what is now the American Occupational Medical Association, assumed the responsibility for the accreditation of over 500 occupational health programs from the American College of Surgeons. Subsequently the Occupational Health Programs Accreditation Commission of OHI was formed. In 1975 the Occupational Health/Safety Programs Accreditation Commission (OHSPAC) became operative with encouragement from NIOSH. A document entitled "Standards, Interpretations and Audit Criteria for Performance of Occupational Health Programs," prepared by the Kettering Laboratory, serves as a basis for the endeavor. The principles of the criteria have equal applicability to large and small employers and should be of help to management in evaluating existing industrial health programs.

17

Setting up an Industrial Health Program

The importance of an industrial health program is such that serious consideration must be given to its concept, its scope, its purposes, and its goals. Obviously, differences would exist from one program to another depending upon such variables as the nature of the operations of the place of employment and the types of hazards involved; whether it has a single location or is multi-sited; its geographic location; and the size of the workforce, its health status, and its demographic, educational, and socioeconomic composition. Preliminary study of all such variables is important, as is consideration of such matters as the organizational structure of the company, the quality of its employee relations, and its financial condition, including its possibilities for growth and expansion.

Preliminary study should also include an examination of all available data on both occupational and nonoccupational injuries and illnesses among the workforce, the incidence of their disabilities, and their records of absenteeism, turnover, and premature deaths. Workmen's compensation and health and disability insurance claims should be scrutinized since they reveal much about the task to be accomplished. The potential hazards to the health and safety of workers inherent in the operations of the workplace should be carefully evaluated. Also to be appraised are the current personnel, medical resources, and safety measures of the company;

the potential of the available community health resources; and the costs of any additional personnel and facilities that might be needed to accomplish the purposes of an adequate industrial health program.

Each of these factors is so important as to require the advice and counsel of professionals in the field of industrial health, as well as of experts within the place of employment. On the basis of the findings of such an examination, and the available financial resources, the industrial health program should then be designed and planned or, if one is already in existence, revised or amplified.

In the establishment or revision of an industrial health program certain essentials are fundamental if the program is to achieve its intended purpose. These are discussed briefly.

TOP MANAGEMENT SUPPORT

It can be said almost with certainty that, as with any other aspect of a business operation, if an industrial health program does not have the support and active interest of top management, it will fail. Management must be convinced of the importance of the program, from both humane and economic considerations. Management must determine the scope of the program, its purposes, and its aims. Management must designate responsibility for the administration of the program and must provide the necessary staff, facilities, and equipment. Above all, management, with active enthusiasm and conviction, must give to the program vision and vitality.

It has been said that a well-planned industrial health program should not only promote the health and safety of workers but should also give those workers tangible evidence of the interest and concern of management in their health and welfare. Management alone, through its approach to an industrial health program, can convey such an interest and concern.

A WRITTEN POLICY STATEMENT

To implement an industrial health program, the establishment and circulation of policy in the form of a written statement is generally considered a necessity so that there will be no doubt in anyone's mind about what is being done, what the purposes are, what is expected to be accomplished, and how it is to be accomplished. The policy position should identify where the responsibility for the execution of the program will lie. Certainly it should make clear that the concern of the program is the health and welfare of the employees. It should be constructive in tone, and the interest of the employer should be made clear. It was shown earlier that for a variety of reasons employees not infrequently have a mistrust of, or lack of confidence in, an industrial health program. The policy statement should make every effort to overcome such reactions.

APPLICABILITY OF THE PROGRAM

The industrial health program should be applicable to every employee at all levels of employment. To be otherwise could appear to be discriminatory and unfair and could so weaken the program as to make it ineffective. Where there are differences in the applicability of certain aspects of the program, such as physical examinations, whether by age categories, sex, type of work hazard involved, personal health status, or job classification, the reasons for such differences should be made understandable.

UNION COOPERATION

In a place of employment where unionized labor is involved, the union should be consulted about the establishment of or changes in an industrial health program, and its cooperation sought. Through such a cooperative approach much can be accomplished to eliminate employee lack of confidence in, or even mistrust of, the program. Union input, furthermore, can strengthen the effectiveness of all aspects of the program, including counseling, health education, and safety efforts. In the instance of an alcoholism control program, for example, joint union-management cooperation is a sine qua non. To ensure the full accomplishment of the program the following should be considered:

Planning and evaluating the program on a mutual basis.
Issuance of a joint policy statement to the employees.
A jointly established training, safety, and health education program for both supervisors and shop stewards.
A jointly established procedure for treatment referral.
Union access to the company health records.
Joint assurance of the confidentiality of records except where otherwise prescribed by law.
Joint consideration and review of occupational health and safety hazards.

Most certainly management-union friction can deter the approach suggested here, particularly when the subject of occupational hazards has entered into collective bargaining negotiations. Yet what is suggested can play a significant role in improving employee relations, in removing any stigma of paternalism that can pervade an industrial health program, and in giving management and labor a cooperative area in which to work together in their mutual self-interest. Such an approach could do much to bring about the fullest compliance with OSHA.

STAFFING AND FACILITIES

The availability of an adequate professional staff is, of course, essential to the conduct of an industrial health program. Both its size and its composition would depend upon the many variables among places of employment noted earlier. They would depend, also, on the purposes and goals of the program and the services to be performed.

A Conference Board study in 1974 noted that unions and other critics of industrial health programs often consider such programs to be inadequately staffed, insufficiently funded, and lacking support by management.[1] These criticisms, the study says, are often voiced by those responsible for the implementation of such programs. This is true, the study comments, even among some large companies with programs of more than average comprehensiveness. The U.S. Public Health Service has also noted such inadequacies as the employment of physicians on a call or emergency basis only, the failure of the physicians to acquaint themselves with general plant sanitation problems, their laxity in making periodic inspections of the plant and its hazards, the failure to maintain adequate health records, and the neglect in providing written standing orders for the nurses.[2]

The Conference Board study further noted that management at times exerts pressure "to interpret work-related disorders as stemming from nonoccupational causes" in order to control workmen's compensation costs, that management and physicians may disagree on health-related hiring and firing policies, and that in some instances the research findings by industrial physicians of occupational hazards are not made known "because management is unwilling to admit to past errors and run the risk of being held liable for their consequences."

Such criticisms, then, present the question: What is adequate professional staffing? The American Medical Association has suggested as a guideline of adequacy two physician-hours a week for the first 100 employees and one additional physician-hour a week for each additional 100 employees. The Metropolitan Life Insurance Company, recognizing that no single standard can prevail because of the variables previously noted, has suggested that when the program includes preemployment, periodic, and special health examinations the usual requirement would be one physician to serve from 2,000 to 3,000 employees and for smaller employers, a full-time nurse to serve from 300 to 750 employees, supplemented by the part-time services of a physician. It is proposed that the small plant with a similar program of physical examinations may require two to three hours of physician services and six to nine hours of nursing service a week for each 100 employees. In large companies, it is noted, industrial hygiene engineers and health educators may supplement the health team.

The National Organization of Public Health Nursing has recommended as a minimum for nursing services a ratio of three hours of physician time to nine hours of nursing time for each 100 workers. For small plants with more than 35 employees, two hours a week of nurses' services is considered minimal. That organization found that in 1951, for fewer than 100 employees, two plants used six to seven hours of nursing time a week and two plants used 13 to 15 hours. In 12 plants with 200 to 250 employees, 4 used one to two hours of nursing time a week, 6 used two to three hours, one used three to four hours, and one used six to seven hours. In nine plants with 300 to 350 employees, one used less than one hour a week of nursing services, three used one to two hours, one used two to three hours, three used three to four hours, and one used 13 to 15 hours of services a week.[3]

The American Public Health Association has said that an occupational health nurse should be available for at least two hours at any one time for a plant with more than 35 employees, that there should be one nurse for up to 300 employees, two or more nurses for up to 600 employees, and three or more nurses for an employee population of 1,000, with one nurse being added for each 1,000 employees up to 5,000, and then one nurse for each additional 2,000 employees. For physician services it is suggested that two hours' time would be necessary for 100 employees and one hour for each additional 100 employees.[4]

Such estimates are by no means definitive, but they do present some indications of the professional staff needs for an industrial health program. The size and composition of the staff of a few actual industrial health programs are shown in Chapter 19. As to the types of health professionals used in an industrial health program, a 1948 survey of 333 industrial establishments in the United States and Canada is of interest since it showed the health personnel employed by those plants, by number of plants (see Table 17-1).[5]

As to the space for the facilities for an industrial health program, this again would be determined by the previously noted variables. Rough estimates have suggested the need for one square foot of floor space per employee for up to 1,000 employees. The American Medical Association has suggested for smaller plants a minimum space of 12 by 20 feet divided

Table 17-1. Types of health professionals used in an industrial health program.

Health Professional	Full Time	Part Time	On Call
Medical directors	78	101	14
Assistant medical directors	22	20	2
Other physicians	26	92	56
Consultants-surgeons and so on	0	18	68
Full-time nurses	189	0	0
Part-time nurses	0	27	0
Oculists	4	27	50
Dentists	4	16	21
Psychiatrists	4	7	18
Industrial hygienists	17	3	4
Psychologists	5	1	5
Podiatrists	0	4	3
X-ray technicians	28	9	13
Medical lab technicians	30	9	7
Physical therapists	10	7	5
Dental hygienists	6	1	0
Nutritionists	9	0	1
Practical nurses	14	5	1
First-aid attendants	41	34	8
Doctor assistants (nonprofessional)	17	3	1

into cubicles, plus an examining room, a waiting room, and a room with a bed. Other estimates range from 175 square feet for fewer than 200 employees to 300 square feet for 500 employees.

The Physician's Role

The role of the physician in the execution of an industrial health program is paramount. It is the physician who has the responsibility for the administration and the functioning of the program. His or her responsibilities would depend upon the purpose, the scope, and the aspects of a particular program, and these should be defined by management. The gamut of the physician's possible concerns and responsibilities are all the facets of an industrial health program discussed earlier.

The responsibilities of the physician in an industrial health program must include the diagnosis and treatment of occupational injuries and illnesses and at least the treatment of minor nonoccupational illnesses and injuries. In instances where more extensive treatment is required, counseling and advice should be given the employee and referral made to his personal physician or to a community health agency. Where preemployment, periodic, or special-purpose (which are required by OSHA) health examinations are involved, these should be conducted by the physician and his staff or under his direction. The physician should provide immunization for certain identifiable work hazards and for diseases that might be prevalent in the community. The physician should provide counsel to management, personnel directors, and supervisors with respect to job placement, job transfer, the employment of handicapped persons, and early retirement based on the health status of the individual employee.

Where certain health- or safety-related responsibilities are multidepartmental, the physician should work cooperatively with the personnel department, supervisors, the safety program, and the industrial hygiene department in all efforts to improve the health and safety of the working environment. This would include an alertness to such potential occupational hazards as the use of chemicals and other toxic materials, the use of tools and machines, noise and vibration, dusts and fumes, and inadequate lighting and ventilation. It would cover plant design and layout, including the adequacy of working space, and the provisions for storage of goods and materials. It would involve sanitation in working areas, lunchrooms, and toilet facilities, including waste disposal and refrigeration. To this end, the physician should participate in all periodic plant inspections. The physician should also be consulted in such matters as the safety of the products being sold or the services being provided, and in the potential effects of the plant operations on the health environment of the community.

As part of his responsibilities the physician should actively cooperate with the local medical society and all health services and facilities in the community. He should have an active interest in the health status of the community since this can have a direct effect on the health of employees and their families. He should play an active role in all industrial health professional organizations.

The physician should also be responsible for the general health education and counseling of employees as respects both occupational and nonoccupational health hazards. Not only can such an effort have a positive effect on the health of employees and their families, but it can make clear to employees management's interest in their health and safety. Of particular importance is the development of an awareness of safety in the home, on the highway, and in the course of recreation. Equally important is the development of an awareness of the various aspects of healthful living, including weight control and diet, exercise and rest, and the avoidance of excessive smoking or the abuse of alcoholic beverages.

Another, and salient, responsibility of the industrial physician is the maintenance and study of the health records on individual employees.

Such records not only reveal the onset of physical or emotional diseases but, when aggregated, serve to identify occupational or community health hazards not otherwise identified. Today such record keeping is required by OSHA, and government inspectors have access to those records on demand. Under OSHA, furthermore, employee representatives can demand to review such health records at any time. The adequacy of health records, therefore, is now a requirement of business operations. A sound system of health records, however, should not be confined to OSHA requirements, since the records can reveal information of value about many nonoccupational health hazards.

The Nurse's Role

Supporting the physician is the industrial health nurse, who should, of course, have special education and training in industrial health. She or he should have a general knowledge of occupational health hazards, of workmen's compensation laws and OSHA, of health resources available in the community, of health education techniques, and of all aspects of preventive medicine.

The nurse should work under the orders of the physician and should report to him. She can provide first aid and certain forms of treatment, and can assist in the conduct of health examinations. She can assist in the preparation of health and safety reports and in the maintenance of health records. She can play a role in the provision of health education information and health counseling, can assist handicapped workers, and can visit disabled employees at their homes. She should remain alert to the identification of mentally ill or emotionally disturbed employees. She should have working relationships with all other departments of the place of employment, particularly those concerned with personnel and safety, and should be a participant in all periodic safety inspections of the plant, and generally should be a participant in all health-related activities at the place of employment.

Facilities

The facilities and equipment of the industrial health program should be as complete as is necessary to accomplish the purposes of the program. This is a responsibility of management. Above all else, the ambiance of the facilities should command the respect and the confidence of the employees. Needed are a waiting room, an examination room, a treatment room, a toilet, and a bed cubicle. In addition, there must be facilities and arrangements for the transportation of injured or seriously ill employees to the hospital. The necessary medical equipment and devices would depend upon the scope of the program.

Shortage of Personnel

There is one distinctly retarding factor to the more rapid development of industrial health programs that must be borne in mind: the shortage of health professionals with training and experience in industrial health. This fact led the National Health Council to comment in 1960: "Many a firm which has seriously planned starting a program for the protection of its employees has had to delay putting it into operation because of an inability to find suitable staff personnel. . . . The company starting a new occupational health program must, therefore, anticipate competing for personnel."[6] While that statement was made almost two decades ago, the situation has not changed to any appreciable extent.

Estimates of the number of available and adequately trained industrial health professionals as well as the extent of the shortages vary among sources. In 1940 the American Medical Association reported that 4,086 physicians devoted some time to industrial practice, but that of these only 1,177 did so on a full-time basis. This total represented about one percent of all physicians. At that same time AMA reported that for every full-time physician in industrial health there were ten full-time registered nurses, some 10,796 nurses in all. Three years later, in 1943, the Social Security Board estimated that there were 113 industrial health programs in existence at that time, 71 of which employed 500 physicians on a full-time basis.[7] By 1960 the National Health Council estimated that there were 5,000 full-time physicians in industrial health and an additional 8,000 who worked in places of employment on a part-time basis. NHC further estimated that over 16,000 nurses were employed in industry.[8]

By 1969 the American Medical Association reported that 10,000 physicians provided services to places of employment, 2,000 of whom did so on a full-time basis. A third of the total were reported to be specialists with membership in the Industrial Medical Association, and 500 more were diplomates of the American Board of Preventive Medicine (board certification did not come into being until 1955).[9] On November 28, 1976, William Abrams reported in *The New York Times* that according to figures published by HEW, there were 4,000 full-time corporate physicians, half of whom were "occupational medicine physicians" concerned with employee health, product safety, and environmental health, and 600 of whom were Board-certified occupational physicians. In addition, 10,000 physicians performed occupational services on a part-time basis, 1,200 practiced in the pharmaceutical field, and 800 in insurance companies. About that same time the American Nurses Association reported that 19,403 nurses were employed in industry, 1,025 Board-certified industrial hygienists were recorded, and an HEW report on health manpower indicated a very uneven distribution of registered occupational therapists

Table 17-2. Shortage of industrial health professionals.

Classification	Present Census	Deficit
Occupational physicians (certified by the American Board of Preventive Medicine)	500	1,200
Other physicians (special training)	2,700	4,200
Industrial hygienists (certified by the American Industrial Hygiene Association)	600	4,000
Safety engineers (certified by the American Society of Safety Engineers)	2,000	4,700
Occupational health nurses (certified by the Association of Occupational Health Nurses)	1,000	8,400
Other nurses	17,000	19,700
Occupational safety and health physicians (not certified)	15,000	8,000

(for example, 362 in Minneapolis, 151 in St. Louis, 76 in Pittsburgh, and 33 in Birmingham). When it is considered that some 12,000 places of employment, excluding government and railroads, have more than 500 employees, that an additional 17,000 have from 250 to 500 employees, that over 53,000 places of employment have from 100 to 250 employees, and that in all there are over 3.5 million places of employment in the United States, the shortage of industrial health professionals is readily evident.

In 1974 NIOSH estimated the present census of various types of industrial health professionals and the shortage for each classification; the figures are shown in Table 17.2

The shortage of the necessary health professionals, then, is severe. Aggravating the situation is the fact that the enrollment of students in occupational medicine and the availability of university faculty in occupational medicine have both receded in recent years, and the number of persons having certificates in occupational medicine has been decreasing. Questionnaires sent to senior medical students in 1970 revealed little evidence of interest in industrial health.[10] While one of the responsibilities of the federal government under OSHA is the support of training for the necessary industrial health manpower, only 4 percent of the NIOSH budget for 1975 was allocated for such training, slightly over $1 million. The picture of the immediate future, then, is not bright.

One contributing factor to the shortage of industrial physicians is their professional status among their peers. Dr. Lorin Kerr of the United

Mine Workers has commented that the industrial physician has "had to struggle with the limitations imposed by his own profession" and that "there were also times when the industrial physician was refused admission in the local medical society." Dr. Kerr recognizes that "there is in reality a marked shortage of industrial physicians" and that "the prospects for training more are slim." He notes that in 1973 there were only eight educational institutions qualified to offer two academic years for residency training in occupational medicine and that many of their positions are unfilled; that only 13 plants offer the third year of residency training; and that only eight medical schools include occupational health in the curriculum.[11]

Much earlier, in 1910, Alice Hamilton had taken cognizance of the problem of peer status when she bluntly stated: "For a surgeon or physician to accept a position in a manufacturing company was to earn the contempt of his colleagues."[12] Fifty years later others commented similarly:

> In terms of his general and colleague status, the industrial physician is not in an enviable position. The stereotype of the company doctor — the incompetent protected from competition — is still to be found. . . . Industrial medicine has low status in medical education, and little encouragement is given to specialization in this area . . . public health, the field to which [industrial medicine] is most clearly allied, also enjoys little prestige.[13]

Apparently other factors than peer status can play a role in the shortages of industrial health physicians. Recently Dr. Gregory Hemingway (son of Ernest Hemingway), who had practiced industrial medicine in New York City, made the decision to join a health clinic in Fort Benton, Montana. His reason for leaving industrial medicine, according to the July 27, 1976, issue of *The New York Times,* was that "the work was necesary but dull."

A comment on this aspect of industrial health contained in the 1974 Conference Board study is also worth noting:

> Many physicians of outstanding qualifications and dedication have been attracted by the professional opportunities that are often present in occupational practice. . . . There is evidence that the calibre of those entering the field has been improving. Nonetheless, it is recognized that a fairly pervasive disinterest and even distaste characterizes the attitude of many physicians toward occupational practice. . . . It reflects . . . the absence of occupational health from the curricula of most medical schools.[14]

The supply, the educational and training facilities, and the motivation of health professionals is an industrial health problem of serious consequence. If industrial health is to progress markedly in the future, as it should and as is warranted, this is a problem deserving the earnest attention of not only the business community, the labor unions, and government, but also of the medical profession and the universities.

CONFIDENTIALITY OF RECORDS

Except where otherwise prescribed by law, the health records of employees gathered by the industrial health program should be confidential to the medical director as a professional matter, and should not be available to management. Of this the employees must be assured. If this is not the case the employee can lack faith in the intent of the program, particularly where nonoccupational injuries and illnesses are concerned. He will be afraid that his records will adversely affect his job advancement or even his employability. Complete confidentiality of health records, then, is a salient element in the full success of an industrial health program.

That such confidentiality does not always prevail was noted in The Conference Board study, which commented: "Management often insists on seeing results of periodic health examinations or on receiving other specific information on the health of employees." Yet despite this, some industrial physicians refuse to make the individual health records available to either management or supervisors.[15] Here it is important to note the position of the American Medical Association, which states without equivocation that "the health records of an individual must be considered a confidential matter between [the employee] and his physician."[16]

THE ROLE OF OTHER SPECIALISTS

Other professional specialists are also important to an industrial health program, functioning either as staff of management or as consultants. In either instance they should function cooperatively and in conjuction with the medical staff. Such specialists include the safety engineer, the industrial hygienist, the sanitation engineer, the health educator, the chemist, the mechanical engineer, the electrical engineer, the ventilation and heating engineer, the illumination engineer, the acoustical engineer, and the statistician.

Jointly their concerns should be to make the working environment as safe and as free of health hazards as is possible. Included in their responsibilities are the safety of tools and machines, the provision of safety devices and guards, and the potential hazards of materials, toxic substances, chemicals, radiation, fumes, gases, and vapors. The adequacy of ventilation and lighting is within their concerns, as is the control of excessive noises and vibrations. Plant sanitation in both working areas and employee facilities is included in their area of accountability, as are such matters as food handling and waste and garbage disposal. They should also have assigned responsibility for the safety of the products sold to the public and for the effect of the plant operations on the health environment of the community, including air and water pollution, waste disposal, noise, odors and fumes, and smoke.

In the execution of such responsibilities these specialists should be required to make periodic inspections of the plant, to study the aggregate health records of employees for what these might reveal, and to evaluate potential health and safety hazards emanating from the operations of a plant as respects both the employees and the community at large. The inspection of plants and the monitoring of potential health hazards is required under the provisions of OSHA. The employees may request, in particular instances, that plant inspections be made, and they must be notified of the presence of any work hazards. The various specialists should be given the responsibility for compliance with these requirements of OSHA. Theirs is an important role in relation to the health and welfare of employees and to the health of the community.

THE SUPERVISOR'S ROLE

The supervisor or the shop steward has a significant role in the operation of an industrial health program. With them rests the responsibility of training the employee for the job, of providing instruction in all respects of safety, and of observing employees on a continuing basis. Furthermore, through daily surveillance they can observe the early onset of illnesses, both physical and mental, as well as such diseases as alcoholism and drug addiction. It is they who can recommend counseling by the medical department.

In the fulfillment of this responsibility the supervisor or the shop steward must have the full cooperation and support of management on the one hand or the union on the other, and this should be supported by written instructions. They should be thoroughly trained for the execution of this responsibility and should be informed about the characteristics of common diseases and their means of identification. In so saying, it should be clear beyond doubt that they are not diagnosticians but that their concern is with such matters as deteriorated job performance, personality changes, friction with fellow employees or customers, faulty judgment, erratic productivity, memory lapses, a failure to meet schedules, absenteeism, and lateness. There should then be clear instructions concerning discussions with the employee and referral to the medical department. The approach must be constructive, impersonal, and clearly one of expressing an interest on the part of management and of a desire to help.

INSURANCE PROTECTION

A necessary element in the establishment of an industrial health program is a review of the adequacy of the health and disability insurance programs. Recommended medical treatment outside the industrial health program can, and very often will, go unheeded if sizable expenditures of personal funds are required or if loss of time from work for a considerable period is necessary. Only through the existence of an adequate insurance program can this obstacle be overcome. Since this subject of insurance benefits is discussed in a later chapter, there is no need to enter into details here, other than to note the important relationship of such benefits to the success of an industrial health program.

REEVALUATION

A final step in the establishment of an industrial health program, and one of importance, is the reevaluation on a continuing basis of all aspects of the program. In this way much can be learned which can lead to improvements and refinements in the program as experience is gained. In this way the program can be altered from time to time in the light of changes that are made in operational procedures, the use of new machinery, and the introduction of new materials, chemicals, or toxic substances. Such reevaluation should be made with full consideration of the health records of the employees. As sound business procedure it should have as necessary background such cost accounting information as the costs of the various elements of the industrial health program as well as its cost-effectiveness in relation to the occurrences of accidents, illnesses and disabilities, and absenteeism and resultant overtime pay; turnover due to health reasons; and the costs of all related insurance coverages and salary-continuance programs. This latter consideration, as will be shown, is too frequently overlooked and yet is a most necessary element in the evaluation and reevaluation of an industrial health program.

OTHER CONSIDERATIONS

Other considerations of an industrial health program include the presence of mental illness and emotional disturbances among employees, as well as alcoholism and drug addiction. These matters are discussed later in Chapter 22. The particular problems of small places of employment in establishing an industrial health program are discussed in Chapter 20.

REFERENCES

1. The Conference Board, "Industry Roles in Health Care," New York, 1974.
2. U.S. Public Health Service, "Industrial Health and Medical Programs," Washington, D.C., Publication No. 15, 1950.
3. National Organization of Public Health Nursing, unpublished data, 1951.
4. American Public Health Association, "Duties of Nurses in Industry," *American Journal of Public Health,* 33:876, 1943.
5. Ethel M. Spears, "Company Medical and Health Programs" (New York: The Conference Board, 1948).
6. National Health Council, "The Health of People Who Work," New York, 1960.
7. M. C. Klem, "Prepayment Medical Care Organizations," Social Security Board, Bureau Memorandum No. 55, 1943.
8. National Health Council, op. cit.
9. H. F. Howe, "Distribution of Occupational Physicians Among American Industries," *Journal of Occupational Medicine,* November 1969.
10. E. M. Dolinsky, "Health Maintenance Organizations and Occupational Medicine," *Bulletin of the New York Academy of Medicine,* Vol. 50, No. 10, 1974.
11. Lorin E. Kerr, M.D., "Occupational Health: A Discipline in Search of a Mission," *American Journal of Public Health,* May 1973.
12. H. B. Selbeck, *Occupational Health in America* (Detroit: Wayne State University Press, 1962).
13. Goldstein, "Medicine in Industry," *Journal of Health and Human Behavior,* ca. 1960.
14. The Conference Board, op. cit.
15. Ibid.
16. American Medical Association, "Guide to the Development of an Industrial Medical Record System," Chicago, undated.

18

What Does It Cost?

The costs of establishing and operating an industrial health program cannot be categorically stated. Quite obviously the total cost would depend upon a number of variables, including the nature of the hazards in the particular place of employment, the number of employees by sex and age, the rate of turnover among employees, the occupational health and safety hazards to which employees are exposed, and the geographic area of the workplace. Most important, though, the cost would be dependent upon the concept of the industrial health program, its scope, the nature of the services performed, the extent of its involvement in the nonoccupational injuries and illnesses of employees, the care provided to the families of employees, the availability of medical services in the community, and whether the program is concerned with potential environmental health hazards to the community or with the safety of products made available to consumers. All such variables can have important bearing on the cost of a program.

In reviewing the cost data that are available, all such influencing factors must be borne in mind lest misjudgments be made. The economies of large-scale operations, or their absence, will also affect the proportionate overall costs of an industrial health program in relation to the size of the employee population. Furthermore, certain costs that are a distinct part of an industrial health program might be allocated among several departments of the place of employment. This is exemplified from correspondence with the medical director of one large corporation who said: "Our total efforts are complex and integrated into our manufacturing, research, sales, and employee relations. No costs are readily definable." Some cost figures, nonetheless, are available and these are shown in the following paragraphs.

The total investment in employee safety and health in 1975 by American business was stated in a May 23, 1975, press release of the Economics Department of the McGraw-Hill Publications Company as follows. Also shown are the preliminary plans for such expenditures in 1978.

"American business plans to invest $3.12 billion for employee safety and health this year, only a one percent rise above 1974, compared with last year's actual gain of 20 percent. After adjusting for inflation, this year's small increase represents a decline of about 10 percent. . . . The rise in job health and safety investment is concentrated primarily in the nondurable goods segment of manufacturing. Manufacturers currently plan to spend $1.64 billion in 1975, a 4 percent increase above last year, with nondurable goods manufacturers increasing spending 23 percent. Durable goods manufacturers are planning a 10 percent cut. . . .

"There are eight major groups planning $125 million or more expenditures on employee safety and health this year. Commerical business plans the largest spending on safety and health facilities for its employees in 1975 ($534 million). Ranking second is the petroleum industry, with $356 million, up sharply from 1974's $216 million. Communications follows with plans for $348 million this year. Electric utilities ($212 million), mining ($195 million), nonelectrical machinery ($166 million), automobiles ($136 million) and rubber ($125 million) complete the list of the big spenders. . . .

"Only 14 out of 26 major industry groups plan gains in their investment in employee safety and health facilities this year. Among manufacturing industries only 10 out of 18 plan rises with 5 out of the 7 nondurable goods industries expecting increases, and only 5 out of 11 durable goods industries planning rises. Four out of eight nonmanufacturing industries plan to lift spending for safety and health in 1975.

"The expected percent changes in safety and health facilities this year run the gamut from a decline of 41 percent for automobiles to a rise of 65 percent for petroleum. The 'other' transportation industry, excluding railroads and airlines, and the 'other' nondurable goods industry — composed of apparel, leather, tobacco, and printing and publishing — expect to raise their spending 60 percent and 52 percent respectively in 1975 from relatively low bases in 1974 to meet safety and health requirements. 'Other' transportation equipment, also up from a relatively low base, follows with a scheduled increase of 50 percent this year.

"Aside from automobiles, nine other industries plan cuts in their safety and health expenditures this year. The largest scheduled declines are in textiles (29 percent), commercial business (16 percent), and instruments (14 percent). The smallest drop is planned by chemicals, off 2 percent. The airlines and gas utilities plan no change this year.

"Planning ahead to 1978, American business now expects to raise its safety and health expenditures by 19 percent to $3.71 billion, about 6 percent per year. By 1978, the nonmanufacturing sector expects to again account for more than 50 percent of health and safety investment. The nonmanufacturing sector plans to increase its spending 26 percent while manufacturing will increase its spending half as fast (13 percent), reversing the pattern of greater growth in manufacturers' expenditures in 1975. Eighteen out of 26 major industries now expect to spend more in 1978 on employee safety and health than in 1975. Fifteen of 18 manufacturing industries expect higher safety and health spending in 1978 than in 1975, with all of the nondurable industries planning increases. The three largest spenders in 1978, however, are expected to be in the nonmanufacturing sector—communications ($525 million), commerical business ($517 million), and electric utilities ($438 million). Electric utilities plan the largest rise in . . . spending for on-the-job safety and health protection, up $226 million to $438 million, a huge 107 percent rise. The communications industry plans the second largest dollar increase in 1978, up $177 million, 51 percent more than its present efforts to upgrade its employees' working conditions. For eight major industries, 1978 spending plans are down from their 1975 spending levels.

"The largest decline is expected by the railroads—down $52 million from the 1975 figure. Automobiles plan a $23 million decline by 1978, the second largest drop, and commercial business plans to cut its 1978 spending for employee safety and health by $17 million.

"U.S. business will slightly slice its share of total capital investment going to eliminate unsafe and unhealthy working conditions this year compared with last year. Planned safety and health expenditures will account for 2.6 percent of 1975 capital investment compared with 2.7 percent in 1974. Business also plans to spend 2.6 percent of its capital spending for upgrading safety and health conditions in 1978. For manufacturing, 3.3 percent of total capital expenditures is currently allocated for the improvement of the working environment. . . . The rubber industry is Number 1 in its 1975 ratio of employee health and safety expenditures to total investment, spending 9.9 percent of every investment dollar to erase work hazards. Instruments (7.5 percent), automobiles (6.2 percent), stone, clay and glass (5.5 percent), textiles (4.9 percent), mining (4.9 percent), and nonferrous metals (4 percent) are the six other industries allocating 4 percent or more of the investment dollar in 1975 to wipe out on-the-job risks. . . . Looking forward to 1978, 17 out of 26 major industries (including the whole nondurable goods group) plan to increase the proportion of their capital spending for safety and health purposes. . . .

"Investment in safety and health is growing considerably slower than investment in pollution control, and it appears that it will continue to grow more slowly in the years ahead to 1978.

"Investment in job health and safety is related, in part, to the present enforcement of the 1970 Occupational Safety and Health Act (OSHA). This is a relatively new area of large-scale capital expenditures which does not result in greater or more modern productive capacity, but it can result in improvement in labor productivity."[1]

A subsequent reporting by the same source showed that in 1976 American industry planned to invest $3.18 billion for employee safety and health, 17 percent above 1975. Industry also planned to spend $1.38 billion on research and development in 1976, whose purpose would be to control pollution, this amount being a 16 percent increase over the 1975 expenditure.

These data present a clear indication of the expenditures by places of employment devoted to the health and safety of employees. They are also revealing as respects the trends in such expenditures.

That the costs of an industrial health program differ with the type of hazards existing in various places of employment is evident from a 1941 survey by the National Association of Manufacturers of 2,064 plants. This survey revealed, for example, that the medical services provided by printing and publishing plants cost less than half the average cost for all plants, while those provided by mining and quarrying operations cost two and a half times the average. The costs in chemical and rubber plants, shoe factories, electrical manufacturers, machine tool makers, and manufacturers of stone, clay, and glass products experienced about average costs. Food processing and textile plants ran below average; and above average costs were found in lumbering, petroleum refining, and the manufacture of metal products. That survey also found that the cost of safety programs was highest in shipbuilding, construction, petroleum refining, mining and quarrying, and steel manufacturing; were average in chemical plants, lumbering, paper manufacturing, and the manufacture of stone, clay, and glass products; and below average in food processing, printing and publishing, public utilities, textile mills, and the manufacture of electrical products and rubber.[2]

A study made by the U.S. Public Health Service of the costs of an industrial health program in 1950 is of more than historical interest.[3] The study noted that the various factors influencing the costs of an industrial health program were as follows.

Type, kind, scope, and extent of services
Basic needs of the plant
    Average number of employees.
    Age and sex of employees.
    Type of industry.
    Expected number of cases to be treated.
    Estimated accident and sickness frequency.
    Estimated labor turnover.
    Indicated type of medical-vigilance program.
    Expected absenteeism.
    Allocated space, facilities, and equipment.
Conceptions of personnel of plant organization
    Predominant activities and hazards.
    Medical, safety, and production personnel.
    Techniques of accounting department in determining the costs of the program.
    Prevailing market for costs of salaries, equipment, and supplies.
Administrative methods for purchase of equipment and supplies
Insurance premiums for health, disability, life, workmen's compensation, and malpractice insurances
Union contract clauses dealing with the industrial health program

The study showed the following average per capita cost of an industrial health program by type of plant in 1950: Paper and its products, $9.66; food and beverages, $11.54; iron and steel, $14.41; and miscellaneous machine operations, $9.81. Of 97 places of employment surveyed the study found that 5 spent $25 or more per capita, 4 spent $20 to $25, 8 spent $15 to $20, 23 spent $10 to $15, 39 spent $5 to $10, 13 spent $1 to $5, and 5 spent under $1 per capita.

Table 18-1 shows the cost distributions on a per capita basis, by type of industry, for 688 companies in 1950. By size of the place of employment these per capita costs for the medical program ranged from $4.37 for those with from 1 to 249 employees to $7.00 for those with over 20,000 employees. For the safety program the range was from $3.76 to $4.00. For the industrial hygiene program the range was from $2.68 to $3.00. The original investment cost for several plants in 1950 were shown as follows:

Number of Employees	Investment Cost
557	$ 3,000
853	1,200
958	1,000
2,329	16,000
3,153	12,500
3,436	11,500
7,621	12,500
10,671	13,500
14,347	15,000

These cost figures are clearly out of date and differences in the costs shown can result from variances in the services provided by the industrial health programs involved. Nonetheless, they present a clear picture of the industrial health efforts a quarter century ago. The study commented that "the costs of industrial medical departments can and should be accurately tabulated and analyzed. ... Cost analyses should be determined by a uniform method to be of value for the purpose of comparison." That statement, unfortunately, is as appropriate today as when it was written.

That the cost of an industrial health program varies in inverse ratio to the size of a plant was revealed in a 1950 survey of 40 plants, which found an average cost of $6.00 per capita in factories employing more than

Table 18-1. Per capita cost for industrial health programs by industry.

Industry	Medical Program	Safety Program	Industrial Hygiene Program
Chemicals	$ 5.42	$ 3.21	$5.33
Electrical products	4.95	1.84	1.98
Food products	2.97	2.54	1.62
Iron and steel	4.51	4.69	6.15
Leather	4.80	4.57	—
Lumber	10.29	3.42	2.09
Machinery	5.35	2.84	3.88
Metal products	6.38	2.84	2.92
Mining and quarrying	12.26	5.27	5.39
Paper	6.05	3.73	3.56
Petroleum refining	6.58	7.87	5.00
Printing and publishing	2.33	1.48	.85
Public utilities	4.83	2.56	—
Rubber	5.51	1.40	1.30
Shipbuilding and construction	3.08	11.83	6.34
Stone, clay, and glass products	4.94	3.72	1.90
Textiles	4.60	1.64	6.77

5,000 employees; an average of $10.15 per employee in plants employing from 2,000 to 5,000 employees; $12.36 in factories having 1,000 to 2,000 employees; and $21.28 in those with 500 to 1,000 employees. Similarly, in 1958 an unidentified large steel company found that with comparable services, the per capita cost was $18 a year in its largest plant and $34 a year in its smallest plant. That same company also found that the cost of industrial health services differed by types of occupation ranging, in 1941, from $13 per capita for the office staff, to in excess of $20 per capita for coal mining and processing employees, to $34 for the steel mill workers.[4]

The Conference Board study released in 1974 also had some interesting findings.[5] For the 828 companies employing more than 500 people in the survey, it was found that in 1971 company expenditures for medical services provided in house or purchased outside, but not including the cost of health or disability insurance, ranged as follows:

Percent of Companies	Expenditure per Employee
17	$35
8	25.01 to 35
13	15.01 to 25
29	5.01 to 15
22	0.01 to 5
11	None

By type of company, the health care expenditures per employee were shown as follows:

	$15 or Less	Over $15
Manufacturing	47%	53%
Transportation and utilities	66	34
Wholesale and retail trade	95	5
Financial	67	33
Other	71	29

By number of employees, the following was shown for the same range of expenditures per employee:

	$15 or Less	Over $15
50,000 and over	33%	67%
10,000 to 49,999	53	47
5,000 to 9,999	67	33
2,500 to 4,999	57	43
1,000 to 2,499	55	45
500 to 999	70	30

That study also revealed that the cost of such programs per employee diminished as the proportion of females in the workforce increased. Thus 53 percent of companies where females were less than 25 percent of the workforce had a per employee cost of $15 or less; where females constituted 75 to 100 percent of the workforce, 85 percent of the companies had a per employee cost of $15 or less for employee health care services.

These data show clearly the effects on the per capita costs of an industrial health program produced by the type of operation at a place of employment and the size and sex composition of the workforce. Granted such cost information is not fully meaningful when the services of the respective programs are not known, but nonetheless, these findings are informative on what was being done, or not done, in the industrial health field a few years ago.

In a certain sense, the costs of industrial health programs as a percentage of payroll can be more meaningful, although again it would be necessary to know the services provided by the program, the type of operation involved, and other relevant factors. One such study was made in 1959 by Dr. Robert J. Kahn.[6] The study included 262 plants and showed that extensive industrial health programs seldom cost more than one percent of payroll and that more usually the cost was .5 or .6 percent of payroll, the percentage increasing somewhat in smaller plants. This study found that for plants with 500 or more employees that provided three or less of the basic industrial health services the cost was $.15 per $100 payroll, for those providing four of the basic services the cost per $100 of payroll was $.23, and for five or more services the cost was $.24. The relative costs per $100 of payroll for smaller plants ranging from 250 to 499 employees were $.28, $.31, and $.34 respectively. (The basic occupational health services were identified as preemployment examinations, pre-job assignment examinations, periodic physical examinations, periodic plant inspection, and the maintenance of health records.) Not only does this report illustrate the difference in the cost of industrial health services as between larger and smaller employers, it also makes clear the fact that as the number of industrial health services increase, proportionately less is the increase in the cost of the program in relation to total payroll.

A few examples of the actual costs of specific industrial health programs at different types of places of employment are of interest. In 1974 the Western Electric Company spent $11,700,000, or about $60 per employee, for strictly medical services (this program is described in some detail in the next chapter).[7] No breakdown was made between occupational and nonoccupational health services, although the treatment provided for occupational illnesses and injuries was almost double the quantity provided for those resulting from nonoccupational causes. These figures do not include the cost of safety or preventive measures, nor do they include the cost of reducing either occupational hazards or environmental health hazards to the community, since these matters are the direct responsibility of divisions of the company other than the industrial health program, and since cost data for those activities were not available.

The Union Carbide Corporation, as discussed in the next chapter, has recently reorganized and expanded its entire industrial health effort. It is estimated that the cost of the program will be $125 per employee.[8]

Another example of the costs of an industrial health program is that of the Mountain States Telephone Company.[9] In 1968 the company had 27,000 employees located in seven mountain states and El Paso, Texas. The program performs the following services by or on behalf of its employees:

Medical examinations
    Preemployment.
    Occupational (periodic).
    Fitness for work.
    Health maintenance (periodic).
Medical care
    Occupational illness or injury.
    Emergencies at work.
    Minor conditions.
Environmental control
    Electromagnetic radiation.
    Systemic toxins.
    Human engineering.
    Skin irritants and sensitizers.
    Nuisance factors.
    Noise.
    Ventilation.
    Illumination.
Health surveys
Research
    Morbidity data.
    Effects of new products or processes.
Medical followup
    Employees with chronic diseases.
    Employees absent due to sickness or injury.
Staff service
    Medical advice to employee benefit committee.
    Assistance in absence control.
    Alcoholism control.
    Safety measures.
    Health education.
    Liaison with medical profession and official health agencies.
    Medical advice to legal department.
    Advice to marketing department.

The use of services at the Denver location of the company with 6,000 employees in 1967 was as shown in Table 18-2, in terms of the number of employee visits. The estimated average annual costs of these services from 1964 to 1967 at the Denver location of the company are as shown in Table 18-3. It will be noted that the cost of treating nonoccupational injuries is about one-third that for the treatment of occupational injuries but that the cost of treating nonoccupational illnesses is more than double the cost of treating occupational injuries.

The total cost of the company's industrial health program in 1967 was as follows:

Medical department operations	$216,023
Payment to private facilities and physicians	63,122
Total	$279,145

This amounted to about $10 per employee per year in 1967. The total cost of operating the medical department in Denver in 1967 was $93,988. These figures included salaries, laboratory and X-ray facilities, supplies, house services (janitor, utilities, postage, printing), and any other expenses directly related to personnel or department activities. Information concerning the cost-effectiveness of this program will be shown subsequently.

A final example is the Seamen's Bank for Savings in New York. The company provided for physical examinations and the care of its employees and their families through the Grace Clinic in Brooklyn. In 1950 the cost of these services averaged 1 to 2 percent of payroll. The costs did not include such indirect expenses as the employees' time away from the job to visit the clinic, nor did they include administrative costs.[10]

Table 18-2. How industrial health services are used.

Reason for Visit	Number of Visits
Occupational injury	662
Occupational disease	9
Nonoccupational injury	794
Nonoccupational illness	7,805
Injections	1,108
Counseling	225
Fitness examinations	472
Applicant examinations	2,108
Periodic examinations	125
Total	13,308

Table 18-3. Estimated average annual costs of industrial health services.

Service	Cost
Occupational injuries	$ 6,560
Nonoccupational injuries	2,148
Nonoccupational illnesses	13,311
Injections	2,697
Applicant exams	17,640
Periodic exams	8,840
	$51,196

258 WHAT IS BEING DONE?

Of interest with respect to the proportionate costs of the various types of employee health services is the following recent hypothetical cost analysis (perhaps based on the experience of the Tennessee Valley Authority) for an industrial health program whose total cost is $1 million:[11]

Interviews	$ 8,958
Immunizations	17,064
Counseling	25,596
Nonoccupational treatment	191,970
Special exams	127,980
Occupational treatment	116,461
Preplacement exam	255,960
Periodic exam	255,960

Further information concerning the costs of industrial health programs, or of specific aspects of such costs, is given in Chapter 26, where the cost-effectiveness of industrial health programs is discussed.

Information on specific aspects of the costs of an industrial health program can also be revealing. While some of this information dates back more than two decades, it nonetheless has relevance.

- In 1956 the earnings of 123 full-time industrial health physicians in 83 different industries were found to range from $6,000 to $30,000, more than half ranging from $10,000 to $15,000. By 1960 the starting salaries were found to range from $10,000 to $12,000, with experienced physicians ranging from $10,000 to $20,000. Such salaries would be supplemented by the customary fringe benefits.[12]

- In 1957 the weekly salaries for registered nurses employed in industrial plants, retail stores, or public utilities in 20 major cities were reported to range from $73 to $89, supplemented by overtime pay, sick leave, holidays, vacations, insurance benefits, and other fringe benefits.[13]

- In 1950 the costs of part-time industrial health services were shown to average $4.00 an hour for physicians and $1.25 an hour for visiting nurses.[14]

- Other costs in 1950 for a plant with 100 employees were shown to be $15 per month for drugs and dressings, $150 for the original costs of dispensary equipment, $75 for the original supply of drugs and dressings, and from $50 to $500 for the original cost of constructing a two-room dispensary, including sink, electrical fixtures, and outlets.[15]

- The costs of physical examinations outside the workplace vary with, among other things, the thoroughness of the examination. In 1968 the cost of such examinations at Better Health Examiners was $50. The cost of a mobile screening program by the Health Testing Services at

Berkeley, California, in 1969 for the California Cannery Workers and the California Processors, involving 6,000 regular employees and 50,000 seasonal employees scattered over 100 plants located in 12 counties, was $36.95 per person.[16] In 1976 the cost of examinations at the Life Extension Institute was $95 for persons under age 35 and $125 for those over age 35, these latter involving more procedures.[17]

- An example of the cost of in-house physical examinations is found in the experience of IBM, which inaugurated a multiphasic screening program for all its employees over age 40 at its 35 locations in the United States and Canada. In 1973 the cost per examination when performed by the industrial health program was $22; for those performed on the outside the cost was $40. The overall average cost was $25 per examination.[18]
- Safety glasses, protective clothing, hard hats, safety belts, respirators, and other safety appliances cost Bethlehem Steel Co. $5 million in 1974, according to a recent company advertisement.
- The Prestolite auto battery plant at Visalia, California, a division of Eltra Corporation of New York, is reported to have spent $750,000 in recent years on industrial hygiene and periodic monitoring to combat lead poisoning and air pollution among its workers.[19]
- The cost of a noise monitoring meter was $700 in 1971, and improved ventilation cost one plant $15,000.[20]
- In 1976 The New York Times improved its fire safety mechanisms with the installment of heat and smoke sensors, firemen's bypass control on all elevators, a two-way public address system, special phones for fire wardens, manual fire alarm boxes, pressurized air in all stairways, and automatic shutoff on the air-conditioning system. The cost was $500,000.[21]
- The cost of maintaining an alcoholism control program at a place of employment has been estimated to average $.35 to $.50 per month per employee depending upon the demographic composition of the workforce, its size, and its geographic location.[22]

In considering such costs it should not pass unnoted that directly related to the expenses of an industrial health program are the costs of certain insurance programs and fringe benefits. These were discussed earlier, but here is a brief summary of what was shown:

- In 1974 group health insurance premiums for employees and their dependents were $27 billion, the larger share of which was paid by employers.
- In that year, disability-insurance premiums were $2.8 billion; a considerable share of this amount was paid by employers.
- In 1974 salary-continuance benefits were $3.6 billion, the entire cost of which was borne by employers.

- In the automobile industry the annual cost of health insurance for employees and their dependents has exceeded $150 per employee.

Of further interest are some available data on the costs to places of employment in their efforts to control potential environmental pollution or other health hazards to the community.

By 1976 the outlay of American industry to overcome environmental pollution was estimated to be $11.3 billion a year. Such costs must be passed on, ultimately, to the consumer. Questions that warrant serious consideration are whether such costs serve to impede plans for industrial expansion or modernization, whether they adversely affect labor productivity, and whether they bring about plant shutdowns or even the closing of plants. In 1976 the U.S. Labor Department maintained that pollution control expenditures were not expected to substantially impede plans for industrial expansion or modernization or to adversely affect labor productivity. It was asserted that between 1970 and 1976 there were only 75 plant closings in which the costs of federally required pollution controls were a factor and that only 13,000 jobs were lost as a result. On the other hand, it was maintained, the federal $18 billion municipal-sewage-improvement grants created an estimated 80,000 jobs for each $1 billion spent.

An example of how pollution control can affect a community is the General Electric plant at Hudson Falls, New York, where 1,350 persons are employed. The unemployment rate in the community had been averaging 11 percent. In 1975 an environmental concern developed over the use of polychlorinated biphenyls (PCB). Control of the potential hazard, for which there is no financially practical substitute, threatened both jobs and the local economy. The gross annual earnings of the GE employees were $14 million. Goods and services purchased locally by the company amounted to $4 million a year. Furthermore, the GE payroll provided $480,000 in state income taxes and the company paid $318,000 in local property, school, and sales taxes in addition to the $1.3 million in secondary local property and sales tax revenues it generated. Thus the workers, the community, and the state were faced with a very considerable problem when attempts were made to control a potentially hazardous industrial operation.[23]

For the control of air pollution alone, the Manufacturing Chemists Association has said that by 1974 industry had invested more than half a billion dollars in air pollution control equipment and that by 1976 this amount would exceed $1 billion. (These amounts were $212 million in 1962 and $288 million in 1967.) In addition, the annual operating and maintenance costs for air pollution control exceed $77 million.[24] The chemical industry alone had spent $85 million on air pollution control by 1969,[25] and industry in Cincinnati is reported to spend $2 million a year to the same end.[26] Other examples of the cost of controlling air pollution are the $15 million spent by Armco Steel Corporation in 1966 to eliminate red smoke from the environment,[27] and the $5 million spent by Consolidated Edison for dust collectors at two of its plants in that same year.[28]

It has been estimated that the cost of adequate noise abatement for 19 major industries would cost $13.5 billion and that further noise abatement would cost $31.6 million. The EPA, however, has estimated that adequate noise abatement would cost U.S. industry $1.2 billion a year for ten years.[29]

With respect to the costs of controlling water pollution, the Manufacturing Chemists Association has reported that up to 1972, 137 of its members had invested $700 million in capital equipment to control water pollution and that by 1976 this investment would be $1.3 billion. (In 1962 the amount of investment in water pollution control by those companies was $264 million.) The annual operating and maintenance expenditures of those companies for the control of water pollution increased from $40 million in 1962 to $121 million in 1972. During that ten years research expenditures for water pollution control grew from $5.5 million to $24.7 million a year. By 1972 the number of full-time personnel engaged in the operation and maintenance of water pollution control in those companies exceeded 3,500.[30] According to recent estimates, the prospective annual outlays for water pollution control have already increased some 70 percent as the 1977 and 1983 deadlines for compliance with federal standards approached.

However viewed, then, industrial health is a costly matter. On the other hand, it is a necessary and humane area for concern on the part of all places of employment, private or public. The gains to be balanced against the costs will be discussed subsequently.

REFERENCES

1. McGraw-Hill Publications Company, Economics Department, press release, May 23, 1975.
2. National Health Council, *The Health of People Who Work,* New York, 1960.
3. Public Health Service Bulletin No. 15, 1950.
4. National Health Council, op. cit.
5. The Conference Board, "Industry Roles in Health Care," New York, 1974.
6. National Health Council, op. cit.
7. Personal correspondence with W. B. Cowen, Medical Administrator, Western Electric Company, June 1976.

8. Information made available to author by John J. Welsh, M.D., Corporate Medical Director, Union Carbide Company, August 1976.
9. Bond, Buckwalter, and Perkins, "An Occupational Health Program," *Archives of Environmental Health*, September 1968.
10. Clarke and Ewing, "New Approach to Employee Health Programs," *Harvard Business Review*, July 1950.
11. James L. Craig, M.D., "A Practical Approach to Cost Analysis of an Occupational Health Program," *Journal of Occupational Medicine*, July 1974.
12. National Health Council, op. cit.
13. Ibid.
14. Public Health Service, op. cit.
15. Ibid.
16. *Employee Benefit Plan Review*, 326-10-70.
17. *The New York Times*, July 31, 1976.
18. F. W. Holcomb, Jr., M.D., "IBM's Health Screening Program and Medical Data System," *Journal of Occupational Medicine*, November 1973.
19. *The New York Times*, June 6, 1976.
20. Susan M. Daum and Jeanne M. Stellman, *Work Is Dangerous to Your Health* (New York: Vintage Books, 1971).
21. Joseph G. O'Connor, "Industrial Health at The New York Times," unpublished paper, May 1976.
22. J. F. Follmann, Jr., *Alcoholics and Business* (New York: AMACOM, 1976).
23. *The New York Times*, February 15, 1975.
24. Manufacturing Chemists Association, "Air Pollution," Washington, D.C., 1974.
25. Wilson and Minnotte, *Journal of the Air Pollution Control Association*, 1969.
26. Ibid.
27. *Environmental Newsletter*, December 26, 1966.
28. *Environmental Newsletter*, January 2, 1967.
29. *The New York Times*, May 2, 1976.
30. Manufacturing Chemists Association, "Water Pollution," Washington, D.C., 1974.

19

Who Is Doing It?

Today a great many places of employment have established industrial health programs. The actual number of such programs is not known, however, but Appendix B lists by type of operation some 250 places of employment having industrial health programs. Certain labor unions that have an active interest in industrial health or have included the subject in their collective bargaining agreements are also shown.

The objectives of such efforts are to protect the health and safety of workers, to prevent to the extent possible injuries and illnesses among the workforce, to create a safe and healthful working environment through the elimination of occupational health hazards, and to provide care to injured or ill workers. Increasingly, these objectives are extended to include a concern for nonoccupational illnesses and injuries. Beyond this, efforts may extend to the protection of the community from potential hazards created by the place of employment and to measures ensuring that products sold to the consuming public are free of health or safety hazards.

That the concept, scope, and services provided by the extant industrial health programs vary quite extensively goes without saying. Some programs are very complete in their services. Others are relatively rudimentary, performing only a few basic functions. Some are limited to only certain aspects of occupational safety and health, while others extend beyond occupational hazards to include concerns for nonoccupational hazards, rehabilitation and employment of the handicapped, the environmental health of the community, and product safety. In part, the differences can emanate from the type of operations engaged in at the place of employment, the hazards involved in its operations, the availability of professionals and health services in the community, the financial status of the business, and the nature of the relationship between management and unions.

The biggest lag in the establishment of industrial health programs occurs in places with a relatively fewer number of employees, businesses dealing primarily with lower-status workers or with a transient or extremely high-turnover workforce, agriculture, and domestic service. In many such cases the employer's commitment to, or even interest in, employee welfare is relatively low. Conditions can be such that it would be virtually impossible to develop any kind of industrial health program of consequence. The specific problems of smaller employers will be discussed in the following chapter.

Unfortunately, meaningful categorical information on the concept, scope, and nature of extant industrial health programs does not exist. Details are available on relatively few programs, and while a broader view emerges from a few surveys that have been on a more encompassing scale, in most instances such information is essentially limited to relatively larger employers. Furthermore, the picture is changing more rapidly today as a result of the increasing employer awareness of the importance of industrial health programs as well as the more recent and intensified activities of government at all levels in occupational safety and health, in product safety, and in the creation of a more healthful environment. For the latter reason alone, information rapidly becomes outdated.

In attempting to give an indication of the present extent and nature of industrial health programs, a review of some relatively early surveys is of interest for comparative purposes. While certain of this information has already been shown, a summarization here is germane.

Earlier, certain of the early beginnings of industrial health programs in the United States, dating into the latter half of the nineteenth century, were noted. Also noted were the subsequent effects of the enactment of workmen's compensation laws by the states. Then, in 1939, the National Safety Council surveyed 213 establishments that were members of the NSC and, therefore, practically by definition, had an active interest in the safety and health of workers. These establishments had a total of 657,937 employees. That they were primarily larger employers is evidenced by the fact that 52 percent of the establishments employed over 1,000 persons, 27 percent employed between 500 and 1,000 persons, and 21 percent had less than 500 employees. The survey found that at that time only 7 percent of those establishments employed full-time physicians, 79 percent had part-time physicians, and 10 percent employed both. Full-time nurses were employed by 78 percent of those establishments and part-time nurses by 14 percent. For certain of the services provided, the following were the percentages of establishments having such services:[1]

Studies of occupational health hazards	86%
Preemployment examinations	87
Periodic examinations	37
Mental tests	13
Treatment of minor illnesses	86
Minor surgery for accidents	91
Treatment of major illnesses	13
Eye examinations	45
Dental services	8

Noteworthy here is the small percentage of establishments at that time having full-time physicians, the fact that one in ten apparently provided no medical treatment at all, and that only a third made available periodic physical examinations.

At that same time the U.S. Public Health Service, in a survey of 16,803 plants in 15 states, found the proportions shown in Table 19-1 having the industrial health services specified.[2]

With the exception of the availability of first-aid kits, the differences between plants with less than 100 employees and those with more than 100 employees were pronounced. The relatively small proportion employing a safety director is also of interest. In comparison with the National Safety Council survey, the lesser degree in the employment of physicians and nurses was marked.

Table 19-1. Percent of 16,803 plants having industrial health services.

Type of Service	More than 100 workers	Fewer than 100 workers
Safety director, full time	33.0%	2.2%
Safety director, part time	29.1	11.9
Shop committees	63.2	12.5
First-aid kit	91.3	91.1
Physician, full time	20.3	.4
Physician, part time	27.6	5.6
Nurse, full time	43.7	.6
Nurse, part time	1.9	.5
Sickness records	55.3	14.1
Accident records	97.4	83.0

About that same time, also, the American Medical Association found that of 611 establishments employing 1,120,000 persons, 24.7 percent employed full-time physicians, 42.4 percent employed full-time nurses, and 7 percent employed part-time nurses. For establishments with fewer than 100 employees these percentages were 0, 2.6, and 5.1 respectively. Of the total establishments, 75.5 percent provided periodic physical examinations, 68.1 percent engaged in accident and disease prevention, 53 percent in plant sanitation, 39.4 percent in health education, 42.1 percent had a clinical laboratory, 45.5 percent provided X-rays, and 39.1 percent provided for physiotherapy.[3] Here, again, the pronounced differences between small and large establishments are evident. Notable also is the high percentage of establishments making available periodic physical examinations in comparison with the National Safety Council findings (75.5 percent compared to 37 percent) and the smaller proportion with full-time nurses (42.4 percent compared to 78 percent).

Ten years later there was a report of a comparative survey of the industrial health programs in 1,570 places of employment between 1930 and 1940 comparing these with the programs in 570 places of employment for the years 1941 to 1948.[4] For the 1930-1940 period, 57 percent of the places surveyed had "approved medical services" as defined by the survey; for the 1941-1948 period this proportion had increased to 77 percent. Between these two periods, however, the proportion of places of employment with full-time physicians increased only 4 percent. Those with part-time physicians increased 13 percent, and those with nurses increased 32 percent. Those with medical supervision of health and sanitation increased 27 percent by 1948. Some progress is revealed by this survey, but most certainly it cannot be called startling.

By 1950 it was estimated that in New York State 19 of every 20 places of employment having 5,000 or more employees had an industrial health program of some type; for those employing from 100 to 200 persons, however, the proportion was only one in 20. In 1957 it was found that of 116 firms with various locations and having 250 or more employees, 74 had some industrial health services. In 1959 Dr. Harold J. Magnuson of the U.S. Public Health Service told the National Health Forum that less than one in ten employees in the United States had preventive health services available. At that same time it was reported that the majority of the 8,000 places of employment having more than 500 employees had some type of industrial health program, but that only relatively few of the businesses with less than 500 employees had such programs.[5] Again, the slow rate of progress is evident, most particularly among smaller places of employment.

Then in 1974 The Conference Board released the findings of a 1972 survey of 858 companies that employed more than 500 persons.[6] That survey showed that only about 12 percent of those companies employed one or more full-time physicians, although the total number of physicians so employed by those companies was about 3,200. In addition, 18 percent of the companies employed part-time physicians, the total being about 10,600 physicians. In those places with more than 50,000 employees, all employed physicians; in those with less than 1,000 employees, only 16 percent employed physicians. Thus the survey found that two-thirds of those working in places that employ more than 500 people were employed by the 30 percent of the companies that used physicians on either a full- or a part-time basis. The following were the proportions of companies, by type of operation, that employed nurses only:

Manufacturing	33%
Transportation and utilities	13
Trade	11
Financial	25
Others	10

Of the companies surveyed, 71 percent provided preemployment examinations and 57 percent made provision for health screening. Only 14 percent of the companies employed an industrial hygiene director and only 34 percent provided any type of health education for their employees. The following were the proportions of companies, by type of operation, that employed no health professionals:[7]

Manufacturing	29%
Transportation and utilities	66
Trade	81
Financial	38
Others	60

In 1974 a release by the Bureau of Labor Statistics reported that 69 percent of employees in manufacturing had at least the services of an industrial nurse, whereas in contract construction the percentage was only 1.2. The release also reported that only 18.2 percent of nonfarm employees had the services of an industrial hygiene director available in their place of employment. At that same time NIOSH surveyed 273 organizations and found that almost two-thirds had written policy statements dealing with occupational safety and health. For the organizations surveyed, 75 percent of those in construction, mining, transportation, communications, and public utilities had such a written policy statement. The proportion in manufacturing was 62 percent, in government 52 percent, and in trades and services 36 percent. At that time NIOSH also reported that only 58.6 percent of employers were providing safety training for their employees.

A final survey of interest is the "Tabulation of the Medical Department Health Services Questionnaire" released by the Akron Medical Center in July 1975, which had been completed by 80 of the industrial health programs at larger places of employment. Of the 80 programs, 18 were in chemical and allied products, 15 were primary metal industries, 13 manufactured transportation equipment, 15 were engaged in petroleum refining, 7 manufactured rubber or plastic products, and 12 were categorized as "other." By size, in terms of numbers of employees, 11.3 percent had over 100,000 employees, 18.7 percent had from 50,000 to 100,000 employees, 41.3 percent ranged from 25,000 to 50,000, 23.7 percent ranged from 10,000 to 25,000, and 5 percent had from 5,000 to 10,000 employees. Of the employers surveyed, 95 percent had in-plant medical facilities at more than one location, 11.2 percent having such facilities at more than 50 locations.

On the average the 80 companies employed 11.17 full-time physicians (.14 per 1,000 employees), 15.67 part-time physicians (.20 per 1,000 employees), 51.43 full-time nurses (.65 per 1,000 employees), and 2.36 part-time nurses (.03 per 1,000 employees). The employment of other types of medical personnel was not shown. Types of equipment used by the industrial health program, in terms of the number of the surveyed companies using each, were shown as follows:

Clinical laboratory	49	Facilities for minor surgery	64
X-ray	61	Facilities for major surgery	1
Sight testing	75	Emergency cart (cardiac)	36
Audiometer	73	Whirlpool	57
Audiometric booth	71	Ultrasound	48
Pulmonary function testing	63	Diathermy	61
EKG	61	Traction	27
Treadmill	20	Exercise modalities	21

By number of companies, the industrial health programs surveyed included the following services:

Preplacement examination	77
Physical examination for disability benefits	63
Physical examination for promotion	27
Physical examinations for transfer overseas	62
Physical examination for management duties	66
Physical examination for detection of effects of environmental health hazards	60

Treatment of occupational injuries and illnesses	73
Limited treatment of nonoccupational illnesses and injuries	70
Unlimited treatment of nonoccupational illnesses and injuries	4
Medical counseling	66
Dispensing of prescription drugs	51
Overseas immunization	71
Visting-nurse service	15
Physical conditioning for management-level personnel	21

With respect to the treatment of nonoccupational illnesses and injuries, it is of interest that 19 of the 80 companies reported providing an unlimited number of treatments, 14 limited the care to two or three treatments, and 33 companies had a limit of one treatment. (The discrepancy between these findings and the foregoing listing could not be explained.) Also of interest is that 38 of the companies use multiphasic screening techniques for the conduct of their physical examinations, four of these using mobile vans for certain of their employees. Concerning the physical examinations for management personnel, it was found that 84 percent of the eligible personnel participated in such examinations. A final note of interest is that 23 of the 80 companies use computerized systems for their medical records, 13 for employee mortality data, 17 for their morbidity data, 16 for OSHA reporting purposes, and 9 for environmental hazard exposures.

While these various surveys, conducted over a period of 35 years, are most certainly noncomparable, a rough picture does nonetheless emerge. Some progress has been made over that time span in the number and scope of the industrial health programs available to employed people. The progress, however viewed, is not as great as one might have expected, and is most pronounced in the relatively large places of employment. The filtering-down process, it seems, will be a slow one. That employers, private or public, are not in all cases adequately coping with occupational health hazards, let alone the other aspects of industrial health, is readily evident from the citations under OSHA noted in an earlier chapter.

A few examples of actual industrial health programs will serve to illustrate what is being done today in relation to different types of places of employment.

Western Electric Company is engaged in manufacturing, installation, distribution, purchasing, and defense activities for the Bell System. It is a subsidiary of AT&T and its subsidiaries, in turn, include Teletype Corporation, Sandia Corporation, and Nassau Smelting and Refining Company. It maintains 14 major plants, 39 distributing houses, and mobile crews of trained technicians. It has more than 145,000 employees located in 29 states. Its manufacturing locations range in size from 1,500 to over 20,000 employees.

The company medical program employs 37 full-time physicians, 99 part-time physicians, 178 nurses, and 28 technicians at 57 company locations. At distributing centers, which range from 120 to 1,100 employees, nurses and part-time physicians are used. The 20,000 employees who constitute the mobile force of the company receive care from the Bell System medical departments or from local private physicians.

The medical program at Western Electric has as its written objective "the conservation and improvement of employees' physical and emotional health . . . under a concept of preserving the dignity of the individual." Emphasis is on preventive medicine. All otherwise acceptable applicants for employment are given a preemployment physical examination, the purpose of which is solely to advise management regarding the degree of the employment risk. Applicants having certain impairments, either correctable or permanent, are given consideration provided that such impairments do not interfere with the job requirements or would not threaten the safety or welfare of other employees. In either case, those employed are given work assignments compatible with their physical and emotional health capacities.

Following employment, employees are given periodic occupational health examinations if they are exposed to occupational hazards. These are provided prior to placement in such jobs, at subsequent designated intervals, and when they leave such jobs. Examinations are also provided for those assigned to work outside the United States or prior to personal leaves of absence. Periodic health examinations are provided annually on a voluntary basis to department chiefs, buyers, research leaders, and those in higher supervisory positions. Examinations are available to section chiefs, professional administrative employees, secretaries, information systems employees, and wage practices associates, as well as to such professional employees as engineers, physicians, nurses, and lawyers on a two-year basis for those age 30 through 49 and annually thereafter. The results of these examinations, along with the health records of all employees, are held in strict confidence by the physicians and are released, except where otherwise required by law, only at the request of the employee. For statistical and study purposes a coding system is used for such records.

Fitness-for-work examinations are provided on the recommendation of a supervisor in cases of occupational injury or disease, disability absences, increased absences from work, decreased job performance, disruptive behavior, or pregnancy (pregnant women are not assigned to jobs involving excessive lifting or straining or exposure to toxic materials or radiation). In all such cases, health counseling is provided if desired by the employee. Examinations are also given prior to intracompany transfer or before retirement.

The medical care provided to employees includes first aid for minor injuries and illnesses, emergency treatment, and care for all injuries and diseases arising out of employment. Care is also provided for nonoccupational illnesses or injuries of a minor nature, followed by referral to a personal physician, if necessary. Immunization is available to employees on a voluntary basis. No charge is made for the care provided in either instance. When absence occurs as a result of illness or injury, home visitation is provided for. First aid and medical treatment are also provided to visitors who become ill or are injured while on company premises.

The Western Electric employee is encouraged to have an interest in health maintenance and safety awareness, both on and off the job. To this end, health and safety instruction is provided through company publications, bulletins, demonstrations, lectures, and conferences on a continuing basis. The company considers the verbal link between its supervisors and its employees paramount in its efforts on behalf of the health and safety of the employees; therefore the supervisor has the responsibility for observing the early development of illnesses, as well as careless safety practices. All employees, furthermore, are encouraged to consult the medical program physician for advice concerning any matters affecting their health. The company also maintains an alcoholism control program.

The health program at Western Electric includes research into and the practical applications of biomechanics, computerized epidemiological studies, lazar hazard evaluation, and the control of all occupational safety and health hazards. Monitoring occupational health hazards, the purpose of which is to reduce or eliminate excessive noise levels, air and water pollution, and harmful effects from toxic chemicals, radiation, and certain raw materials, proceeds at Western Electric on a continuing basis. Standards are set for the control of such hazards, and employees are trained and educated in safety precautions. Unphysiologic, repetitive activities are arrested and there is emphasis on engineering design for safety in the use of tools and machinery. Such activities involve the medical director, the design engineer, the factory engineer, the safety supervisor, the industrial hygienist, the machine and tool designer, the industrial engineer, and the functional supervisor.

Beyond this, the company also engages in basic research to develop knowledge of such employee health problems as heart disease, respiratory diseases, and on- and off-the-job injuries, as well as the effectiveness of various safety practices. The effects of smoking and poor diet are also studied, as are the personality factors that affect health and safety.

The company also supports the participation of its industrial health staff in professional organizations and encourages participation of its staff in community and voluntary health efforts and services. Of all gifts made by the company, 17 percent are donated to health-related endeavors.[8]

The industrial health program of the *Union Carbide Corporation* is of particular interest since it has recently been completely reorganized. Under the reorganization plan its industrial and environmental health efforts are combined under a single department of Health, Safety, and Environmental Safety. The reorganized program will include in its responsibilities both the occupational and nonoccupational health and safety of employees, product safety, and environmental health both within and outside company plants.

In 1976 Union Carbide employed 120,000 persons, 70,000 of whom are located in the United States. The company maintains 200 plants within the United States, 100 of which are quite small in terms of the number of employees. The industrial health program functions under a written corporate policy statement. Within that policy statement, however, each division of the company is free to select those aspects of the program that it will implement.

The program encompasses mandatory preplacement examinations, special examinations for specific occupational hazards and for service out of the United States, and voluntary in-plant health-screening examinations on a yearly basis for those age 45 and older and on a biennial basis for employees between ages 30 and 45. Special examinations are available to company executives and management personnel on an in-plant or out-of-plant basis, and some 85 of these employees take advantage of such examinations. The records of all voluntary examinations are confidential. Immunization is available on request and is provided for all employees with duties outside the United States. Treatment is provided for all minor conditions; for other conditions treatment is recommended and referral made to the personal physician. Prescribed drugs are available. Health counseling is also available to all employees, and health education proceeds on a continuing basis. An alcoholism control program is fully operational.

The program is concerned with the elimination or reduction of all types of occupational hazards, including those resulting from the use of machines, tools, toxic materials, chemicals, and nuclear materials, for which severe standards of exposure are applied. It is also responsible for controlling excessive noise and vibration, for correcting any inadequacies in lighting and ventilation, and for controlling fumes, gases, or vapors. All elements of employee safety are within its scope, including that of plant design and layout. Every aspect of industrial hygiene is within the responsibility of the program. Records on the medical history, mortality, and morbidity of all employees are maintained on a computer, and these are studied for clues in the detection of both occupational and nonoccupational health hazards, with emphasis on potential chemical and nuclear hazards.

The industrial health program at Union Carbide is staffed by 30 full-time physicians, more than 100 part-time physicians, and 6 half-time physicians. These are supplemented by 101 full-time nurses, 4 part-time nurses, and 2 physician assistants. The company employs 10 industrial hygienists, one biostatistician, and 2 health-related technicians. Various types of engineers complete the staffing. The facilities, in addition to examination rooms, surgeries, and other necessary space allotments, include clinical laboratories, facilities for minor surgery, and such equipment as EKGs, X-rays, audiometers, sight testers, treadmills, emergency carts, whirlpools, sound testers, pulmonary function testers, and the means for physical therapy and diathermy.[9]

At *Eastman Kodak,* with some 32,000 employees, some elements of the industrial health program are still under development or in prospect. Basic to the evolving concept is a teamwork approach involving all the related disciplines: physicians, nurses, industrial hygienists, safety and design engineers, chemists and toxicologists, human factor specialists, research experts, physicists, and psychologists. Because of the size and diversification of the plant operation, 14 teams have been developed, each being responsible for health and safety in a particular area of the plant. The program is viewed as having four aspects.

1. *Clinical medicine,* including first aid, a plant ambulance, a 24-hour-a-day dispensary, ambulatory care for illnesses and injuries, diagnostic studies, referrals to personal physicians or community services, followup of chronic illnesses and disabilities, health records maintenance, laboratory services, X-rays, and EKGs.

2. *Health maintenance,* including health education, counseling, immunization, physical examinations, selective job placement, occupational hazard surveillance, the development of employee health baselines and risk profiling, analysis of health records, the development of supervisory responsibilities for health and safety, improved job methods, hazards control, safe design, medical intelligence concerning illnesses in the workforce, and employee training.

3. *Rehabilitation,* including advice on the management of disabilities, home visitation, and a rehabilitation plan.

4. *Environmental health,* including the recognition, identification, and evaluation of health and safety hazards in the workplace and in the community, and their elimination, substitution, or control; foreseeing and predicting accidents; the safety of the products produced; and the safety of outside maintenance personnel, salesmen, and visitors to the plant.[10]

At *Exxon* an industrial health program was started in 1920 and has been expanded very considerably since that time. For example, in 1947 an industrial hygiene program was instituted. In 1958 a medical research and engineering department was founded. In 1972 a physical fitness laboratory was established, as was a program for alcoholism control. Exxon has 137,000 employees in the United States and abroad. It employs 125 physicians, 181 nurses, 28 industrial hygienists, and 418 other professionals and supportive personnel, including laboratory technicians and those engaged in record keeping. At the New York City office, with 2,200 employees, the company has seven full-time physicians, three nurses, five laboratory technicians, three physiologists, and a supporting staff. Here it maintains two consultation rooms, three laboratories, an X-ray room, a physical fitness laboratory, a records room, a library, and necessary space for office, equipment, and clerical staff.

In its entirely the industrial health program at Exxon is concerned with the health of its employees, its customers, and the public. Its goal is the elimination of potential health hazards that might be inherent in its processes and in its products. Accordingly, the program is concerned with any potential environmental health hazards that might result from its operations, either at the place of work or within the community. Included here are such potential environmental hazards as air pollution, fumes, noise, vibration, and heat, and those that might be caused by the use of such materials as lead, benzene, vinyl chloride, nickel, sulfides, or carbonyl.

A unique aspect of the Exxon industrial health program is its physical fitness laboratory established in 1972 (although the National Cash Register Company established a "recreation" program in 1904). Its intent was to provide a "highly organized physical and cardiovascular conditioning system." It makes exercise accessible to participants through a schedule of activities supervised by the medical department. The facility has 2,900 square feet of space adjacent to the medical department. Equipment includes an electromechanical pacing device, medicine balls, wall-pulley weights, knee-thigh weights, a rowing machine, a stationary bicycle ergometer, a treadmill, dumbbells, and a punching bag. All clothing is provided except sneakers.

A special staff conducts orientation meetings, plans and supervises activities, and records the evaluative data. The unit is available before, during, and after working hours. Three hours of activity a week is suggested. Initially the program was limited to a specific group of company executives. Of the 422 who participated, 309 completed the evaluation procedure, their mean age being 46.4 years. The 113 who did not participate were, on the average, older, in poorer physical condition, smoked more, and had a greater prevalence of hypertension and coronary heart disease. After six months the original participants were in better physical condition, 75 percent had an increased sense of well-being, and 90 percent had an improved capacity for physical effort. Of the 54 who were smokers, 27 percent stopped or decreased their tobacco consumption. About 28 percent lost weight; 8 percent had a weight increase. Only 15 percent increased their degree of outside physical activity. Of the original participants, 75 percent have remained active in the program.[11]

At the *Olin Mathieson Company* plant in North Carolina, located in open country 30 miles from the nearest large city, an industrial health program was established in 1950 with well-defined objectives. It was recognized that any such program is doomed to failure unless it has the support of all levels of management, the confidence and acceptance of employees, and the cooperation of the local medical profession. A medical center was established with adequate facilities, and was placed under the direction of a medical director. The program includes complete annual physical examinations for all employees (more frequently for older employees and those with physical abnormalities). The medical records are strictly confidential, and the program is closely correlated with the plant safety program. Special services provided by the program emphasize the early detection and prevention of such chronic conditions as hypertension, diabetes, cancer, arthritis, and emotional disorders.[12] The cost-effectiveness of this program is discussed in a later chapter.

For a quite different type of employer operation, the industrial health program at *The New York Times* is of interest. The medical staff is composed of a medical director, two internists, a surgeon, and six nurses. Some of the physicians are on a part-time basis. The program provides preemployment examinations, emergency care, and basic medical care. More extensive care is provided in a hospital. Testing includes blood pressure, vision, hearing, urinalysis, blood sampling, X-rays, and EKGs.

A separate safety department at the Times has as its purpose the promotion of safety and healthful working conditions. It reviews all cases of injury to determine if additional safety precautions are warranted and, where necessary, machinery or procedures are altered. Safety training is provided by the shop foremen, and the American Newspaper Guild has a safety committee that works in close conjunction with the safety department. That the effort has been effective is evidenced by the fact that in recent years the incidence of injuries at the Times has been reduced 52 percent.

Recently a program was instituted by the safety department to protect employees in the composing, photoengraving, and remelt rooms, and the master job area against inorganic lead poisoning. In cooperation with the engineering department a system of testing was installed, the amount of lead dust and fumes was reduced, and protective respirators and filters were provided.

The safety department is also responsible for compliance with the applicable OSHA standards. While the Times has had several OSHA inspections, it has received only one citation: for excessive noise levels in the pressrooms. Since redesign of the presses was not feasible, protective hearing equipment was provided all pressmen. A hearing-conservation program was also inaugurated for all employees exposed to excessive noise. Under this program periodic audiometric testing is provided all pressmen twice a year; cases of hearing deficiency are referred for additional testing.

The approaches to fire prevention taken at the Times have been mentioned in the preceding chapter, and recently its alcoholism control program was revised under a department separate from the medical department.[13]

Of special interest here is the health-screening and medical data system inaugurated by *IBM* in 1968 for all employees at its 35 locations in the United States and Canada. The purpose of the program is the early detection of disease and the accumulation of medical data. The program is voluntary and the results are confidential. Over a six-year period 72 percent of the eligible employees participated in the program (42,700). Of these, 73 percent were found to have some physical defect, and in 12 percent of the cases such defects were unknown to the employee. Undetected conditions included tachycardia, diseases of the nervous system, vision and hearing defects, circulatory, endocrine, and respiratory diseases, blood diseases, diseases of the digestive system, diabetes, hypertension, and cardiovascular diseases. These findings have redirected the focus of the company's health education program, with emphasis on heart disease. The cost of inaugurating the program could not be determined. The cost per examination in 1973 averaged $25: $22 for those conducted in plant and $40 for those done by outside services.[14]

An example of a union program is also of interest. In 1967 the *California Cannery Workers* and the *California Processors* instituted a health-screening program provided by Health Testing Services in Berkeley, California. Involved were 6,000 regular employees and 50,000 seasonal employees (15,000 of whom had three or more years of plant seniority). These workers were scattered over 100 plants ranging in size from 20 to 2,000 employees and located in 12 countries. Consequently, mobile units were used for the testing. The 29 tests performed included the detection of heart and blood vessel diseases, lung and chest diseases, visual defects, anemia, nutritional status, bone diseases, diabetes, gout and arthritis, venereal disease, kidney and bladder diseases, liver diseases, and cancer of the cervix and breast.

In 1969, of 21,000 workers examined, 41 percent had one or more abnormalities, the proportion being higher for those over 45 years of age.

Of the abnormalities found, 49 percent were previously known, 36.7 percent had not been known, 5 percent were not considered significant, and 9.3 percent were not considered valid. Discovery of abnormalities that were not previously known resulted principally from the Pap smear (71.4 percent), ocular tension (65.3 percent), hemoglobin (51.4 percent), uric acid (49 percent), and cholesterol (44.5 percent). The findings were sent to the personal physician. Cost of the program was $36.93 per examinee, including not alone the testing but also personnel, rent, other expenses of the central office, and insurance. This cost could have been reduced to $29.96 had the program functioned on a year-round basis. No cost-effectiveness or cost-benefit analysis was made, recognizing that more exploratory work had to be done as respects such matters as reduced morbidity and mortality and changing patterns of medical care utilization in relation to multiphasic screening tests.[15] The applicability of this program to the problems of smaller employers, to be discussed in the following chapter, is evident.

Two examples of employer safety programs are also of interest. At the *Baltimore Gas and Electric Company* it is considered part of the job of every company officer, every department head, every foreman, and every worker to prevent or avoid accidents. Management actively supports and participates in the program. There is a clear policy statement on safety, and responsibility is placed in the operating departments. The program is supervised by a safety staff. Safe plants, equipment, and tools are provided and workers are given job training, safety education, and safety training. That the program has been successful is evidenced by the fact that in 1928 there were 263 employees who suffered disabling injuries on the job. By 1973, with double the 1928 workforce, there were 20 such injuries.

The other example of an employer safety program is that of *Norris Industries* with headquarters in Los Angeles. Under this program, responsibility for safety rests in the industrial relations managers and in the plant foremen. The program functions under a corporate safety director. Monthly safety talks are given employees on specific accident problems, and in 1974 a safety contest was instituted at all company locations. Each location is judged by the number of accidents requiring physician care, the number of cases involving loss of time from work, and the number of workdays lost as a result of on-the-job accidents. The company considers it a good program.[16]

These examples of what is being done in the various aspects of industrial health are indeed encouraging. That considerably more should be done has been noted previously. That more will be done, if only as a consequence of the demands of government agencies, would seem to be self-evident. Today an increasing number of organizations stand ready to assist employers in the development, revision, or expansion of industrial health programs.

REFERENCES

1. National Safety Council, "Health Services in Industry," Chicago, 1939.
2. Public Health Service Bulletin No. 259, 1940.
3. *Journal of the American Medical Association,* February 17, 1940.
4. Gaylord R. Hess, "Basic Approval of Medical Services in Industry," *Journal of Industrial Medicine,* July 1949.
5. National Health Council, "The Health of People Who Work," New York, 1960.
6. The Conference Board, "Industry Roles in Health Care," New York, 1974.
7. Ibid.
8. Materials sent to the author by Western Electric.
9. Information furnished the author by John J. Welsh, M.D., Corporate Medical Director, Union Carbide Corporation.
10. W. D. Hoskin, M.D., "The Practice of Occupational Health," paper presented before the American Medical Association conference, September 21, 1976.
11. Patricia M. Yarvote, M.D., et al., *Journal of Occupational Medicine,* September 1974.
12. N. H. Collisson, "Management and an Occupational Health Program," *Archives of Environmental Health,* Februrary 1961.
13. Joseph G. O'Connor, "Industrial Health at The New York Times," unpublished paper, May 1976.
14. F. W. Holcomb, Jr., M.D., "IBM's Health Screening Program and Medical Data System," *Journal of Occupational Medicine,* November 1973.
15. *Employee Benefit Plan Review,* 326-10-70.
16. "How Two Companies Are Promoting Safety," *Supervisory Management,* March 1975.

20

The Problems of Small Employers

As was evident in Chapter 19, there is a lag in the development of industrial health programs in smaller places of employment. The problems of smaller businesses, however defined, in coping with industrial health problems are difficult. Certainly much of what has been said here about the establishment of an industrial health program could be of only limited applicability to very small places of employment and could easily appear to be "fancy talk." Two factors are significant in considering industrial health programs for small businesses. One is that there is no established medical department at the place of employment. The second is that the costs involved would not have the advantage of being spread over large numbers.

Dr. Logan T. Robertson, an occupational health consultant in Asheville, North Carolina, told the 1959 National Health Forum that the small employer is hesitant to embark upon a well-rounded occupational health program as a result of "lack of knowledge of its probable costs and possible savings."[1] Another discouraging factor was noted by Dr. Donald F. Buchan of The Prudential Insurance Company, who cited an instance in New England, where ten plants, each employing fewer than 90 persons, wanted to set up an industrial health program on a cooperative basis but "could find no doctor interested in undertaking this challenging task."[2]

Other reasons for the nonestablishment of industrial health programs in small places of employment have been identified as being:

The cost.
Other expenditures that take priority.
The lack of space for a dispensary.
The absence of a health problem.
Employees' use of their own physician, the cost of which is covered partly by insurance.
A lack of interest because of the type of employee involved.
A concern that such a program would interfere with the present close relationship between management and employees, or would be considered paternalistic.
The seasonal nature of a business.
The high rate of labor turnover.
Unfortunate previous experiences.

Still, small business, no less than others, encounters the same problems of both occupational and nonoccupational illnesses and injuries, and the costs to the enterprise can be relatively greater, if only because they cannot be spread over large numbers. In addition, the smaller place of employment frequently has greater difficulty in competing in the labor market to find an adequate replacement for a disabled employee. More to the point, the accident-frequency rate in smaller plants has been found to be two and a half times as high as in larger places of employment, 25 percent of workers in small plants have been found to have an occupationally derived disease, and absenteeism due to illness is also found to be greater in smaller places of employment.[3]

Statistics released by the Bureau of Labor Statistics in 1972 of the incidence of occupational injury and illness by type of industry and size of place of employment are shown in Table 20-1. Notable here, and contradicting what has been shown earlier, is the quite low rate of incidence of occupational illnesses and injuries for the very small places of employment, those with fewer than 20 employees and, in some instances, those having from 20 to 49 employees. No explanation for this is given. It might lie in the manner in which records are maintained. A factor that might be influencing, however, is the very nature of such small enterprises, with an awareness on the part of each employee of the significance of the role he or she plays, coupled with the fact that the employer is most usually personally present and intimately aware of the health problems of the employees. In such cases the employer can be subject to the same health hazards as are employees and therefore can have a personal interest in their elimination. These data, nonetheless, have led the National Safety Council to comment: "These figures indicate that although smaller busi-

Table 20-1. Incidence of occupational injury and illness by industry and size of workplace.

Industry	Number of Employees							
	1–19	20–49	50–99	100–249	250–499	500–999	1,000–2499	2,500+
Construction	14.3	19.8	22.8	24.9	24.2	19.7	15.1	12.7
Manufacturing	11.8	16.5	19.5	20.2	17.3	14.3	11.9	12.4
Transportation and public utilities	8.0	12.7	12.6	11.6	9.0	10.0	11.3	11.4
Wholesale and retail trade	4.6	8.9	11.1	12.1	11.5	12.4	11.9	10.2
Services	2.7	4.6	7.6	7.9	8.4	9.4	8.9	6.4
Finance, insurance, and real estate	2.1	1.9	2.9	2.8	3.0	2.9	2.8	1.8

ness employers may not be able to afford full-time safety and health professionals, they have on-the-job work injury problems which need attention. When added to consumer product safety and health, environmental safety and health, and energy problems, a much stronger case for management concern and attention to safety and health responsibilities is made."[4]

The importance of smaller business in the entire spectrum of business organizations is evident in Table 20-2, which shows the number of companies in the United States and the number of employees in the workforce by size of company (excluding the self-employed, government employees, and railroad employees).[5] It can readily be seen that half the workforce shown are employed in almost three and a half million places of business that employ fewer than 100 people. Furthermore, a quarter of the workforce are employed with companies that employ fewer than 20

Table 20-2. Number of companies in the United States and number of employees in the workforce by company size.

Size of Company by Number of Employees	Number of Companies	Number of Employees
1–3	1,698,585	2,985,453
4–7	763,516	3,954,231
8–19	618,509	7,377,848
20–49	286,670	8,719,647
50–99	92,213	6,325,646
100–249	53,061	8,057,863
250–499	16,779	5,764,232
500+	11,513	14,830,984
Total	3,540,846	58,015,904

employees, and almost 3 million people work in places that employ from one to three employees. At the other extreme, a quarter of the workforce are employed in places having over 500 employees.

It has been said that with relatively few exceptions, workers in small industries do not have the benefit of occupational health services. Many receive only emergency care for injuries, and unfortunately even this service is often inadequate. Small businessmen may recognize the value of occupational health services but find that individual programs are expensive and that other employers are hesitant to engage in cooperative projects. The smaller the work group, the more difficult it is to arrange for health services. These industries could use to advantage part-time nursing in conjunction with periodic visits to the plant by a physician. When services available through voluntary and public health agencies in the community are also used, small industries can provide health benefits that compare favorably with those of larger establishments.[6] Thus a 1957 study in Michigan found that in 34 places of employment, each of which had from 50 to 249 employees, only one employed a full-time nurse; the other 33 had neither a regular doctor nor a nurse on their staff.[7]

Where small plants have set up any type of industrial health program they most usually employ a physician part time to provide care for work injuries and perhaps to perform preemployment or periodic physical examinations. Where dispensaries have been set up they are most frequently served by a full-time or part-time nurse.

That industrial health services in smaller places of employment are less complete than those made available by larger employers is evident from the findings of two studies. A 1950 survey by the National Association of Manufacturers showed that at that time the percentages of companies shown in Table 20-3 had the type of health program services shown:[8]

Table 20-3. Percent of companies having health program services.

	Size of Company		
Type of Service	1–250 Employees	251–500 Employees	Over 2,500 Employees
In-plant medical services	33.9	69.5	96.5
Out-of-plant medical services	84.0	86.2	83.2
Health education	21.1	54.3	80.2
Accident prevention	58.8	85.4	98.0
Safety committee	6.9	30.4	47.8
Mass X-ray	33.8	66.5	71.4
Cancer test	4.0	5.3	19.8
Blood test	4.9	11.5	21.1
Syphilis	4.5	11.6	35.2

Eight years later a survey by the U.S. Public Health Service found that in manufacturing plants with less than 100 employees only .5 percent had adequate physician services and only 1 percent had adequate nursing services. For plants with 100 to 250 employees these percentages were 5.2 and 13.3 respectively. On the other hand, for plants with 5,000 or more employees the percentages were 100 in both instances. In plants with fewer than 100 employees, the following were the proportion of employees eligible for certain services:[9]

Preemployment examinations	23.5%
Periodic physical examinations	8.0
Medical advice	7.2
Treatment of plant injuries	84.2
Treatment of minor on-job illnesses	39.1
General medical care	13.1
General medical care of employees' families	3.9
Health education	2.8
Plant health inspections	3.9
Accident records	74.4
Illness records	21.7

It is evident from what has been shown that smaller places of employment provide considerably less in the way of industrial health services than is the case with larger employers. The problem of smaller businesses is not, however, impossible of solution. What is primarily needed is a concern and awareness by small employers of both the costs and the effects of nonoccupational as well as occupational health and safety hazards on their business operations, and a decision to take a point of action that will lead to positive results.

Essentially, the approach to occupational and nonoccupational illnesses and injuries in small places of employment is the same as that in the large places: observance and removal of work hazards, job and safety precautions and training, observation of the work performance of the individual employee, awareness of his or her health status, and the motivation of the person needing treatment to seek such treatment at an early stage. Where large organizations must rely on supervisors or shop stewards for such observation, in the smaller plant the owner or manager can easily perform this function.

Beyond this generality, each of the size groupings shown earlier can have its particular type of difficulty in controlling both occupational and nonoccupational health hazards. Each, furthermore, can differ from the other depending upon such variables as the type of operation involved, the occupational hazards presented, turnover among the workforce, and the age, sex, and socioeconomic composition of the workforce. All such differences can present their own unique type of problem.

Regardless of such problems and difficulties, hundreds of smaller places of employment have developed at least some aspects of an industrial health program. The approaches taken vary considerably and, it must be noted, do not always prove to be successful or lasting.

The most usual approach taken, of course, is the employment of a part-time physician or nurse, perhaps in conjunction with the use of community health resources. An alternative to this approach would be the cooperative use of an industrial nurse by a few places of employment. In either instance, periodic health examinations can be provided through the use of community resources. A local physician can be used for the conduct of preemployment examinations and for the treatment of occupational injuries. An industrial health physician can be used on a monthly or semimonthly basis to counsel management in the handling of personnel and job placement problems and in the development of safety measures. State and local government agencies can provide guidance and counseling on environmental health hazards.[10]

A few examples of such approaches were noted by the U.S. Department of Health, Education, and Welfare some years ago.[11] The Ketterlinus Lithographic Manufacturing Company in Philadelphia, with from 180 to 300 employees, in 1951 engaged the services of a part-time physician for three days a week and a visiting nurse who served two hours a day, five days a week.

The Allen Manufacturing Company in Hartford, with 595 employees in 1951, employed a full-time nurse and a part-time physician who served eight hours a week. The program included preemployment, special, and periodic physical examinations, immunization, chest X-ray, health counseling, the maintenance of health records, periodic plant inspection, and safety glasses.

The Frank H. Fleer Corp. in Philadelphia, with 265 employees in 1951, established an in-plant dispensary, the cost of which was $30.30 per employee. And the A.C. Horn Company in New York City, a paint manufacturer with 230 employees in 1951, established an in-plant dispensary that provided preemployment and periodic physical examinations, treatment for occupational injuries and illnesses, treatment for nonoccupationally caused emergencies, nutritional and health counseling, and periodic plant inspections.

Other approaches taken by smaller employers can involve some type of cooperative endeavor. This might entail the joint use of existing services or facilities available in the community. It might require the establishment of services and facilities on a cooperative or consortium basis. In either instance, the catalyst can be the smaller employers in the community, the trade organizations to which employers belong (which are frequently used to enable small employers to purchase group insurance coverages), a group of independent physicians providing services on either a contract basis or a fee-for-service basis, or a local hospital or HMO. Required would be adequate staffing, facilities, and equipment. This, in turn, would require adequate financial support, both at the outset and once the program became operational. Basic to such approaches would be a consensus as to the services to be provided. The advantages are essentially the avoidance of the need for each employer to make his own arrangements, and the spreading of costs over larger numbers. While more experimentation, experience, and information are sorely needed, it is noteworthy that today consortia in various parts of the country are proving valuable for employers in coping with the problem of alcoholism among their employees.[12]

Yet another approach that would seem to have possibilities for smaller employers is where union employees are involved. An industrial health program could be established under the aegis of the union, or of several unions in a community acting cooperatively. As has been noted earlier, many unions operate their own health centers. Quite frequently such centers are developed for union members employed by small firms and, in a sense, take the place of an industrial health program. It is not inconceivable that such centers, working cooperatively with small employers, could form the basis for an industrial health program that would be economically feasible since the services would be spread over large numbers. Requisite, of course, would be a cooperative attitude between employers and the union, which admittedly does not always exist.

Following are some examples of different types of cooperative approaches which have been developed for the benefit of small employers and their employees. It will be noted that not all of these have proved to be successful.

In 1946 the Hartford Plan was inaugurated by six employers and was jointly sponsored by the Joint Committee on Industrial Health of the Manufacturers Association of Connecticut and the Connecticut State Medical Society. The original employers had between 165 and 746 employees each, totaling 3,561 persons. Each establishment had its own in-plant dispensary and employed one or more full-time nurses. They, in turn, functioned under the direction of a full-time physician, employed by the plan, who spent specified periods of time at each of the places of employment at least semiweekly and was on call for emergencies. The services provided by the physician included preplacement and periodic physical examinations; termination examinations; private consultation with employees on plant premises; treatment of occupational illnesses and minor occupational injuries; laboratory tests; consultation with management regarding safety precautions, sanitation, nutrition, and rehabilitation; supervision of the health aspects of job placement; the evaluation of health complaints; supervision of the nurse; and referral of employees to their personal physician. After the formation of the plan a number of other employers in the area became participants.[13]

Another example of a joint approach to an industrial health program, albeit a more limited effort, is that of the Industrial Health Council of Birmingham, Alabama. Brought into being in 1947 through the combined efforts of business executives, physicians, civic leaders, and public health officials, the Council had a dual purpose: (1) the advancement of health education and (2) the conduct of multiphasic health screening examinations. The program commenced with 9 employers, but by 1958 had grown to include 166 places of employment involving over 17,000 employees. The Council is headed by a medical director. Employee participation in the health screening is voluntary and the examinations are conducted during working hours. The program does not perform other aspects of an industrial health program.[14]

Yet another cooperative approach was the Philadelphia Medical Society–Chamber of Commerce Small Plant Program organized in 1943. The program established a central dispensary to serve 101 places of employment, each of which had from 250 to 500 workers. In that same city the Philadelphia Health Council had been formed in 1924 to serve 31 places of employment, each of which had from 25 to 560 employees. The businesses involved were 11 confectioners, 2 cake bakers, 2 pork packers, 1 root beer manufacturer, 3 lithographers and printers, 6 textile firms, 2 paper manufacturers, 1 tannery, 1 cigarette factory, 1 iron and steel mill, and 1 lead and alloy foundry. The services provided were health examinations, accident prevention, occupational hazards surveys, sanitation surveys, the treatment of occupational accidents, the provision of emergency care, the diagnosis of chronic conditions, and health education. The program, however, was discontinued in 1932.[15]

Several attempts to serve the needs of small employers have also been made by clinics or hospitals in various parts of the country. By 1966 there were reported to be over 20 private clinics serving 9,100 places of employment. One such is the Northwest Industrial Clinic in Minneapolis. The clinic provides diagnostic procedures in toxicology, job evaluation and preplacement examinations, periodic physical examinations, executive examinations, immunization, industrial ophthalmology, audiometric testing, orthopedic surgery and care of trauma, physical therapy and rehabilitation, in-plant medical services, and surveys of the occupational environment. In-plant visits by the physicians are made on a scheduled basis, and services can be tailored to the needs of a particular plant based upon a preliminary survey.[16]

Another such clinic is the Portland (Oregon) Industrial Clinic, which provides industrial health services for a variety of clients including a department store, a foundry, a scrap metal yard, and a wood manufacturer. Among its services are preemployment examinations, consultation, immunization, and the supervison of in-plant dispensaries. Half the financing comes from the charges made to employers and half comes from the care provided employees injured at work and treated on a fee-for-service basis.[17]

A third example is the Petrie Clinic in Atlanta, Georgia, which provides services for several employers having from 100 to 500 employees. The clinic maintains a health center and provides part-time physician and full-time nursing services on an in-plant basis. The services include preemployment examinations, periodic physical examinations for all employees, emergency treatment, nursing services for ill employees, home-nursing visits, the maintenance of health records, health education for employees, and the study of occupational hazards. Two other examples of industrial clinics are the Manufacturers' Health Clinic at Winder, Georgia, serving six places of employment through a central clinic and a program in Tacoma, Washington, which provides complete medical care for over 7,000 industrial groups employing more than 22,000 men and 8,000 women.[18] In other instances clinics or hospitals might provide certain industrial health-related services on a contract or fee-for-service basis. Several of these are listed in Appendix C.

It is generally recognized that industrial health clinics should be designed specifically to serve places of employment, that their services should be tailored to meet the needs of the individual place of employment, and that they should provide adequate care for occupational injuries and illnesses. The degree to which a particular clinic can perform the necessary industrial health services will depend on the available facilities, the number and type of firms being served, the adequacy of its range of services, the qualifications of its staff, and the fees it can charge.

A critical view of industrial clinics was taken by Dr. D. John Lauer of the Jones and Laughlin Steel Corporation before the 1959 National Health Forum when he said: "The traditional industrial clinics that exist in so many of our large cities are often no more than physicians' offices gussied up to take care of large volumes of compensation cases and examinations. From such clinics, no one ever goes near a plant or attempts to practice preventive medicine."[19] That this is a sweeping statement to which there are exceptions is evident, but it strikes a cautionary note which is not unwarranted.

One more approach to making industrial health services available to smaller employers is to extend the services of large employers to the small employers in the community. The program established by the Cummings Engine Company, Columbus, Indiana, for its 9,500 employees is an excellent example of such an approach. In 1960 the company expanded the program to form the Columbus Occupational Health Association servicing 15,000 employees of 13 other employers on a fee-for-service basis. The services include preemployment examinations, periodic physical examinations, special examinations for employees with chronic conditions or who drive motor vehicles, and the treatment of occupational illnesses and injuries. Another example of such an approach is the Gates Rubber Company in Denver, which developed a program for its 10,000 U.S. employees and then expanded the program, as the Gates Medical Clinic, to provide service on an HMO basis to the community. Care is provided for both occupational and nonoccupational illnesses and injuries by 50 physicians, 17 nurses, 10 dentists, 6 pharmacists, 5 technicians, and 2 physiotherapists.[20]

A final possibility exists for smaller employers, as suggested by Edward M. Dolinsky of the Metropolitan Life Insurance Company. He believes that the way to bridge the gulf that separates industrial medical care from family medical care is through the use of HMOs and that in this way the HMO could bring the concepts and concerns of occupational medicine into the family health center. Noting the $375 million available through the HMO Act of 1973 to support the determination of feasibility and planning, implementation, and early operating costs of new HMOs, Mr. Dolinsky maintains that such planning could include departments of occupational medicine in those new HMOs or could provide for the expansion of extant HMOs into the industrial health field. He says: "We need to develop an expanded concept of HMOs which would, logically, incorporate departments of occupational medicine as integrated services of the HMO where industrial families have reasonable access to existing or proposed HMO health centers." He feels that the combination of OSHA and the HMO legislation "offers a potential means for regenerating occupational medicine."

Noting that the conceptual emphasis of an HMO lies in preventive medicine, Mr. Dolinsky points out that the prevention of illnesses and accidents is also the goal of an industrial health program and that this provides an obvious intellectual tie between the two. To make his concept a reality, Mr. Dolinsky suggests that a department of occupational medicine within an HMO could contract with employers to provide emergency services at the place of employment and to provide preemployment and periodic health examinations at the HMO. Such a department could assume responsibility for the study and identification of occupational health hazards, and the HMO could maintain and review a complete health record of each employee.[21] Such a concept for industrial health would indeed provide a source to which small employers could turn.

That smaller places of employment are called on to develop industrial health programs under OSHA has been made evident. That small businesses can receive federal loans for developing such programs was also shown. The Department of Labor has available a pamphlet, "OSHA Fact Sheet for Small Businesses on Obtaining Compliance Loans," which contains information on where to obtain such loans, the required collateral, interest and fees to be paid, and the terms of maturity. Such loans may be for the purpose of constructing new buildings, moving to a new location, additions to or alterations in equipment, or otherwise to meet the standards established by OSHA. These loans are made through the Small Business Administration. By October 1974, 115 such loans had been made at $6\frac{1}{2}$ percent interest.[22]

To assist small businesses in evaluating their environmental and physical conditions to determine whether they meet OSHA standards, the National Safety Council has prepared a comprehensive checklist, "OSHA Standards Handbook for Small Business." The checklist covers available medical services and first aid; general environmental controls such as sanitation, occupational health and environmental hazards (such as ventilation, noise, and radiation), and the use of hazardous materials (including gases, liquids, toxic substances, and explosives); the availability of protective equipment for workers; the means for fire protection; hazardous machinery and machine guards; power tools; electrical equipment; welding and cutting processes; materials handling and storage; and working surfaces.[23]

Other sources of assistance to small employers include the American Medical Association's "Guide to Small Plant Occupational Health Programs," published in 1973; the 1973 recommended safe practices and procedures of the Manufacturing Chemists Association, "Safety Inspection Committee for a Small Plant"; and certain of the sources mentioned in the following chapter and in Appendix C. These sources can provide considerable guidance to smaller employers.

REFERENCES

1. National Health Council, *The Health of People Who Work,* New York, 1960.
2. Ibid.
3. Ibid.

4. National Safety Council, *Handbook of Occupational Safety and Health*, Chicago, 1976.
5. "County Business Patterns," 1972; and *U.S. Statistical Abstract*," 1973.
6. HEW, *Community Health Nursing for Working People*, 1970.
7. Institute of Social Research, University of Michigan, 1957.
8. George W. Bachman, Brookings Institution, 1952.
9. HEW, *Small Plant Health and Medical Programs*, 1958.
10. National Health Council, op. cit.
11. HEW, op. cit.
12. J. F. Follmann, Jr., *Alcoholics and Business*, AMACOM, 1976.
13. National Health Council, op. cit.
14. Ibid.
15. HEW, op. cit.
16. Leonard Arling, M.D., "Industrial Clinics and Occupational Health," *Archives of Environmental Health*, May 1966.
17. Forest E. Rieke, M.D., "Industrial Clinic Services to Small Industries," *American Journal of Public Health*, January 1972.
18. HEW, op. cit.
19. National Health Council, op. cit.
20. The Conference Board. "Industry Roles in Health Care," New York, 1974.
21. Edward M. Dolinsky, "Health Maintenance Organizations and Occupational Medicine," *Bulletin of the New York Academy of Medicine*, Vol. 50, No. 10, 1974.
22. Virginia Reinhart, "Small Business Grabs a Lifeline," *Job Safety and Health*, March 1975.
23. National Safety Council, op. cit.

21

Where to Receive Help

There are many sources of counsel, help, guidance, and supportive materials available to employers or unions in the establishment, enlargement, improvement, or reevaluation of industrial health and safety programs, in their concerns for nonoccupational health hazards, in the surveillance of product safety, and in dealing with the problems of environmental health hazards. Of particular significance are the sources of help that are available to the smaller places of employment, some of which have been mentioned in the preceding chapter.

The areas of concern in which places of employment of all sizes frequently need assistance run a broad gamut, depending in part on the type of operation and the hazards involved. The enactment and enforcement of OSHA, the activities of the FDA and consumer organizations concerned with product safety and the protection of the consumer, and the authority vested in the Environmental Protection Agency and its state and municipal counterparts in recent years have heightened and intensified this need for guidance and assistance. The areas of need require a broad spectrum of specialists, ranging from industrial physicians, nurses, and hygienists; to engineers trained in safety, accident prevention, and protection against fire and explosion; to engineers and architects skilled in such matters as plant construction and layout, machine design, ventilation, protection against extreme temperatures, illumination, the control of vibration, waste disposal, sanitation, housekeeping and storage, and the control of smoke, dust, fumes, and odors; to chemists knowledgeable in such areas as toxic materials and substances, plastics, solvents, metals, and radiation; to psychiatrists and psychologists versed in the control of stress, monotony, and other influences that affect mental and emotional stability; to those trained in such diverse matters as the control of alcoholism and drug abuse, rehabilitation, health education, laboratory techniques, record keeping, and health-related research. This is a formidable need, particularly when it is generally recognized that the entire industrial health field is in serious short supply of many of the needed professions and skills.

Despite this, employers must fulfill their responsibility to their employees, to the consuming public, and to the community in which they function. To do so is humane, is a constructive policy of operation, is a fulfillment of corporate responsibility, is in the best interest of the economy and the welfare of the nation, and in their own self-interest from every standpoint. Where, then, can they turn for guidance and assistance?

There are many professional organizations that can provide assistance in obtaining industrial physicians, nurses, hygienists, and supportive personnel on either a full-time or part-time basis. Sources for industrial physicians include the Industrial Medical Association, the American Association of Industrial Physicians, the Council on Industrial Health of the American Medical Association, and the state and local medical societies. Assistance in obtaining nurses can be received from the American Association of Industrial Nurses, the American Nurses Association, the National League for Nursing, and the state or local visiting nurse associations, which often have specially trained and supervised nurses available to industrial health programs on a salary or contract basis. Assistance in obtaining the services of an industrial hygienist is available from such organizations as the American Industrial Hygiene Association and its local affiliates, and the American Conference of Government Industrial Hygienists. All such organizations, as well as the Occupational Health Institute, can provide valuable counsel and assistance in the establishment or revision of an industrial health program.

There are many professional organizations through which competent technical personnel can be obtained to cope with specific types of hazards and problems. These organizations are highly specialized in such areas as illumination, noise, vibration, temperature control, ventilation, gas and fumes control, fire and explosion protection, waste disposal, safety, and other engineering concerns. Others are specialized in such matters as chemicals, toxic substances, petroleum, plastics, radiation, and air and water pollution. Such organizations are of inestimable value to an industrial health program, depending upon the needs of a particular program and the potential hazards involved. Among such organizations are the American Society of Mechanical Engineers, Inc., the Manufacturing Chemists Association, the National Safety Council, the American Society for Safety Engineers, and many others of equal importance, including several government agencies, noted in Appendix C.

Of very considerable importance are those organizations that develop and make available standards concerned with the health and safety of workers, with the prevention of nonoccupational illnesses and injuries, with product safety, with environmental control, and with the maintenance of health and safety records. Such organizations include committees of the American National Standards Institute and the National Safety Council, the Occupational Safety and Health Administration, the National Institute of Occupational Safety and Health, and many of the technical organizations in specialized fields previously mentioned.

The efforts of the American National Standards Institute (ANSI), in cooperation with many other organizations, in the development of voluntary consensus standards cover a broad range of concerns, including occupational safety (such as protective equipment, lighting, construction and layout, conveyances, electrical equipment, ventilation, insulation, and fire and explosion prevention), certain industrial processes (such as chemicals and materials handling, vibration, radiation, packaging, shipping, storage, noise, ventilation, temperature control, and toxic dusts and acids), certain employee health matters (such as sanitation, hygiene, medical devices, and first aid), consumer product safety, highway safety, information systems, and record keeping. Ten years ago there were 2,900 American National Standards. Today there are more than 6,000. A high percentage of these are used by industry to protect the workforce, to protect the public, to provide safer and more satisfactory consumer products, and to reduce health and safety hazards in the home, on the highway, in public places, and in the course of recreation. For places of employment, they assist in compliance with government regulations, reduce workmen's compensation costs and the costs of health and disability insurance, reduce operating costs (including absenteeism and labor turnover), and increase production. In 1976, ANSI approved and published the first safety standard for refuse collection, identified by the National Safety Council as one of the most hazardous occupations, having eight times the incidence of accidents as industry as a whole.

The National Safety Council (NSC) also has developed safety standards, working in conjunction with ANSI. Such standards are concerned with chemicals and gases, noise, eye protection, electricity, fires and burns, falls, machine guards, tools, lifting, materials handling, vehicles, office accidents, home accidents, industrial housekeeping, recreational hazards, hygiene, clothing, first aid, the protection of limbs, and the role of training and supervisors.

Both ANSI and NSC work cooperatively with NIOSH and OSHA in the development of standards. All such standards are of importance in the operation of an industrial health program, in the concern of employers over nonoccupational illnesses and injuries, in the assurance of the safety of products offered the consuming public, and in concerns over potential adverse effects on the environment.

Business organizations also provide considerable assistance to places of employment in the establishment and maintenance of industrial health programs. Certainly many of the technical, professional, and standard-setting organizations mentioned previously are within this category since they are supported by employers. In addition, the National Association of Manufacturers and the U.S. Chamber of Commerce are valuable sources of information and assistance to industrial health programs. Beyond such organizations, it cannot be overlooked that those programs established by individual places of employment, some of which have been in operation for many years and have gained considerable experience, can be of inestimable value in setting up, revising, or enlarging a program. Their practical knowledge and data gathering can be a primary source of help. Individual employers also form and finance a great many foundations with various purposes. Many of these are devoted partially or solely to health matters and conduct or support research in many areas related to industrial health. Beyond this, many corporations manufacture safety devices, testing instruments, and other equipment that can be of use not only in operating an industrial health and safety program but also in coping with potential environmental hazards. Manufacturers of these devices also offer valuable consultation.

Substantial aid is provided by insurance organizations, such as the American Insurance Association and the American Mutual Insurance Alliance, as well as individual companies. These organizations have had a long-standing interest in both occupational and nonoccupational injuries and illnesses. Frequently they maintain safety engineers and industrial hygiene personnel and laboratories for the study of health hazards and the means for their control. They furthermore maintain useful data. In some instances they provide rehabilitation services. Many insurers also stand ready to assist in the establishment of an alcoholism control program for employees. With their long experience in workmen's compensation, disability, and health insurance, insurers are quite willing to assist in the establishment of an industrial health program. For example, insurance companies have some 10,000 safety engineers available to their policyholders to provide consultation in such areas as environmental health, monitoring and inspection, loss control, and record keeping. Many have available helpful manuals, literature, and films, and some give courses of instruction in safety. The demands on these services have increased over 30 percent since the enactment of OSHA.[1]

Organized labor can be a source of assistance, particularly in the establishment of union health programs. Sources can be found in the AFL-CIO as well as in particular unions with a long-standing interest in industrial accidents and diseases (such as the UAW, the UMW, the United Steel Workers, and the Chemical Workers). They can also be instrumental in bringing about employee cooperation with an industrial health program. The United Paperworkers International, for example, conducts courses in safety.

In many localities there are available consultants and clinics of various types that can be of direct assistance, particularly to smaller employers, in the establishment or conduct of an industrial health program. Some provide counsel for setting up a program, others provide some or all of the necessary services. In many instances the services provided are essentially the performance of preemployment and periodic physical or screening examinations for the early detection of disease. These services can be of equal value to the conduct of union health programs.

A great many organizations, in addition to those previously mentioned, are sources of information on aspects of specific diseases and hazards and on the prevention of illnesses and accidents. In particular, the materials made available by these organizations can be of great help in the conduct of health education and safety programs at places of employment or by labor unions. The available materials include posters, pamphlets, materials for journals, lecture guides, slides, and films. The specific diseases of concern include heart diseases, cancer, diabetes, respiratory diseases, hypertension, and arthritis. Health education materials also deal with such matters as excessive smoking, alcohol and drug abuse, diet and obesity, exercise, and rest. These are of value in the prevention of nonoccupational illnesses. The National Safety Council, for example, has a complete catalog of slides and films dealing with many hazards to safety and health, both on and off the job, including instruction to supervisors. NSC also conducts an employee safety award program, which has been very effective in reducing the incidence of accidents. There are many other sources of films available dealing with specific aspects of both occupational and nonoccupational health and safety hazards.

Several colleges and universities also play a significant role with respect to industrial health programs. They train the health professionals, physicians, nurses, industrial hygienists, technical professionals, engineers, and chemists. More related to the concerns here, in some instances they provide courses of instruction in industrial health for the representatives of employers and unions. Their research projects are also of inestimable value to an industrial health program.

Finally, government agencies at the local, state, and federal levels provide significant guidance and consultation in the establishment and operation of an industrial health program. These include city and state health or labor departments, OSHA, NIOSH, the FDA, and the Environmental Protection Agency. The Small Business Administration can be of particular assistance to smaller employers. Industrial hygiene is one of the areas where such public agencies play a supportive role through the inspection of workplaces, improved sanitation and housekeeping, and the elimination of health and safety hazards. Such agencies frequently provide consultation in starting an industrial health program, which can be of particular value to smaller places of employment. One problem with many such programs, however, is the inadequacy of their staffs. Thus a 1958 study commented that "because of limited personnel, [state] agencies can reach no more than 10 percent of the nation's workers in any one year."[2] It is evident, then, that employers and unions cannot expect too much from this source.

There are, then, a great many sources to which employers and unions can turn for assistance in their concerns over industrial health and related hazards. Which source is most valuable in a particular instance can be determined by such differentials as the number of employees involved, the type of operation and hazards presented, the particular geographic location, and the discerned needs and problems to be solved.

Many, but by no means all, of these sources of guidance and assistance are listed in Appendix C.

REFERENCES

1. Doris Baldwin, "Ask What Your Insurance Company Can Do For You," *Job Safety and Health*, September 1974.
2. HEW, "State Occupational Health Programs," 1958.

22

Some Special Problems

Three problems warrant special mention in relation to industrial health programs since, to a certain extent, they call for approaches different from the prevention and treatment of physical illnesses or injuries, whether occupationally induced or otherwise. These three problems are mental illness, alcoholism, and drug abuse.

MENTAL ILLNESS

Mental or emotional illness is the most vexing problem facing places of employment today. It is estimated that 25 percent of the working population suffers from some degree of emotional illness. Half of these cases can be expected to have a good prognosis with proper identification and treatment. The remainder are likely to be chronically ill, requiring long-term psychotherapy, and being subject to recurrences of the disease.[1]

The mentally ill or emotionally disturbed employee has a greater degree of absenteeism, a higher incidence of accidents, and an undue use of health and disability insurance benefits. Turnover is greater among such employees. They can be a cause of friction among employees, adverse customer relations, and harmful public relations. It is generally recognized that personality problems are the predominant and most common cause of discharge. Mental illness and emotional disorders, then, are problems demanding the thoughtful attention of places of employment.

Identification of the mentally ill or emotional disturbed employee can best be accomplished by the supervisor or the shop steward in the place of employment. To be observed are the work performance and the functional effectiveness of the worker, his or her competence, sense of well-being, and job satisfaction, any failure to grow on the job, tenseness and irritability, resentment, a nonextension of effort, poor relations with other employees, and certainly absences from work and accident proneness. It is also important to develop an awareness of such highly personal matters as marital or family troubles or the accumulation of debts. To make such efforts effective, managers, supervisors, and shop stewards should be informed and trained in the nature of mental illness and emotional disturbances, the identifying symptoms of such diseases, and the forms of treatment available in the community. By no means should such personnel attempt diagnosis: Once a supervisor suspects that a mental problem exists, referral should be made to the medical staff for consultation and diagnosis. The employee should be informed of the professional opinion and, if he wants it, referred for care. Subsequent progress evaluation and followup should then be made, and this should continue after the employee has returned to work. Every effort should be made to help the employee in the rehabilitation process, including the possibility of limited duty or responsibility for a time, or even the possibility of a change of job assignment.

At the same time, supervisors and shop stewards should be trained to detect the effects of various types of working conditions on the mental or emotional stability of employees. Such factors as inadequate lighting, poor ventilation, offensive odors, and excessive noise and vibration can take their toll. So can such little-understood matters as job monotony and working under stress. Such matters as a low level of employee morale, inadequate wage levels, faulty supervision, or lack of recognition and advancement can affect the disturbed employee more than his fellow workers. Unfortunately too little is known about the relationship of the job and working conditions to the problem of mental illness. There is a need for much more research and knowledge of this important subject.[2]

Beyond this, any health education program at a place of employment should include information about mental illness and emotional disorders, explaining their nature and the types of treatment available. Principally, the purpose of such information should be to overcome the stigma that frequently attaches to such diseases.

The cost-effectiveness of the Illinois Bell Telephone Company program for mental illnesses and emotional disturbances among its employees is shown subsequently.

ALCOHOLISM

In two preceding chapters the considerable incidence of alcoholism in places of employment and the resultant sizable costs have been shown. Since half of all alcoholic persons are employed and since they have a greater incidence of illnesses, accidents, and disabilities, a marked degree of deteriorated work performance, and a higher rate of labor turnover, alcoholism is a problem that places of employment must face squarely. Particularly is this so since alcoholic employees are most usually between ages 35 to 50, have been employed with the same employer for more than ten years, and are therefore experienced employees of value. The successes of those alcoholism control programs that have been established, experiencing from 60 to more than 80 percent recoveries, with marked savings in operational costs, argue for the establishment of such programs in all places of employment.

Today over 600 places of employment have alcoholism control programs of various types. Granted, the definition of alcoholism is not a resolved matter, too little is known of its causes (as is the case with many diseases), its diagnosis can be vague or misleading (due, in part, to the frequent presence of physical manifestations and to the stigma that attaches to alcoholism), and identification of the alcoholic employee in the early stages of the disease is not by any means a simple matter. Nonetheless, much is being accomplished in places of employment today to control the disease.

Many aspects of establishing an alcoholism control program are similar to those involved in developing an industrial health program, even though in some instances the alcoholism control program functions as a more or less separate entity. For example, the Morgan Guaranty Trust Company has a Special Health Services unit, and the New York Times alcoholism control program functions as a separate entity.

Of paramount importance is the full, active support of top management, a formal written statement of policy, and the active cooperation of any union involved. The program must be applicable to employees at all levels and without discrimination. All health records must be maintained on a confidential basis. Treatment recommended by the medical department must be optional and should dovetail with the advice of the employee's personal physician.

Alcoholism control programs differ from other aspects of an industrial health program in essentially three areas. First, the managers, supervisors, and shop stewards must have special training in coping with alcoholism. Identification of the alcoholic employee, particularly in the early stages of the disease, is difficult, among other reasons because the alcoholic is secretive and evasive. This calls for intelligent and sympathetic understanding of the employee's problem. In all instances, identification should rest not in the physiological symptoms of the disease but in the presence of deteriorated work performance documented with facts. Confrontation on this basis should lead to referral to the medical department.

The second differing aspect of an alcoholism control program lies in the types of treatment modalities to which the medical department can refer the alcoholic employee. Available facilities differ widely from community to community, and their level of efficacy can also cover a rather broad range. It is a responsibility of the medical department to be aware of the available facilities and of their value in leading to recovery. Furthermore, the type of facility or facilities suggested can vary with the state of the disease in an individual case, as well as the presence or absence of physical complications. For example, in some cases immediate referral to Alcoholics Anonymous might be sufficient. In other instances the use of detoxification centers, alcoholism treatment centers, or hospitals might be necessary, with long-term followup therapy in AA.

A third difference occurs in those instances where suggested treatment is refused or where treatment is not continued to completion and recovery. The motivation of an alcoholic person to seek treatment is at best difficult, since he or she finds it next to impossible to conceive of functioning without the support of alcohol. It must therefore be made clear from the start that refusal to seek treatment will result in disciplinary action. Since the job is important to the alcoholic employee, perhaps much more so than to other employees, often being the last real substantial base left in the person's existence and one with which he is familiar and comfortable, such threat most usually will prove effective. The disciplinary action, dependent upon the individual circumstances, could be demotion, suspension without pay, early retirement, or dismissal. This approach is most usually effective.[3]

DRUG ABUSE

Drug abuse may take three forms: (1) the abuse of drugs and medications commonly purchased, such as barbiturates and antihistamines; (2) the misuse of prescribed drugs by overuse; or (3) the illegal use of mind-altering drugs such as heroin or marijuana. This latter form of abuse is of concern here.

That drug abuse can be a problem to places of employment is evident from an estimate that one-third of all drug addicts are employed. This has become particularly so with the increasing degree of employment of the disadvantaged. It has also been shown that the majority of addicts, unlike alcoholic employees, are in the age bracket 20 to 25, and that three-quarters of such employees have had less than four years of service with their employer. Drug users are generally recognized to have undue absenteeism and labor turnover, a greater incidence of accidents, decreased productivity, and a proclivity for theft.

Despite the scope of the problem it has been noted that "most private employers are not doing very much in this area.... Some changes, however, are being attempted. Many major employers are taking a nonpunitive approach to drug abuse among current employees."[4] Many employers today will hire addicts who have undergone treatment and rehabilitation if they are recommended by a responsible treatment agency. Increasingly, employers are coming to recognize that they have a social responsibility to attempt to control drug abuse among their employees. To this end, the subject of drug abuse is now frequently included in health education programs at places of employment.

The approach to drug abuse in places of employment is quite similar to that taken in coping with alcoholism. Once management policy has been established and supervisors and shop stewards have been adequately trained, the drug abuser is identified through unsatisfactory job performance, diagnosis is made by the medical department or consultant, the employee is referred for treatment, and diligent followup is made. Refusal of treatment calls for disciplinary action.

A few examples of the approaches being taken by employers today will suffice. The Weyerhaeuser Company is one of the companies that has a management policy dealing with the abuse of drugs. Identification of the addict is based on deteriorated job performance, and the education of supervisors to this end is provided. Treatment is then recommended and there is followup on the rehabilitative process. The buying and selling of illicit drugs on company premises is not condoned. Known addicts are not employed.[5]

AT&T also has a management policy concerned with drug abuse. The company makes available educational materials to its employees and conducts meetings on the use of drugs. The program is generally similar to its alcoholism control program. The illegal sale, purchase, transfer, use, or possession of illicit drugs by employees on company premises or while on company business is prohibited.[6]

A final example is the drug abuse program at the Illinois Bell Telephone Company. A physician with a background in psychiatry administers the rehabilitation program for drug abusers. A full-time qualified drug counselor conducts individual counseling and group therapy sessions. In 1974, 89 employees were referred to the program, 62 of whom were males and 27 females. Referrals come principally from supervisors

(63 percent); the remainder are made by the medical department (24 percent), by the individual (2 percent), and by relatives (11 percent). The addicts are young and have had a relatively short period of service. Of those referred, 71 percent have consented to treatment. It is considered too early to date, to attempt to demonstrate the positive effects of the program.[7]

With all such special problems of employee health, as well as other aspects of an industrial health program, it is important that the health and disability-insurance benefits available to employees be adequate to provide the necessary financial support for the recommended treatment. In practice, unfortunately, the insurance benefits to cover the cost for the period of such treatment can be less than adequate, and at times such conditions are entirely exempt from the coverage. When this is the case, the effectiveness of the industrial health program will be placed in jeopardy.

REFERENCES

1. Robert R. J. Hilker, M.D., "Behavioral Problems in Industry," paper presented before the Pennsylvania Hospital conference, October 1975.
2. Stanislaw V. Kasl, "Mental Health and Work Environment," *Journal of Occupational Medicine,* June 1973.
3. J. F. Follmann, Jr., *Alcoholics and Business,* AMACOM, 1976.
4. Drug Abuse Council, Inc., "Employment and Addiction: Overview of Issues," Washington, D.C., 1973.
5. John T. Redfield, M.D., "Drugs in the Workplace," *American Journal of Public Health,* December 1973.
6. Metropolitan Life Insurance Company. "Drug Abuse in Industry," New York, 1970.
7. Hilker, op. cit.

Part VI

Relationship of Financing Programs for Health Care and Loss of Income

The relationship of an industrial health program to programs for financing health care and loss of income resulting from disability is an extremely close one indeed. No industrial health program, no matter how comprehensive, could be complete in the absence of adequate provision for the financing of health care and for the protection of workers against income lost as a result of disability.

The incidence of accidents, illnesses, and disability in the workforce and their economic costs were shown earlier. These costs have a direct relationship to the financing programs to be discussed here. As will be shown, an effective industrial health program, in turn, can affect the costs of financing programs.

In Chapter 23 the publicly established programs for financing medical care and loss of income for employed people will be discussed. Chapter 24 explores the relationship of private insurance coverages and employee benefit programs to industrial health.

23

Publicly Established Programs

Several types of publicly established programs for financing health care and/or loss of income resulting from illness or injury can have a direct or indirect relationship to industrial health programs. Of direct relationship, since they pertain only to employed people, are the workmen's compensation programs in all states and the disability benefits and medical care benefits established as part of the federal Social Security program (OASDI). Indirectly related, since some employed persons might become eligible for their benefits, are the Medicaid program, the categorical public assistance programs, CHAMPUS, the VA program, and the publicly provided forms of health care. Each of these will be discussed briefly. The question of national health insurance in the United States and its relationship to industrial health programs will then be explored.

WORKMEN'S COMPENSATION

Without question, the workmen's compensation system has the most direct relationship to industrial health programs of all the publicly established programs for financing health care and loss of income. As was shown in Chapter 2, it was the enactment of workmen's compensation legislation by the states that provided the first significant impulse for the development of industrial health and safety programs. A review of the present status of workmen's compensation in terms of its scope, its benefits, and its costs is, therefore, directly relevant here. This being so, criticisms of workmen's compensation and proposals being made for changes in the present system are also important to note.

Commencing in 1911, all states and the federal government (for longshoremen, harbor workers, employees in the District of Columbia, and railroad workers) enacted workmen's compensation legislation. In nine states the coverage is elective on the part of employers; in the remainder it is compulsory. Some 84 percent of the civilian workforce today are protected by such legislation. There are differences among the states as respects the types of employees covered, the benefits provided, and the types of insuring mechanism employed.

None of the workmen's compensation laws covers all employees. Usually exempt from the coverage are farm workers (who are only covered in about half the states), domestic employees (who are only covered in a few states), casual workers, and those working for small employers (14 states—most usually those employing less than three employees).

The benefits pertain to occupational accidents and occupational diseases, although in four states the coverage is applicable to only certain specified diseases. The benefits provided for are of basically the following three types.

1. *Disability benefits* to partially replace income lost as a result of occupational injury or disease. The benefits differ for (a) temporary total disability (the majority of cases), (b) temporary partial disability, (c) partial impairment of a permanent nature (usually providing for a flat-sum payment, the sum depending upon the nature of the impairment and varying among the states), and (d) permanent disability. In cases involving disability the benefit differs among the states, most usually being 60 or 67 percent of average wages after a one-week waiting period. For permanent disability benefits the states vary in limiting the payment of the benefit to a specified number of weeks (usually ranging from 330 to 550 weeks); in addition some states have established a maximum overall dollar amount of benefits to be paid (ranging from $21,000 to $40,000). The majority of the states, however, provide for the payment of permanent total disability benefits for life. By 1973 only 12 states had made any adjustment in the disability benefits to cover the effects of economic inflation.

2. *Medical care benefits* to defray the cost of the treatment of occupational injuries and diseases and the costs of physical rehabilitation. Today most states provide for the payment of all reasonable and necessary costs of such medical care without limitations or exceptions. In a minority of the states (five), however, a variety of limitations to such

benefits is provided in terms of the maximum amount of benefit payable or the duration of time such benefits shall be paid.

3. *Survivors benefits* in the case of fatal injuries. These benefits include burial allowances and the payment of a proportion of the worker's former weekly wages to the spouse until remarriage and to the children to a specified age. Some states stipulate a maximum period of time for which such benefit is paid (most usually ten years, with a maximum dollar amount payable ranging from $16,000 to $45,000).

Most workmen's compensation laws limit employer liability in second-injury cases (where a preexisting injury combines with a second injury to produce disability greater than that caused by the latter alone). To cover such cases second-injury funds are created, thereby encouraging the employment of physically handicapped persons and more equitably allocating the cost of providing benefits for such employees. Under this arrangement "second injury" employers pay compensation related primarily to the disability caused by the second injury alone, the difference being compensated from the second-injury fund.

The insuring mechanism used by employers might be a state fund, insurance companies, or self-insurance. In 6 states the state fund is used exclusively (although 3 of these permit self-insurance). Insurance companies provide the coverage in 44 states, in 12 of these in competition with a state fund. In 43 states self-insurance is permitted. Insurance company premiums rates are subject to state regulation. For larger employers the premiums are experience-rated, thereby encouraging the development of safety programs. Workmen's compensation benefits today exceed $5 billion yearly. Until recently 64 percent of the benefits were paid by insurance companies, 23 percent by the state funds, and 13 percent by self-insured programs. With the enactment of the federal black lung legislation in 1971, these proportions changed, with insurance company benefits dropping to 58 percent of the total and state fund payments increasing to 29 percent. The cost of workmen's compensation is borne entirely by employers. In 1973 this cost was $4.9 billion, or 1.2 percent of payroll.[1,2]

The relationship of workmen's compensation to the rehabilitation of disabled workers is to be noted. The goal is the restoration of the disabled worker to some useful and remunerative occupation to the greatest degree possible through prompt treatment and therapy, the provision of social and psychiatric services, and the provision of vocational counseling and training through sheltered workshops or otherwise. Commencing in Massachusetts in 1918, state workmen's compensation laws came to recognize the importance of rehabilitation and required payment for physical rehabilitation services under the medical payments benefit structure.

Vocational training is almost universally provided by the states today. One of the results has been the creation of a new medical specialty: physical medicine.[3] Today there are in excess of 1,700 rehabilitation facilities in the United States, many of these maintained by insurance companies.

Many organizations have a direct interest in the rehabilitation of workmen's compensation cases and many of these, in turn, have adopted formal resolutions on the subject. Included here are the National Association of Manufacturers, Chamber of Commerce of the United States, American Mutual Insurance Alliance, American Insurance Association, National Council on Compensation Insurance, Compensation Rating Board, AFL–CIO, National Rehabilitation Association, American Medical Association, American College of Surgeons, International Association of Industrial Accident Boards and Commissions, International Conference on Employee Benefit Plans, U.S. Department of Labor, and state departments of labor.

Among the benefits of rehabilitation efforts, in addition to the clear and obvious benefit to the worker and his family and the reduction in workmen's compensation claims, is the encouragement given to the employment of physically handicapped persons. Here it is of interest that as early as 1944 studies by the Labor Department found that 87 percent of employed handicapped persons were just as efficient as other workers, 8 percent were more efficient, and only 5 percent were found by employers to be less efficient.[4] It was also found that 56 percent of the employed handicapped persons had a lower accident rate than other employees, 42 percent had about the same incidence of accidents, and only 2 percent of the handicapped had a higher accident rate.[5]

It is generally recognized, however, that a deterrent to the rehabilitation process can be the absence of motivation on the part of the disabled worker. Lack of motivation can be psychological in nature; it can also have its roots in economic considerations in cases where excessive benefits are available to the disabled worker for so long as he is disabled, where he is nearing retirement age, or where he is faced with reduced earnings on return to work, or even with unemployment.[6] Other deterrents are the shortage of available professional personnel, the processes of litigation, the payment of lump sum settlements, the failure of some cash benefit structuring to provide an incentive to rehabilitation, the arbitrary limits on medical benefits under the laws of some states, and the frequent absence of adequate reporting and records of workmen's compensation experience.

Practically from its inception the workmen's compensation system has been subjected to a host of criticisms by labor unions, employers, insurers, health economists, academicians, and politicians. One consequence is that the state laws have been frequently amended. A few examples of such criticisms will suffice.

In 1958 Jerome Pollack, then with the UAW, criticized the inadequacy of the cash disability benefits received by injured workers. He was also highly critical of the nature of the medical expense benefits, particularly those for occupational diseases, many of which go "unrecognized and uncompensated," the absence of free choice for the worker in the selection of a physician, and the attitude of physicians in minimizing the work origin of a disease or injury.[6] Subsequent to the extension of Social Security benefits to totally and presumably permanently disabled workers, Congressman A. Sydney Herlong, Jr. and the American Mutual Insurance Alliance were critical of the effects of duplicate benefits under the new DI program and under workmen's compensation, maintaining that the effect would be to discourage workers from undergoing rehabilitation and returning to work by the removal of the financial incentive. In 1960 at a discussion of risk management conducted by the American Management Associations, Hiram L. Kennicott, Jr. of the Lumbermen's Mututal Casualty Company was critical of the concept of self-insurance by employers for workmen's compensation coverage, which was defended by Everett Parks of Marsh and McLennan.[7] In 1971 the AFL-CIO adopted Workmen's Compensation Resolution No. 131. The resolution supported a Federal Workmen's Compensation Act but, in the interim, called for the following program of benefits under workmen's compensation:

Compulsory coverage with no numerical exceptions for smaller employers.
Coverage of agricultural workers.
Benefit levels not less than two-thirds the injured worker's average wage.
Time limits of at least one year for the filing of an occupational disease claim.
Full statutory coverage for all occupational diseases.
Full medical expenses.
Free choice of a qualified physician by the worker.
Inclusion of rehabilitation in the workmen's compensation agency.
Maintenance benefits during rehabilitation.
Benefits for the totally disabled for the period of disability.
Exclusive administration by a state compensation fund.
Widow benefits to be paid for life or until remarriage.
Double benefits to minor employees if illegally employed.
Prohibition of lump sum settlements of claims.

It has been noted that one of the provisions of the Occupational Safety and Health Act of 1970 established a National Commission on State Workmen's Compensation Laws. That Commission issued its report on July 31, 1972. It concluded that in general the present state programs were "neither adequate nor equitable" but that the states should continue to have primary responsibility for workmen's compensation and that they should be given the opportunity to appropriately amend their then existing laws before any federal mandatory standards should be adopted. The principal recommendations made by the Commission were as follows:

1. Coverage should be compulsory rather than elective and should eventually be extended to cover farm workers, household workers, and casual workers. There should be no exemptions for any class of employees (for example, professional athletes and employees of charitable organizations or small employers).

2. There should be full coverage of work-related diseases.

3. There should be full coverage for medical and physical rehabilitation services without limits as to dollar amounts or length of time.

4. There should be adequate weekly cash benefits for temporary total disability, permanent total disability, and death. The amount paid should be at least two-thirds of the worker's gross weekly wage.

5. There should be no arbitrary limits on the amount or duration of benefits for permanent total disability or death, with benefits paid for the duration of disability or life and for the lifetime of a widow unless remarried.

6. The employee should be given the choice of jurisdiction for filing interstate claims.

7. OASDI benefits should continue to be reduced in the presence of workmen's compensation benefits.

8. Permanent total disability and death benefits should be increased through time in recognition of economic inflation.

9. The worker should be permitted initial selection of the physician.

10. Each state should establish a second-injury fund with broad coverage for preexisting impairments.

11. Each state should establish a medical rehabilitation division.

12. Time limits for filing claims should be liberalized to recognize the time lag inherent in certain occupational diseases.

The Commission left to each state the type of insurance arrangement to be used but encouraged the extension of the concept of experience rating as a further incentive to safety programs. It recommended that each insurance carrier be required to provide accident prevention services.[8]

The challenge that this report made to the states did not pass unheeded. Since that time many states have liberalized their programs in several respects.

SOCIAL SECURITY DISABILITY BENEFITS

Another publicly established financing program directly related to industrial health is the Disability Insurance (DI) program of the Social Security system (OASDI). In this instance the relationship is with respect to income lost as a result of total and presumably permanent disability and to the costs of hospital and medical services under the Medicare program.

In 1956 the Disability Program was established by Congress to provide OASI (Social Security) income benefits for the seriously disabled age 50 or over and until age 65. In 1960 the age-50 limitation was removed. In 1965 the definition of disability was changed to the "inability to engage in any substantial gainful activity by reason of any medically determinable physical or mental impairment which can be expected to result in death or has lasted or can be expected to last for a continuous period of not less than 12 months." Monthly payments are payable after an elimination period of five months. For any month in which a worker receives a workmen's compensation benefit the DI benefits are reduced to the extent that the total benefits payable to him or his dependents under both programs do not exceed 80 percent of his average monthly earnings prior to the disability. In 1973 the DI beneficiaries became eligible for the Medicare benefits for hospital and medical care expenses to the same degree as persons age 65 and older.

Eligibility for benefits under the DI program depends on work experience, but certain employed people (such as domestic workers, farm workers, employees of nonprofit organizations, certain self-employed persons, and public employees) are not included under the program. Trial work periods are permitted without interference with benefits provided that the work is part of a rehabilitation program. Each year the cases determined to be eligible for benefits are reexamined (in the main those disabled as a result of mental illness, tuberculosis, or heart disease) and some 25 percent of these have their benefits terminated, most usually because they have returned to work.

The program is financed in the same manner as are the OASI benefits: specifically designated taxes on employers, employees, and the self-employed. The program is administered by the Social Security Administration.

Almost 90 percent of those in paid employment are covered by the

DI program. In 1974 the number of beneficiaries under the program was about 3.5 million workers and their dependents.

The DI program makes provision for vocational rehabilitation of eligible beneficiaries through the appropriate state agencies, with the federal government reimbursing the states. In 1973 the number of wage earners reported to have been rehabilitated was 13,919.[9]

MEDICAID

The Medicaid program (Title XIX of the Social Security Act) is a federally assisted state program established by Congress in 1965 to finance the medical care for "medically indigent" persons. The federal government makes funds available to the states in varying proportions (from 50 to 78 percent of the costs). By July 1977 all states and the District of Columbia had Medicaid programs, Arizona establishing its program in 1977. The states determine the eligibility of recipients (although recipients of the categorical-assistance programs for the aged, the blind, the totally and permanently disabled, and families with dependent children must be included), the types of medical care for which payment is to be made (although inpatient and outpatient hospital care, laboratory and X-ray services, skilled nursing home care, physician services, and home health services must be included), and the amounts to be paid for such services as well as their duration. Variations among the states are considerable. Providers of care are free to accept or reject Medicaid patients. By 1976, ten years after the Medicaid legislation became effective, many criticisms were being made of the program and of its administration, including such matters as the escalating costs, operational weaknesses, and abuse of the program by some providers of care. Some states have turned over the administration of the program to private health insurers.

The Medicaid program is primarily designed for the benefit of needy persons who require medical care. It might, therefore, be looked upon as being of secondary importance to those in the workforce needing medical services. This is said in the sense that employed persons, in addition to having recourse to their wages and savings, most usually have private health insurance available to them. If disabled in the course of their work they have available to them the workmen's compensation benefits described earlier and, in some cases, the benefits of the DI program and Medicare. In most cases it would be only when these various resources had become exhausted that workers or disabled workers requiring medical services would turn to the Medicaid program. While the number of workers or disabled workers who are Medicaid recipients is not known, it is reported that 13 percent of the 23 million Medicaid recipients, or about two million persons, are permanently and totally disabled (54.4 percent of these are dependent children and 24.4 percent are aged). These disabled persons account for $210 million of the $14 billion cost of the Medicaid program in 1976.[10] To workers or disabled workers as a group, therefore, the Medicaid program is not a matter of great moment.

CHAMPUS

Another public program of secondary importance to workers or disabled workers is the CHAMPUS program begun in 1946 to finance the medical care for dependents of the active-duty and retired uniformed services personnel (including the army, navy, air force, coast guard, Geodetic Survey, and Public Health Service). The medical care is provided by civilian physicians and other professionals or civilian hospitals (when not provided at government facilities) for some 6 million such dependents. The covered care includes semiprivate hospital rooms, surgery, acute medical care, obstetrical and maternity care, treatment of acute mental disorders, and diagnostic tests. Private insurers administer the program.

It is conceivable that in some instances an employed spouse who did not otherwise have health insurance protection would have recourse to the CHAMPUS program benefits upon becoming injured or ill. The degree to which this is so is not known. It probably is not great.

VETERANS ADMINISTRATION

Yet another publicly established program of secondary importance to disabled workers is the Veterans Administration program. Under this program medical care and disability compensation are provided for veterans with service-connected disabilities and for veterans in need who are permanently and totally disabled by virtue of disease or injuries without regard to service origin. It is this latter category that could have applicability to disabled workers who are veterans. The disability payments in such cases vary, depending upon income and resources, but they are minimal. Apparently some 165,000 disabled workers who are veterans benefit from this program.[11]

PUBLIC ASSISTANCE PROGRAMS

Public assistance programs of various types conceivably could play some role as respects disabled workers, although the extent of this is not known. The medical care aspects of public assistance programs have already been discussed. Beyond that, public assistance programs provide income at the subsistence level. These programs are categorized as Old Age Assistance, Aid to the Blind, Aid to the Permanently and Totally Disabled, Aid to Families with Dependent Children, and General Assistance. In all cases the intent is to help the needy. Unquestionably such programs would benefit disabled workers who had been employed in occupations where no protection is offered by workmen's compensation or the federal Disability Insurance program.

It is reported that 53 percent of the severely disabled on public assistance were employed at the onset of disability.[12] It is also reported that the majority of public assistance recipients, including those severely disabled, had never applied for disabled-worker benefits and that most of them did not have the required work experience under OASDI to be insured for disability.[13] It has also been reported that employment injury or disease accounted for less than 2 percent of APTD (Assistance to Permanently and Totally Disabled) recipients.[14] It would appear, therefore, that the public assistance programs do not play a highly significant role for disabled workers but that such programs are of unquestionable value for disabled workers who do qualify.

PUBLIC PROVISION OF CARE

There are many types of publicly provided medical care programs established by the federal, state, county, and municipal levels of government. Included are hospitals, mental hospitals, provision for immunization and physical examinations, and several types of special clinics for mental illness, alcoholism, drug addiction, venereal disease, maternal care, and pre- and postnatal care. Conceivably public provision of care could play a role with respect to disabled workers whose other benefits have expired. It is not known to what degree such care is of value to disabled workers.

TEMPORARY DISABILITY BENEFITS

The temporary disability benefits compulsory in five states are not discussed here because, while they are applicable to certain employed people and provide fixed cash amounts when a worker is temporarily disabled for more than 7 days and up to 26 weeks, they exclude disability resulting from occupational accidents or diseases. Such programs are quite related to the private short-term disability benefits discussed in Chapter 24.

THE QUESTION OF NATIONAL HEALTH INSURANCE

The issue of national health insurance has been discussed in the United States since 1912. In the 1940s the Wagner-Murray-Dingell bills received considerable public attention and were the subject of many hearings before Senate and House committees. In no instance was enactment recommended, and during the next two decades interest in the subject waned. These early proposals, it should be noted, would have made provision for financing both medical care and loss of income resulting from disability. Their principal focus of attention was on the working population and their families.[15]

It is of interest that on January 30, 1948, President Truman requested the administrator of the Federal Security Agency "to undertake a comprehensive study of the possibilities of raising health levels." On September 2, 1948, Oscar R. Ewing, administrator, tendered his report. The report, in advocacy of a federal government insurance program, stated: "Adequate health insurance is, by its very nature, a great assistance in preventive medicine. It places people in a position in which they can get medical care more promptly and more readily than otherwise.... Concerted effort to encourage prevention of disease ... would become possible because lack of finances would no longer stand in the way."

It is also of interest that this report stated: "This examination of the facts makes it clear that, at a maximum, only about half the families in the United States can afford even a moderately comprehensive health insurance plan on a voluntary private basis." Today some 87 percent of people under age 65 have private health insurance protection. While "moderately comprehensive" was not defined in the report, it might reasonably be assumed that most of the private health insurance in force today would meet this description and therefore that the report simply failed to anticipate developments in the years subsequent to its preparation. In any instance, the report also said of voluntary insurance that it "probably will never be able to cover more than half the population." Here its hypothesis was grossly in error.

Of further interest is the estimate in the report that: "Our analysis of the costs of the national health program laid down in the report shows that we may reasonably expect that an annual investment of $4 billion by state, local, and federal governments can ultimately produce an annual return—in national wealth—of several times that amount ... we can expect to reduce deaths from preventable causes ... by almost one-fourth in the next 10 years."[16] One need only review the cost estimates for the national health insurance proposals presently before Congress to note how grossly in error this cost estimate was, even in constant dollars.

In the course of preparing this report, Oscar Ewing convened a National Health Assembly in May 1948 "to secure a firm basis for my report." The report of the Assembly was published in 1949.[17] It recommended, among other things, that:

- On-the-job health-maintenance services should be available to employees with chronic diseases, whether or not they are disabled, as a means of keeping them in productive employment.
- A comprehensive program of nutrition service and nutrition education should be an essential part of all health, education, and welfare programs. . . .
- Accident prevention should be considered a major responsibility of state and local health departments.

No actions resulted from these efforts as respects national health insurance. Then, on December 29, 1951, President Truman appointed a President's Commission on the Health Needs of the Nation. The report of that Commission in December 1952 enunciated certain health principles, including the following:

- Access to the means for the attainment and preservation of health is a basic human right.
- Effort of the individual himself is a vitally important factor in attaining and maintaining health.
- Comprehensive health service includes the positive promotion of health, the prevention of disease. . . .

The Commission then recommended greatly increased expenditures for the prevention of disease to speed the eradication of tuberculosis, syphilis, typhoid fever, diphtheria, and other communicable diseases. It also recommended programs for disease detection by expanded health departments and other health agencies in cooperation with local medical groups and other health agencies.

The report recognized that disease prevention can never be entirely delegated to others: "This kind of [personal] disease prevention cannot be done for the individual. It demands individual initiative." The report further stated: "Nowhere does money invested in human welfare pay greater returns than in disease prevention . . . [it] is extremely profitable. It pays the individual to be healthy; it pays industry to have healthy employees; it pays the nation to have a sound and healthy citizenry." The report took cognizance of the fact that: "The chief drawback to complete medical examinations for everybody is that there are not physicians available to examine 150 million people each year and still have time left over for the care of the sick. Another objection is the excessively high cost." There were also recommendations concerning water supply, sewage disposal, pure milk and other foods, insect and rodent control, and both the living and the working environment.[18]

Until 1967 there was no active interest in the subject of national health insurance. Then in that year Walter Reuther of the UAW appointed what is now known as the Committee on National Health Insurance. This action immediately created a broad interest in the subject. In 1968 and 1969 a host of proposals for a system of national health insurance were placed before the Congress sponsored by such diverse groups as the AFL-CIO, the American Medical Association, the American Hospital Association, the Health Insurance Association of America, and the National Association of Manufacturers. In addition, Senator Long made a proposal for a federal system of catastrophic health insurance benefits, and the administration under President Nixon proposed a system of national health insurance.

Since that time each session of Congress has received proposals for national health insurance from these same sources, each time amending their previous proposals. By the 1972 presidential election, however, it was quite evident that public interest in the subject had again waned. Other public concerns took precedence. As a consequence, estimates of the enactment of any such legislation were extended from 1975 to 1976 to 1978 to 1980. Then in 1976 Governor Carter, in running for the presidency, reintroduced the subject of national health insurance. At this writing (mid-1977) it is not possible to judge whether the subject of national health insurance will become a vital issue in the immediate future or whether Congress will see fit to act on the subject in the years immediately ahead.

This being the case, it would avail little here to enter into a comparative discussion of the various proposals for national health insurance that have been placed before the Congress, particularly since each year these proposals are subject to amendments by their sponsors. Each would establish a system of national health insurance to cover practically all persons under age 65 (or possibly to supersede the Medicare program for persons age 65 and older). Each would cover, to varying degrees, most medical care expenditures as specified and to the degree specified. Under some, administration would be by the federal government; under others private insurers would administer the program. Under most proposals, private insurers could supplement the benefits of the compulsory program. Under some, the financing would be by federal taxation on employers, employees, and the general taxpayer; under others financing would be by employer–employee premium payments, supplmented to some degree by federal taxation.

It is of pertinence here, however, to note the relationship of these proposals to preventive medicine. Most would cover certain of the costs for well-child care, prenatal care, immunization, eye and teeth examinations, and periodic health examinations at specified intervals, all subject to any cost-sharing or other specified limits. In no instance is there emphasis on preventive medicine, nor has any meaningful interest in this aspect of a national health insurance scheme been evidenced. The emphasis, rather, has been on the payment of the costs of curative care and the possibilities of containing the costs of health care through various proposals and of improving the availability and distribution of health care services. The importance of the relationship of a system of national health insurance to preventive medicine has not been overlooked, however, by some thoughtful students of the subject.

Kindig and Sidel have commented:

> Special and heavy emphasis should be placed on health maintenance programs, particularly those which focus on health education and formation of good health habits (e.g., accident prevention, smoking deterrence), on early reportings of medical symptomatic disease (e.g., glaucoma, hypertension) and screening for—and prompt treatment of —diseases apparently better treated when diagnosed in the asymptomatic stages. It is not at all clear at present which diseases fall into the last category.... Emphasis should be placed on these elements of health maintenance which have clearly beneficial effects, and carefully controlled experimentation should proceed on the other elements.[19]

Dr. Paul Ellwood has noted:

> Multiphasic screening and testing devices ... are examples of the various kinds of technology now being developed to increase the productive capacity of the health care system. However, most of these are still in the experimental and demonstration stages. Indications are that such technology can indeed expand capacity and improve quality, although at present their use would considerably increase the per unit costs of service.... The two best ways or providing cost control are ignored and perhaps even aggravated—that is, supply is not increased in numbers or efficiency, and emphasis on preventive medicine is not encouraged.[20]

Victor Fuchs, an economist, in discussing the costs of national health insurance, has said:

At first, health status will be the same regardless of the health plan adopted. However, to the extent the different plans have different effects on health, this could become a relevant variable in the long run. For instance, if different plans make different provisions for the use of preventive services, and if preventive services change health status, utilization could be affected. . . . The principal ways that health status might be altered are through change in environment, advances in medical science, preventive health services, and health education. . . . The plans do differ with respect to provision for preventive services. . . . None of the plans makes any special provision for health education. . . . None of the plans contemplates any direct attempts to alter individual health practices as, for example, by offering lower premiums or other rewards to non-smokers or to those practicing preventive dental care at home.[21]

Dr. Ernst Wynder of the American Health Foundation has said:

> Whatever type of insurance scheme this nation will select, periodic health surveillance examinations must be an integral part.[22]

The Committee for Economic Development has noted the importance of prevention:

> To achieve a national health-care system that stresses prevention as well as cure, we recommend. . . .[23]

Rashi Fein has taken the position that:

> National health insurance lies at the heart of the solution to a number of other medical care issues. Solving the barrier to prevention requires more than the elimination of the economic barrier alone. . . . The major barrier to a better distribution of preventive care is not a shortage of funds.[24]

REFERENCES

1. Chamber of Commerce of the U.S., "Analysis of Workmen's Compensation Laws," Washington, D.C., 1974.
2. *Social Security Bulletin,* October 1974.
3. Dr. A. A. Dun, "Rhyme and Reason in Rehabilitation," paper presented at the National Conference on Health, Welfare, and Pension Plans, Inc. (now the International Foundation of Employee Benefit Plans), Brookfield, Wis., 1963.
4. *Monthly Labor Review,* October 1944.

5. Annual Report of the Office of Vocational Rehabilitation, 1944.
6. Jerome Pollack, paper presented at the National Conference on Labor Health Services, American Labor Health Association, June 16, 1958.
7. *Risk Management Today: Problems, Trends, and Practices,* American Management Associations, Report No. 47, 1960.
8. Report of the National Commission on State Workmen's Compensation Laws, OSHA, 1972.
9. Research and Statistics Notes, Social Security Administration.
10. *Social Security Bulletin,* February 1975; Weikel and Leamond, "A Decade of Medicaid," *Public Health Reports,* July-August 1976.
11. HEW Research and Statistics Note No. 7, 1975.
12. HEW, "Social Security Survey of the Disabled," Report No. 9, June 1970.
13. Henry P. Brehm, "The Disabled on Public Assistance," *Social Security Bulletin,* October 1970.
14. Preliminary draft of confidential interoffice memorandum made available to the author.
15. J. F. Follmann, Jr., *Medical Care and Health Insurance* (Homewood, Ill.: Richard D. Irwin, Inc., 1963).
16. HEW, "The Nation's Health: A Ten-Year Program," September 2, 1948.
17. *America's Health: A Report to the Nation* (New York: Harper & Row, 1949).
18. The President's Commission on the Health Needs of the Nation, *Building America's Health,* December 18, 1952.
19. Eilers and Moyerman, "National Health Insurance," proceedings of a conference at the University of Pennsylvania, 1971 (Homewood, Ill.: Richard D. Irwin, Inc., 1971).
20. Ibid.
21. Ibid.
22. Ernst L. Wynder, M.D., paper presented before the American Health Foundation, September 25, 1970.
23. Committee for Economic Development, "Building a National Health-Care System," New York, 1973.
24. Rashi Fein, paper presented before the New York Academy of Medicine, April 25, 1974.

24

Private Insurance Programs and Employee Benefits

Several types of private insurance and employee benefit programs have a direct relationship to industrial health programs. If those in the workforce are to be in reasonably good health it is necessary that they, in significant measure, be relieved of the economic burdens that can result from the costs of health care, either therapeutic or preventive, and the loss of income that often accompanies medical treatment, adequate recuperation, and disability. Provision must also be made for their dependents in the event of premature death.

That the costs of such employee programs are borne largely by employers was shown earlier. Significant is a recent report in the *Washington Post* that at General Motors the annual costs of health benefits exceed $825 million, an amount that adds $175 to the cost of every car and truck manufactured. For the big three automakers the cost of health care benefits was $1.04 billion in 1975, or some $1,800 for each employee.[1]

The types of programs to be discussed here are health insurance protection against the costs of medical care for either the employee or his or her dependents, loss-of-income protection in the event the employee becomes disabled, salary-continuance programs during the employee's illness, and life insurance. The role of insurers in the prevention of illnesses and injuries and in rehabilitation will also be discussed.

HEALTH INSURANCE

In the United States there are essentially three types of private health insurance plans that provide protection against the costs of medical care: (1) insurance companies, of which some 1,000 provide such protection on a cash indemnity basis to about 108 million persons; (2) 74 Blue Cross plans and 70 Blue Shield plans, which have agreements with the majority of hospitals and physicians and make payments directly to such providers of care on behalf of some 78 million covered persons; and (3) the some 400 prepaid group practice plans (HMOs) and union labor health centers, which provide the health care directly on a prepayment basis for about 8 million subscribers or members (of this number the Kaiser Foundation Health Plan covers 2.7 million persons and the Health Insurance Plan of Greater New York covers .75 million persons). In all, some 88 percent of the civilian population have some form of private health insurance protection, the benefits of which exceed $29 billion yearly.

Of the covered persons, insurance companies provide protection for 58 percent, Blue Cross and Blue Shield plans for 38 percent, and the plans that provide the care directly do so for 4 percent of all covered persons.[2] Some employers, it should be noted, finance medical care on a self-insurance or self-administered basis acting, as it were, in the role of an insurance company for their employees and their dependents. In addition, of course, there is the direct provision of care by employers on an in-plant or contract basis, as discussed in earlier chapters. (In the interest of completeness it should be noted that private health insurance is also available for specific forms of care such as prescribed drugs, dental care, and vision care, and that protection is also made available to certain population groups such as members of some fraternal societies, some rural or consumer health cooperatives, and those eligible for college health programs.)

With respect to insurance company coverages it should be noted that 80 percent are written on groups of people, predominantly employed groups. The remainder are individually written policies for the benefit of persons with no group affiliations, self-employed persons, or professional people, or for those who wish to supplement other insurance protection. Blue Cross and Blue Shield plan enrollments are predominantly on a group basis, as are the enrollments in the prepaid group practice plans or HMOs.

The forms of protection offered by insurance companies and Blue

Cross and Blue Shield plans assume essentially two types: basic coverages and major medical expense insurance. Under the basic coverages the costs of care in hospital, the costs of surgery, the costs of physician care out of hospital, and the costs of out-of-hospital diagnostic and X-ray services can be covered. The limits for hospital care are for the cost of semiprivate care or a fixed amount per diem for a limit ranging from 21 days to one year or longer. The costs of surgery can be on a schedule basis for each type of procedure or can cover the reasonable and customary cost of such care. Physician charges for out-of-hospital care follow this same pattern. Major medical expense coverage, on the other hand, covers practically all types of care in or out of hospital after the payment by the insured of a deductible amount, which can range from $25 or $50 to $300 (although this amount may be zero for hospital care), and up to a maximum amount which can range from $5,000 to $250,000. For all covered medical expenses the insured bears a share of the cost, called coinsurance, which most usually is 20 percent of the costs of care (although this can be decreased on a graduated basis where large maxima amounts are payable). Today more than 90 million persons have this form of health insurance protection.

Prepaid group practice plans, or HMOs, which have attracted considerable attention and some federal government financial support in recent years, provide or arrange for the provision of health care on a prepayment basis. The care provided is quite complete, although with a few exceptions does not include the treatment of mental illness. Such plans place emphasis, as will be noted subsequently, on preventive medicine.

Some HMOs have experienced difficulties in becoming established and accepted by large numbers of employed people and their dependents. One example of this is found in the case of the three HMOs in Rochester, New York (the Genesee Valley Group Health Association, Health Watch, and the Rochester Health Network). While 40 percent of the employees in the Rochester area have the option of membership in one of these groups, with the employers paying a sizable proportion of the cost, less than 5 percent of those workers have taken the option.[3] (Reasons for workers not opting for HMO care can include such matters as the inconvenience of the HMO to the residences of employees or their families, satisfaction with present health care services, or a lack of public acceptance of the concept.)

Regardless, it is of interest that insurance companies and Blue Cross plans have developed an active involvement in HMOs. By mid-1976, 22 insurance companies were involved in 50 operational HMOs in 25 states. Such involvement might include financial support, administrative services, consultation, marketing assistance, or the provision of hospital insurance, out-of-area coverage, or reinsurance. In addition, 31 Blue Cross plans have an essentially similar involvement. Today HMOs are located in 35 states.

As respects employed people, a study by The Conference Board in 1974 presents a good indication of the scope and degree of health insurance protection afforded employees today.[4] The study was based on a survey of the employee benefits of 1,600 companies of various types and sizes. That study found that the vast majority of the employers in the survey had basic hospital-surgical-medical coverage for all employees and their dependents. Major medical expense insurance coverage for office employees was in effect in 95 percent of the places of employment, and for nonoffice employees in 85 percent of the places in the survey. In the majority of cases (65 percent) the major medical expense coverages were supplemental to basic health insurance coverages. The majority of the programs were insured with insurance companies. Only a small minority were self-insured. The benefits paid under these employment-centered programs, according to the report, account for about 85 percent of all medical expense benefits paid by private insurers.

The major medical coverages most usually provided for a $100 deductible amount and 20 percent coinsurance for the employee's share of the costs of care. The maximum amount payable was most usually $20,000 for office employees and $15,000 for nonoffice employees. In a few cases the maximum amount was $250,000, although 90 percent of the cases had a maximum amount of $50,000 or less. Under the basic coverages the hospital benefit was most usually the full cost of semiprivate care for 120 days (although over 20 percent of the cases provided such coverage for one year); the surgical benefit was on a schedule basis (usually up to $500) in two-thirds of the cases, with the remainder covering the reasonable and customary charges; the reasonable and customary physician charge for out-of-hospital care was covered in 21 percent of the cases; and maternity benefits had special limits in 70 percent of the cases.

In 90 percent of the cases the costs of the treatment of mental illness were covered, but usually with limitations on the coverage, such as 50 percent of the costs of outpatient psychiatric treatment and a limit on the number of such episodes of treatment, usually 20 per year. In 75 percent of the cases the treatment of alcoholism on an in-hospital basis was covered, and in 63 percent of the cases the treatment of drug addiction was covered. In 13 percent of the cases nonoffice workers (9 percent for office workers) were covered for the costs of dental care.

As respects the option of employees to enroll in an HMO for the receipt of medical care, only a bare minority of the employers in the survey made such option available. Of these, 28 were large industrial establishments, 12 were insurance companies, 8 were banks, 2 were smaller manufacturers, and one was a utility. These figures, of course, are not indicative of any attitude on the part of employers or unions, since it is quite possible that the majority of the places of employment surveyed are not located in a geographic area where the services of an HMO are available. Unfortunately, the survey did not show the proportion of employees who exercised the option in those cases where it was available.

Concerning the continuity of health insurance coverages during a plant layoff, the survey showed that 42 percent of the cases made such provision. Also of interest is that in 29 percent of the cases the coverage was continued in instances of strike.

In the majority of these places of employment the employee makes no contribution to the cost of the health insurance protection. The proportion of employers not requiring contributions, by type of employee and type of coverage, was shown to be: nonoffice employees, 74 percent for basic coverages and 69 percent for major medical coverage; office employees, 62 percent for basic coverages and 61 percent for major medical coverage. In some instances the employee shares half the cost of the coverage for dependents. These contributions on the part of the employer constitute, in effect, wages that are exempt from income taxes from the standpoint of both the employee and the employer.

That health insurance benefits have an important relationship to an industrial health program, particularly as respects the care of nonoccupational illnesses and injuries, as well as the care of dependents of employed persons, is apparent from what has been shown.

The adequacy of these employee health care benefits has never been satisfactorily or meaningfully measured. There is no question that the health insurance coverage on some workers is less than adequate, having restrictive amounts of benefits or time limits on the coverage. In other instances the types of medical services covered are less than complete.

Beyond this there is a sizable segment of workers who do not have health insurance benefits available to them at all through the place of employment. The Congressional Subcommittee on Health reported in 1974 that 29 percent of all full-time workers did not have such insurance protection, the proportion being highest among farm workers (81 percent), service workers (47 percent), and sales personnel (37 percent). The lowest proportion of uninsured workers was found among blue-collar operatives (19 percent). By industries, the Subcommittee found the following proportions of places of employment with no provision for health insurance: agriculture, 79 percent; retail trade, 45 percent; construction, 41 percent; mining, 11 percent; manufacture of durable goods, 10 percent;

and communications and public utilities, 8 percent.[5] The Subcommittee also found that in 1975 of those workers who were unemployed and who previously had health insurance protection at the place of employment, 70 percent had no provision made for the continuation of the coverage after employment ceased. Whether or not such unemployed workers had the right to convert their group insurance coverage to an individual policy was not shown, although the payment of the premium for such converted coverage can be a problem to an unemployed worker.

Unquestionably small employers would constitute the principal problem in these findings, although today small employers have available to them group insurance programs through their associations or through trustee groups. In many instances, however, group health insurance coverages are available through the labor unions where unionized workers are involved. Beyond this, many workers not covered through their place of employment purchase protection on an individual policy basis, most usually through insurance companies. These coverages are also used to supplement the insurance available through the place of employment.

DISABILITY INSURANCE

What has been discussed until this point has had to do with group insurance for health care costs. Of equal importance are the coverages that protect employed people against loss of income resulting from disability. It is an important form of protection for employees, self-employed persons, and professional people who become disabled as a result of illness or injury. Because of the compulsory workmen's compensation benefits for loss of income resulting from occupational hazards, disability insurance benefits are most usually provided for disabilities resulting from nonoccupational hazards, although there are exceptions to this, as will be shown, particularly for long-term disability benefits. When occupational hazards are covered by disability insurance, an offset is provided as respects the workmen's compensation benefits and, in the case of long-term disability insurance, against the Social Security Disability Insurance benefits.

Essentially there are two types of group disability insurance: short term and long term. In either instance they might supplement salary-continuance or paid-sick-leave programs, workmen's compensation benefits, compulsory temporary disability benefit programs, or the Social Security DI program. Also available to many employees are disability pension programs and individually purchased disability-income insurance policies.

The majority of group disability insurance is written for *short-term disability,* most usually nonoccupational. At one time these coverages provided protection for disabilities lasting up to 13 weeks. Today, however, the predominant period covered is 26 weeks, with many plans providing protection for 52 weeks. The benefit, which is always wage-related, is most usually for 50 to 80 percent of wages up to a maximum dollar amount, which can range from $50 to $125 a week. In some cases the benefit is offset by the benefits under the compulsory temporary disability laws in five states.

The Conference Board study mentioned earlier showed that of the employers surveyed, two-thirds had such short-term disability coverages for their nonoffice employees, and 39 percent had such coverages for office workers. One reason for this difference is the fact that in 85 percent of the places of employment salary continuance or paid sick leave is provided, primarily for office employees. Interestingly, the proportions of employers providing these coverages in manufacturing and large industry were 79 percent for nonoffice employees and 53 percent for office workers. That study also found that 55 percent of the employers surveyed provided short-term disability benefits that continued for 26 weeks. In 17 percent of the plans coverage was for 13 weeks, and in 23 percent of the plans it was for 52 weeks. Benefits under these plans ranged from 50 to 70 percent of wages to maxima of $50, $75, $100, or $125 a week.[6]

During 1974 and 1975 the Labor Department prepared a digest of the employee insurance benefits for 89 individual employers and 42 industrywide (multi-employer) programs, of which 7 were national in scope for such industries as the railroads, the upholstering trade, the clothing industry, the bituminous coal industry, the distillery industry, the furniture industry, and the luggage and leather goods industry (the remaining 35 were local or regional in nature and included trucking, shipyards, laundry, lumber, maritime, leather, metal manufacture and repair, taxicabs, milk dealers, millinery, shipping, forest products, upholstery, printing, publishing, construction, retail food, building service, personal services, hotels, jewelry, furniture, furs, brewers, restaurants, retail trade, dolls, dresses, retail drugs, and hospitals.)[7]

That study found that the majority of single-employer programs for short-term disability insurance and a large majority of such programs under collectively bargained multi-employer programs were limited to nonoccupational disabilities. Many, however, covered both nonoccupational and occupational disabilities, the benefits in the latter case being offset by workmen's compensation benefits.

The digest showed the following with respect to the waiting period before short-term disability benefits commence. The large majority of the programs commence benefit payments on the eighth day for disability resulting from sickness and the first day for disability resulting from accidents. In some cases the waiting period is made retroactive once disability is established. However, some plans commence disability payments for sickness on the fourth day, and a few have no waiting period for either accident or sickness. In several instances the eight-day waiting period for sickness disabilities is waived if hospitalization is required. With respect to the period of time of disability covered, that digest showed that short-term disability benefits most usually cover a period of 26 weeks. Several, however, cover 52 weeks of disability.

The amounts of the benefits payable for short-term disabilities were shown in the digest to vary in almost every instance. Some pay stated dollar amounts with a minimum and maximum amount depending upon the wage level of the individual. In many cases the stated dollar amounts are flat weekly sums. In a few instances the short-term disability benefit is stated as a flat percentage of wages. In other instances the benefit is stated as a percentage of wages but with a dollar maximum amount.

With respect to these benefits for short-term disability it is to be borne in mind, as will be shown in the following section, that many employees receive salary-continuance or paid-sick-leave benefits that may be in lieu of short-term disability insurance benefits or may be supplemented by the insurance benefits.

Long-term disability (LTD) insurance has been a rapidly growing form of employee protection over the past decade. Under this form of protection benefits are provided for loss of income as a result of long-term or permanent disability, in the vast majority of cases whether the cause is from nonoccupational or occupational hazards (although some are limited to nonoccupational causes only). Where occupational disabilities are covered, an offset against workmen's compensation payments is most usually specified. In practically all LTD cases an offset is provided against the Social Security DI benefits.

Some employee benefit programs do not provide for LTD, and some exclude hourly employees. Here the Conference Board 1974 study is of interest. Of the employers surveyed it was found that 72 percent provided LTD benefits for managerial personnel, 62 percent for office employees, and 28 percent for nonoffice employees.[8]

Under LTD programs benefits commence in almost all cases either after six months of disability or after the cessation of short-term disability or salary-continuance benefits. The Conference Board study found that the period was one month in 5 percent of the places of employment surveyed, 2 months in 4 percent of the cases, 3 months in 23 percent of the cases, 6 months in 62 percent of the cases, and 12 months in 6 percent of the cases.

LTD benefits most usually continue at age 65, the Conference Board study showing that this was the case in 88 percent of the places of employment surveyed, with 8 percent of the places providing such benefits for life. In 4 percent of the places a specified period of time for the payment of such benefits was established. The Conference Board found that of the employers surveyed, 46 percent had LTD plans paying 60 percent of salary to a maximum amount of $1,000 to $2,000 a month and that 29 percent had plans paying 50 percent of salary up to similar maxima.[9]

A related long-term disability benefit frequently provided is that of *disability retirement provisions under pension plans.* Typically these plans provide disability coverage on the completion of a length of service requirement, such as 10 to 15 years, and/or attainment of a specified age, such as 45, 50, or 55. The benefits in such cases are generally equal to the accrued normal retirement pension benefit, but with some minimum benefit provided for, which might be a flat dollar amount or might be a percentage of the wage level at the time of disability. The Conference Board study in 1974 showed that of the employers surveyed, 47 percent had such pension provisions for office employees and 62 percent made such provision for nonoffice employees. In union-negotiated plans, 71 percent of the cases had such disability pension provision.[10]

Individually purchased disability income insurance policies offer a wide range of benefits, usually up to half or three-quarters of earnings, the choice resting with the purchaser. Benefits might be continued for anywhere from two years to age 65, or even for life, again depending upon the purchaser's choice. The unit cost is greater than for group insurance coverage as a result of increased merchandising, underwriting, and claims administration costs. Most such coverages today are written on a noncancelable basis, but the prospective purchaser must give evidence of good health and sound credit standing. In cases of impaired health, substandard insurance is available at increased premium rates or reduced benefits.

Today it is estimated that 30 million employed persons are protected by group insurance for loss of income resulting from short-term disabilities, that 11 million employees have long-term disability protection, and that 16 million income earners have individual policies for disability-income protection. However, the annual estimates of the Health Insurance Association of America indicate that 63 million workers have short-term disability-income protection, 41 million of which is through insurance companies, and that 18 million have long-term disability protection. In all, the benefits exceed $2.5 billion yearly. Of interest is the fact that the number of persons with long-term disability-insurance protection increased 91.8 percent between 1969 and 1975, showing the rapidly growing interest in this form of protection.[11] Today it is estimated that 74 percent of the workforce has some type of disability insurance protection: In 1946 this percentage was 48 percent. Insurance company coverages are estimated to provide the protection for half the workforce. Noticeable here, of course, is that about one-quarter of the workforce does not have such protection. Also to be noted is that 70 percent of workers today are estimated to be covered by pension plans that have a provision for disability retirement.[12]

In all, it was estimated by the Social Security Administration that in 1972, $6.6 billion was paid for short-term nonoccupational disabilities through insured, self-insured, and salary-continuance programs.[13] With respect to such benefits it is to be noted that they are, in considerable part, tax-free under federal, state, and local income tax laws. An effect of this is that a benefit providing for two-thirds of wages rises to an actual value of from 72 to 91 percent of wages, depending upon the income level of the individual involved.

Disability-insurance benefits are most frequently noncontributory on the part of the employee; that is, the employer pays the entire cost. The Conference Board study in 1974 found that for short-term disability insurance the coverage for nonoffice workers was noncontributory in 75 percent of the places of employment surveyed. This proportion was 84 percent of the places of employment where collective bargaining agreements were in effect. For office workers the proportion was 66 percent. For long-term disability these proportions were in both instances 48 percent. For those programs where the employee contributes to the cost it most usually is half the total cost, the employer paying the other half.[14] Another study, in 1970, found that in 30 percent of LTD cases the employee made no contribution to the cost, in 40 percent of the cases the employee shared in the cost, and in 30 percent of the cases the employee paid the entire cost.[15]

With respect to these insurances for disability income, it has been seen that there can be differences from one plan to the next concerning the waiting period before benefits commence, the percentage of earnings paid in benefits and the maximum dollar amount of such benefits, and the time period for which benefits are paid. There can be strong differences of opinion with respect to each of these. Should the waiting period be retroactive after disability is once established and how long is a reasonable waiting period? Should the definition of disability be the inability to perform "his occupation" or should it be "any occupation"? What is a reasonable or adequate proportion of wages to be paid in benefits and what should reasonable dollar maxima amounts be? At what point do benefit levels impede rehabilitation? For what time period should benefits be payable? It is noticeable from the previous descriptions of these benefits that opinions vary considerably.

SALARY-CONTINUANCE PROGRAMS

Another form of benefit for employees that replaces, in whole or in part, income lost as a result of illness or injury is provided by salary-continuance or paid-sick-leave programs. Such programs might be on a formal basis (in writing) or on an informal basis (not in writing, with the benefits determined by individual situations). In some instances these benefits stand alone in the absence of any disability-insurance benefits, in some cases they precede the long-term disability-insurance benefits, and in some they supplement short-term disability-insurance benefits or workmen's compensation benefits if the salary continuance is applicable to occupationally caused disabilities.

Not all places of employment make provision for salary continuance in the event of illness or injury, particularly if a disability-insurance program has been established or if the employees are in a state where short-term disability insurance is compulsory under a temporary-disability-benefits law (California, Hawaii, New Jersey, New York, and Rhode island). Thus it is estimated that over 19 million employed people have protection under formal salary-continuance programs and that perhaps an additional 2 million have some protection under informal programs.[16] This latter estimate is necessarily with little foundation and is probably ultraconservative when one considers the large number of very small places of employment.

The variations among salary-continuance programs are seemingly infinite. Differences can pertain to such matters as the types of employees covered or not covered, the amount of the benefit, the duration of the benefit, the waiting period before benefits commence, the distinctions between occupational and nonoccupational causes of disability, and the relationship to such compulsory programs as temporary disability benefits and workmen's compensation. While complete delineation of these differences would be too lengthy and serve little purpose here, the following will serve to illustrate the differences and variations among the formal salary-continuance programs. The information shown derives from two principal recent sources: the Conference Board *Profile of Employee Benefits* and the U.S. Labor Department *Digest of Health and Insurance Plans*.

One basic difference among salary-continuance programs concerns which employees are to be covered by the program and which are to be excluded. As a generality, such programs are limited to regular full-time employees, often with some period of established service such as six months or a year. It would be extremely difficult to administer a program for casual workers, short-term workers, or part-time workers. Hence the employees' length of service permeates such programs, not only as respects eligibility for benefits but also the extent of the benefits. The Conference Board survey has shown that of the employers surveyed, 45 percent required one year or less of service, the remainder requiring from one to ten years of service for eligibility. The U.S. Labor Department *Digest* shows that the benefits under the vast majority of programs increase as the length of service of the employee increases.

Beyond this, other differences among covered and noncovered employees occur. The Conference Board survey showed that 85 percent of the employers surveyed made such benefits available to office or salaried employees. This proportion dropped severely to 46 percent for nonoffice employees and further to 35 percent for employees under labor union collective bargaining agreements.

The benefits under such programs vary widely. In the vast majority of cases the employer provides full pay for a specified period, ranging from five days to six months. In many instances, the period of time is related to the length of service of the employee, up to some maximum limit. In some, the period of full-pay benefit is supplemented by a further period of partial pay. In some instances the salary-continuance benefits are cumulative; in the majority they are not. In the large majority of reported cases the benefits are payable from the first day of sickness or injury. In the remainder there can be a waiting period before benefits commence, although in some of these the benefits can be retroactive to the first day, and in others the benefits begin on the first day if hospitalization occurs. The waiting period in such cases can be from one day to eight days. Under 25 percent of the employer programs, absence from work due to maternity is also covered at full pay, most usually for a period up to six weeks.

In a minority of cases where both occupational and nonoccupational disabilities are covered by the salary-continuance program, a distinction is made in the benefits between the two types of disability.

From what has been shown it is evident that salary-continuance benefits or paid-sick-leave programs differ considerably among employers. Their relationship to industrial health programs at places of employment is most pertinent as respects nonoccupational disabilities not involving long-term periods of disability. The reasons for the large variety of differences among the various programs, aside from their relationship to disability-insurance programs, the processes of collective bargaining where these are applicable, or the availability of funds, are unexplainable. Certainly they have little, if any, relationship to the needs of working people for adequate protection.

LIFE INSURANCE

Life insurance benefits are of particular importance to the dependents of an employed person who suffers premature death. Today life insurance companies have insurance in varying amounts in force on over 140 million people in the United States and each year pay over $16 billion in death benefits and annuities. Certainly the majority of persons with life insurance protection are employed or formerly employed people. A significant proportion of these have life insurance benefits at their place of employment, although the first policy of group life insurance was not written until 1911 when the Equitable Life Assurance Society of the United States insured the employees of Montgomery Ward. The Conference Board study of employee benefits in 1974 found that virtually all the employers surveyed provided group life insurance for their employees, and three of four of these also provided accidental death and dismemberment insurance benefits. The trend with respect to these benefits is increasingly toward a noncontributory basis of premium payment on the part of the employee, with 66 percent of employers paying the full cost for nonoffice employees and 55 percent for office employees. Accidental death and dismemberment benefits (AD and D) were noncontributory in 80 percent of the places of employment surveyed for nonoffice employees and in 76 percent of the places for office employees. The U.S. Labor Department *Digest* similarly showed that the large majority of group life insurance programs, including those on collectively bargained multi-employer groups, required no contribution on the part of the employee: In only about 13 percent of the cases, including a few multi-employer cases, was some degree of employee contribution required. It should also be noted that many employers offer the employee the option of purchasing additional life insurance protection at group insurance rates. Group life insurance benefits in force today have been estimated to be some $3 billion.[17]

Beyond these life insurance benefits, many plans provide for a *survivor or dependent benefit,* most usually paid in monthly amounts for a specified number of months, such as 24 months. Many plans also give the employee the *option to purchase additional life insurance* at group rates on a contributory basis. Most usually this option is related to wages or to the basic amount of life insurance.

In certain of these life insurance programs provision is made for the payment of a benefit in the event of *total and permanent disability.* In the large majority of cases the life insurance benefit is paid in installments upon the establishment of total and permanent disability. In several plans the life insurance benefit is maintained even though employment has ceased, and in some plans part of the life insurance benefit is paid in installments and part is maintained as a death benefit.

In a great many of the programs an *accidental death and dismemberment* benefit is also provided for in addition to the life insurance benefit, and in some cases the employee can purchase supplemental amounts of this benefit. In the vast majority of cases the coverage pertains to either nonoccupational or occupational accidents, although in some the coverage is limited to nonoccupational accidents, and in some it is limited to occupational accidents, at times being limited to accidents occurring while the insured is traveling on company business.

The benefit structure under these coverages varies considerably. In the majority of cases the benefit doubles the amount of the life insurance benefit. In a few it is half the life insurance benefit. In some cases the benefit is a flat stated amount, the amount most usually varying with salary level. In quite a few cases the benefit is stated as a multiple of the annual salary of the employee. Not all these plans provide for a benefit in the event of a single dismemberment. Where such provision is made it is one-half the accidental death benefit.

THE ROLE OF INSURERS IN PREVENTION AND REHABILITATION

A quarter of a century ago insurance companies and Blue Cross–Blue Shield plans made no coverages available for the various aspects of preventive medicine in the absence of a disabling illness or injury. The rationale was that the costs of periodic health examinations and other preventive measures were predictable, that they were subject to personal budgeting, and therefore that insurance was not needed for such costs. Since that time thinking on the subject has changed.

To determine the degree to which insurance companies today provide, or make available, insurance for the cost of preventive services, I sent a questionnaire to 18 of the larger insurance companies. These companies write over half the health insurance coverages written by all insurance companies, so that the findings might be considered indicative. Of the 18 companies, 14 today make available group medical-expense insurance coverages for the costs of preventive medicine without evidence of the presence of a disability sickness or injury. The preventive services covered include physical examinations, multiphasic screening, diagnostic tests, immunization, and well-baby care.

The amounts of the benefits under these coverages vary with the wishes of the purchaser in most cases, the variants being from limited amounts on a scheduled basis to comprehensive coverage with substantial maximum dollar amounts and little or no internal limits on a "reasonable and customary" cost basis. Coverages for physical examinations differing from this pattern are as follows:

> $50 or more per calendar year (2 companies).
> Full cost: One examination per year.
> Schedule basis up to $25 or $50 per examination.
> Adults, $75 per examination; children, $10 each to age 6 months, $25 each to age 5, $50 each thereafter until age 23.
> 50% of expenses, $75 maximum in any 12 months.

These benefits, in some instances, are subject to the deductible amounts applicable to the coverage in its entirety.

The limitations on the coverages vary considerably, depending on the negotiations with the purchaser, as well as the nature and extent of the benefits. Some cases contain age limits for the benefit, others do not. In all cases there is a limit on the number of covered examinations over a stated period of time. For the most part, provision is made for one examination a year. The following show some variations:

> Up to age 6 months, 6 visits.
> 6 months to 2 years, 4 a year.
> 2 to 5 years, 1 per year.
> 5 to 40 years, 1 every 5 years.
> 40 to 65 years, 1 every 2 years.

Another example of a variant is the following:

> Up to age 40, 1 examination every 2 years.
> Age 40 and over, 1 every year.

The monthly premium costs for adding physical examinations to the health insurance coverage for insured groups of people vary considerably, depending on the extent and frequency of the covered examinations, the level of benefit, and whether a deductible amount is applicable. Some general indication of the monthly cost per person, however, is revealed by information gained from the previously referred to survey. This showed a range of $1.25 to $2.64 per employee, $1.25 to $3.00 per spouse, and $1.52 to $5.83 per dependent child. In the interest of completeness, it should be noted that at least three insurance companies now have insurance in force that covers the costs of multiphasic screening examinations.

Generally speaking, the insurance coverages available today do not cover primary care in clinics attempting to alter deleterious lifestyles of individuals. An exception is the now generally available coverage under group insurance for the treatment of alcoholism, the availability of which is now required by law in 15 states. This area of primary prevention is one which warrants the reconsideration of insurers, since the potential here appears to be appreciable.

From what has been said, it is apparent that many of the major insurance companies do, today, make available health insurance coverages for the cost of physical examinations. Unfortunately, there is very little buyer demand for this coverage, with the result that very little of it is sold. The survey of the major insurance companies, for example, revealed the numbers of persons having such group coverage in 1974: One company has 287,000 persons so covered, another has 18,000, another 28,000, another 200,000 to 400,000, and yet another has 14,000 so insured. These are amazingly small numbers when compared to the total number of persons in the United States having private health insurance protection. Unfortunately, it is not yet clear to insurers how much market there is for insurance coverage for preventive health services.

The details of coverages for preventive medicine by the Blue Cross–Blue Shield plans are not known. It is a supposition that some plans do not make available such coverages, and certainly that where such coverages are available they are not purchased to any appreciable degree. It is known, however, that many Blue Cross–Blue Shield plans have added outpatient diagnostic X-ray and laboratory services to their coverages, such services to be performed in the physician's office or in the outpatient department of a hospital. It is noteworthy, however, that Walter J. McNerney, president of Blue Cross Association, has written: "Blue Cross plans, among others, are now moving into the area of preventive services realizing that successful programs will depend on both imagination on the part of the provider and adequate financing."[18] John van Steenwick, in a study of preventive care, has commented that one of the organizational needs is: "Prepayment for all the services relevant to a health protection program. This includes the screening tests, physical examinations, related health education, and referral facilities; and it should also include prepayment for followup care, for confining tests, and so on. It would be less than desirable if a disease were to be discovered in an individual who would then be left to rely strictly on his own resources in financing appropriate followup care."[19]

Of interest is a five-year study commenced in August 1976 by the Blue Cross Association supported by the National Cancer Institute to determine whether the early detection and treatment of cancer would lead to improved survival rates and lower treatment costs and, if so, whether it would be economically feasible to offer the new benefit to cover the costs of examinations on a nationwide basis. The study will try to determine which screening tests would prove the most effective in detecting which types of cancer before symptoms develop on a cost-effective basis for large population groups. The study is aimed at seeking the answer to such questions as: How will the cost of treatment and the decrease in deaths from cancer for patients being screened compare with the costs of treatment without early screening? Will savings, if gained, balance the cost of administering the program and providing the benefit? What educational programs will be necessary to encourage the public to use early screening?[20]

Prepaid group practice plans and HMOs enjoy a unique position with respect to the prepayment of periodic physical examinations. One of the original and underlying purposes of these plans has been that of health maintenance and the early detection of unknown disease conditions. Such plans as the Kaiser Foundation Health Plan, the Health Insurance Plan of Greater New York, the Group Health Association in Washington, D.C., the Group Health Cooperative of Puget Sound, and the Community Health Association of Detroit built into their benefit design the provision of prepaid physical examinations or multiphasic screening for presumably well persons.

The Health Insurance Plan of Greater New York encourages enrollees to have periodic physical examinations. The history and examination are specially oriented to preventive medicine and early detection of illnesses and include a urinalysis, hemoglobin or hematocrit, serological test for syphilis, X-ray or photofluorogram of the chest, stool for parasites and ova if the individual has lived in or visited any subtropical area outside the Continental United States, and a cytological cervical smear. In addition, every type of X-ray and test that may be indicated by the examination (provided that it is not of an experimental nature) is covered by the HIP contract. Early specialist consultations are available and encouraged (specialists give a little over 50 percent of all services, not including laboratory services). Although treatment by psychiatrists is not provided, HIP provides one or two psychiatric visits for diagnosis and evaluation of treatment needs. Preventive services also include monthly examinations of infants, immunization, and pre- and postnatal care. The difficulty in persuading subscribers to take advantage of these services, however, has been quite apparent.

Since the inception of the Kaiser Health Plan Foundation three decades ago preventive medicine has been a basic principle. Today the Kaiser program includes physical examinations, immunization, pre- and postnatal care, well-baby clinics, and pediatric checkups as a regular part of its prepaid benefits. Emphasis is placed on the early detection of disease, employing a quite complete system of multiphasic screening and any necessary follow-through. Such examinations are available every three years for persons under age 30, every two and a half years for those in their 30s, every one and a half years for those in their 40s, and annually for those age 50 and over. The Kaiser Medical Center in Oakland has established ten preventive health maintenance offices for this purpose. The Kaiser Plan also conducts health education seminars and clinics, as well as prepares publications on the various aspects of health education.[21] With respect to the multiphasic screening examinations, each is followed by a return visit to an internist for any necessary followup and further examination.

The cost to the member at the time of examination varies according to the type of prepayment coverage. Since most members have coverage in which laboratory and X-ray outpatient services are prepaid, only the doctor office visits, at a cost of $1 each, would remain as a charge to the member. The difficulty in having members take advantage of these services is evident, although the proportion of members using such services at Kaiser appears to be greater than at the other prepaid plans. Approximately half the patients who take the examinations are found to have a clinically significant abnormality requiring followup medical care.[22]

The Group Health Association (GHA) of Washington, D.C., has made physical examinations available for years but no organized effort was made to encourage such examinations. The content of the examinations varied from physician to physician, as well as by the age and sex of the individual. In 1963 a study of these examinations revealed that the rate of their use was 33 percent of all adult members. The approximate cost of an examination (in 1963) was $12.08, which, when converted to a monthly premium charge, was $1.01 per patient. For the entire membership, the premium would be one-third that amount, reflecting the relatively low degree of utilization.[23] At GHA, 26.7 percent of the examinations were given to persons with no previous physician's services during the year. On the other hand, about 50 percent were received by individuals with one or more physician's contacts within the preceding six months. GHA members receiving physical examinations in general did not appear to be high users of medical services. In late 1973 GHA put into effect a Health Assessment Program, a screening service believed to be a more efficient and time-saving alternative to the routine physical examinations that had been provided. Complete health assessments are available every year for members over age 60, every other year for those between 50 and 60, every three years between ages 40 and 50, once in every five years for those between 30 and 40, and once in ten years for adults under age 30. Health assessment consists of a defined and limited number of routine screening procedures and laboratory tests, varied according to the age and sex of the individual. Only those procedures believed to be effective in diagnosing diseases in which early discovery has been demonstrated to lead to cure or more successful treatment are included. Members are encouraged to take the examination.

Of interest is the 1962 report that the Group Health Cooperative of Puget Sound, which had included preventive services in its program, advised its members at that time that after ten years of recommending annual physicals, "with relatively little knowledge as to their true effectiveness in improving health . . . it is the conclusion of both the medical staff and the management that annual physical examinations will neither improve the health of our members nor provide the best medical care. Indeed, in some instances the member might suffer from a false sense of security."[24] Instead, the plan believed that more can be accomplished through an intensive health education program and an encouragement of immunization, prenatal care, well-baby supervision, chest X-rays every two years for all adults, and tests for cervical cancer for women.

One of the roles of employers as respects industrial health is that of making available to employees an adequate program of health insurance benefits. Certainly if employers or unions demand insurance coverages for physical examinations and other preventive measures, insurers would willingly oblige. The fact is that, by and large, such purchaser demand has not been made. In part, of course, this results from the fact that many employers provide physical examinations on an in-plant basis through their industrial health services or arrange for such examinations on a contractual basis. Thus the Conference Board study released in 1974 showed that of the employers surveyed, 57 percent provide or make provision for such examinations.[25] In such cases it may well be considered that health insurance benefits for such services would not be necessary.

Insurances against loss of income resulting from disability provide valuable aid during the period of *rehabilitation* through maintenance of a portion of the lost income. In this way a financial deterrent to obtaining rehabilitation services is relieved, and the benefits might be used, in part, to pay for necessary services. It, furthermore, has become a quite general, although not universal, practice for insurance companies to continue paying loss-of-income benefits in whole or in part even though the insured individual has returned to some form of employment. In such cases the prognosis must be one of prolonged disability and the individual must participate in undergoing a formal rehabilitation program in which the employment plays a definite part. This practice is followed for carefully selected cases with the approval of the attending physician. It might be done informally by specific agreement in individual cases, or it might be a provision stipulated in the contract. The employment for rehabilitative purposes might be on a full-time or part-time basis or might be subject to earnings limitations, such as 15 percent of the disability-income benefit. There might be a time limit for which the income benefits are payable under such circumstances. This might be three months, six months, or twelve months, but usually some extension is possible in meritorious cases. The income benefit paid might be in full, or might be the full benefit less the wages earned. In a few instances, under loss-of-income coverages, the company will also assume some or all of the costs of physical or vocational rehabilitation even though such costs are not covered by the policy. With respect to these extensions of the loss-of-income coverages, some companies have experienced difficulty in the determination of cases that are suitable for rehabilitation. Problems have been occasioned also in obtaining the cooperation of the physician, of the patient, or of the family of the patient. Since all three are indispensable, any failure in this direction presents a serious handicap to bringing about rehabilitation.

Medical expense programs, as they are presently conceived, provide considerable benefits for expenditures for rehabilitation, although these are not identified in the contract as such. Coverages for in-hospital care generally include all the medical services provided while hospitalized. Psychotherapy administered by a qualified physician, vocational therapy, and social services provided in the hospital do not usually appear to be covered, although a few insurance companies, by applying a liberal attitude to the definition of covered medical services, do pay benefits for these services in certain instances. Speech therapy generally does not appear to be covered. Under these coverages, the cost of rehabilitation services provided outside the hospital would not be covered, since the conceptual nature and purpose of the coverages pertain to hospital-centered care.

Under major or comprehensive medical expense coverages, where both in- and out-of-hospital medical expenditures are covered, the foregoing pattern for services provided in hospital, in the main, pertains, although some companies do not cover physiotherapy, and none appear to cover vocational therapy. On the other hand, some companies include psychotherapy administered by a qualified physician. With respect to services provided out of hospital, physician services, surgery, and the purchase and fitting of artificial limbs and other prosthetic appliances are generally covered; some companies cover physiotherapy and psychotherapy, some do not; a few will consider vocational rehabilitation services, most will not; and a few cover speech therapy.

Generally, under this form of insurance, social services are not covered, although some insurers will consider payment for such services where necessary. With respect to care provided in rehabilitation centers or clinics, practices differ. Some insurers do not cover such care, some do if the facility is recognized as a hospital, and some provide coverage if the care is warranted even though such care is not covered under the terms of the contract. Rehabilitation services provided by publicly financed agencies are covered by some insurers and not by others in those instances where the patient who can pay is, under the law, charged for the services regardless of the presence of insurance.

As is readily apparent, the clear categorization of insurance company practices with respect to rehabilitation services under their medical expense coverages is complicated by variations of practices among companies. It also is complicated by conditions that pertain to certain rehabilitation services, facilities, and agencies. Professional qualifications are not always clearly discernible. Licensure practices not only set but minimal qualitative standards, they frequently obscure the picture through outmoded concepts and the creation of artificial categorizations. The absence of an accreditation program for rehabilitation facilities leaves insurers in doubt with respect to the worthiness of some. This situation makes the definition of rehabilitation services in the insurance contract difficult if not impossible. In addition, the dearth of factual information with respect to both the cost and utilization of rehabilitation services of various types means that insurers have no data available upon which to calculate premium charges for coverage of these services.

A further question also faces health insurers: To what degree should they assume the initiative in bringing about the application of rehabilitation services? In the main, health insurers do not presently assume the initiative, although in a few instances companies have established formal rehabilitation programs and do take positive steps. Rather, their present role for the most part varies from that of doing nothing to that of encouraging or suggesting the rehabilitation process wherever feasible, of cooperating with the group policyholder where interest is expressed, of providing assistance where this is called for, of directing the patient to the proper agencies, and of paying the benefits contracted for under their policies with a liberal attitude as respects nonmedical rehabilitation services. In some instances the insurance company can be a factor in case finding, although here the cooperation of the patient and the employer or labor union is usually a prerequisite, and the confidential nature of certain of the medical and diagnostic information in the claims records must be respected on behalf of the patient. However, those companies with a positive rehabilitation program do, through their claims personnel and agents, actively engage in case finding.

In order to effectively assume the initiative in bringing about rehabilitation, an insurance company would have to consider the following: the establishment of a definite organizational plan through a specified unit within the company; the development of knowledge of the services available, including tax-supported vocational services provided at nominal cost to the patient; the promotion of consciousness of the subject on the part of claims personnel in the selection of claimants with rehabilitation potential; the establishment of a system of home office screening and evaluation by claims personnel, specially designated rehabilitation coordinators, or medical personnel; the maintenance of accurate records; the establishment of close cooperation with the patient and his physician; and the establishment of cooperation with employers and labor unions.

It would seem that several other accomplishments are necessary before there can be any resolution of the question of what the role of health insurers should be in relation to rehabilitation efforts and the nature of and degree to which health insurance coverages should provide benefits for rehabilitation efforts to a greater extent than is now the case. These might include:

> Clearer delineation of the purposes of rehabilitation and of the types of rehabilitation services.
>
> Based on the foregoing, a more precise estimate of the needs for rehabilitation by types of cases and types of services, and a measurement of the degree to which these needs are unmet.
>
> Education of physicians and the allied health care professions in the principles, processes, and potential of rehabilitation.
>
> More definitive examination of the factors that assist or retard the rehabilitative processes.
>
> Consideration by employers and labor unions of the role they should play with respect to rehabilitation.
>
> Further research that might lead to continuing developments in the field of rehabilitation.
>
> Maintenance of a listing of rehabilitation agencies and centers of credible standing, and the development of an accreditation program for such facilities.
>
> A more precise estimate of the costs of rehabilitation services, including the development of reliable utilization and cost data.
>
> Development of a uniform pricing system, based upon cost accounting principles, which would determine the charge to be made for a particular service, regardless of how paid for, and the establishment of charging practices by public agencies that do not discriminate against insurers.[26]

The situation described here for insurance companies writing health insurance coverages is essentially similar to that which prevails for the Blue Cross-Blue Shield plans. In a few cases Blue Cross plans have specific contractual arrangements with rehabilitation centers (for example, Harmarville Rehabilitation Center in the Pittsburgh area).

Certain of the prepaid group practice plans, or HMOs, also directly provide rehabilitation services as part of their prepaid arrangement. For example, the Health Insurance Plan of Greater New York provides physiotherapy and a number of other rehabilitative services, and the Community Health Association of Detroit includes physical therapy as well as medical rehabilitation services.

That such disability and health insurance benefits are a significant element in the rehabilitative process for employed people could hardly be argued, provided proper management of the rehabilitation services is a part of the procedure. In the absence of such insurance protection it must be expected that some patients who could otherwise be rehabilitated will continue to be disabled.

REFERENCES

1. *The Washington Post,* March 16, 1976.
2. Health Insurance Institute. "Source Book of Health Insurance Data," New York, 1975-1976.
3. Dennis Brev, "HMO Skirmish Heats Up in Rochester, N.Y.," *American Medical News,* October 27, 1975.
4. The Conference Board, *Profile of Employee Benefits,* New York, 1974.
5. Subcommittee on Health, Committee on Ways and Means, U.S. House of Representatives, "Basic Facts on Health Insurance Coverage of the Unemployed," April 1974.
6. The Conference Board, op. cit.
7. U.S. Department of Labor, *Digest of Health and Insurance Plans,* 1974; Supplement, 1975.
8. The Conference Board, op. cit.
9. Ibid.
10. Ibid.
11. Health Insurance Institute, op. cit.
12. Subcommittee on Health, op. cit.
13. HEW Research and Statistics, Note No. 23, 1973.
14. The Conference Board, op. cit.
15. Dorothy R. Kittner, "Changes in Health and Insurance Plans for Salaried Employees," *Monthly Labor Review,* February 1970.
16. Health Insurance Institute, op. cit.
17. The Conference Board, op. cit.

18. John Van Steenwick, "Implementing Programs for Preventive Health Care Services" (Chicago: Blue Cross Association, 1970).
19. Ibid.
20. *The New York Times,* July 20, 1976.
21. Kaiser Foundation Medical Care Program, "Preventive Medicine," Oakland, Calif., 1974.
22. Morris F. Collen, M.D., "Multiphasic Screening for Periodic Health Examinations," undated.
23. Scott I. Allen, M.D., "Five Hundred Consecutive Physical Examinations at GHA," paper presented before the Group Health Institute, May 10, 1963.
24. *Seattle Times,* February 11, 1962.
25. The Conference Board, op. cit.
26. J. F. Follmann, Jr., "Prevention of Loss and Rehabilitation" (Washington, D.C.: Health Insurance Association of America, 1959).

Part VII

The Cost-Effectiveness of Industrial Health

In relatively recent years considerable interest has developed in the cost-effectiveness or the cost benefits that might and with seeming logic should be inherent in various approaches to preventive medicine, including industrial health programs. That there is an assumption that preventive medicine, in its broad sense, produces economic savings is evident from any review of the literature of the subject; it pervades the field with a continuum. For example, Morton S. Hilbert as president of the American Public Health Association said in 1975, in addressing the annual meeting of that Association: "It is time to redirect the prevailing local, state, and national focus of health care to an emphasis on prevention and health promotion. Both the direct and indirect costs of illness prevention are substantially less than the costs of sickness care. Thus, prevention programs not only prevent illness, disability, and premature death, they reduce the burden on the medical care system so it can more effectively handle unpreventable ills."

That such statements are reasonable is evident from the potential of preventive medicine in reducing the incidence and severity of certain diseases and accidents, as shown earlier. That the assumptions made can often be exaggerated is evidenced by the British National Health Service and the drastic adverse economic effects that resulted from overrating the potential of preventive medicine. Among other things, in the making of such assumptions, there can be a disinclination to remain aware of the deterrents to the maintenance of good health.

To be discussed in Chapter 25 is the matter of evaluating the cost-effectiveness of preventive medicine — its concept, the cost-effectiveness of certain specific programs, and the effectiveness of physical examinations and health-screening techniques. In Chapter 26 the cost-effectiveness of industrial health programs will be discussed.

25

Evaluating the Cost-Effectiveness of Preventive Medicine

The raison d'être of preventive medicine, including industrial health programs, is of course the improved health and well-being of people. Economically, however, the not infrequent speculation as to the cost-effectiveness of preventive measures, including the reductions they might bring about in the utilization of and expenditures for health services, in disability and absenteeism, in the costs of the various types of programs, and in premature deaths can be based on faulty or incomplete reasoning. Beyond that, the methods of arriving at such evaluations, as will be shown, are not simple matters and there are differences of opinion as to the methodology to be employed.

To some it perhaps seems unbecoming to speak of the life and health of people in economic terms. It must be recognized, however, that wherever one looks around the world, regardless of the degree of national wealth, varying political ideologies, the nature of the economic structure, prevailing religious beliefs, or the manner in which health care is financed, there is always present the consciousness that resources are not limitless and that the wishes of a people for improvement in their welfare take a great many forms, the cost of any one of which has to be balanced against all the others. In this process, preventive medicine has never assumed a position of priority. Mankind has always placed, as respects its health, primary emphasis on curative care, displaying a willingness to pay rather dearly for the choice.

In a certain sense, then, any type of prevention program is called upon to justify its existence in economic terms, as illogical and perhaps inhumane as this might appear to some. But this is not unreasonable once it is considered that people must be remunerated for whatever work they do or services they perform, and that materials and facilities all have an economic value. Since funds, whatever their source, are never unlimited, choices must constantly be made as to how they will be spent and what for. Granting that preventive medicine does not enjoy a top priority, the demand for its economic justification is therefore greater than is the case for certain other human needs or desires.

Complicating this process is the unfortunate fact that there is as yet too little information available to aid the process of such decision making. While many cost-effectiveness analyses in the health field have been made in relatively recent years, the extant studies to date are quite scattered and fragmentary and often limited to small samples or to isolated aspects of health. It is an important field in which considerably more research is needed and one that could benefit from considerably more interest and effort on the part of the providers of health care, the universities, government, and most certainly employers, labor unions, and insurers.

Beyond this, difficulties are presented by the lack of agreement among economists as to which factors are to be included in or excluded from the considerations in determining the cost-effectiveness of preventive measures, nor is there agreement as to the value to be placed on each of the factors involved. To say this is by no means to be critical; it is simply to take cognizance of the situation as it is today. The determination of cost-effectiveness or cost benefit in the field of health is a relatively new field of exploration and one that will continue in an evolutionary status for some time to come.

THE CONCEPT

An economic cost approach to the value of health has a relatively long history, dating as far back as the concepts of Sir William Petty, in England, in 1667 and those of Adam Smith in the following century. Burton Weisbrod commented over 15 years ago that "estimates of losses from disease involve questionable, misleading, or simply incorrect procedures . . . costs, real costs, that is, and money expenditures are not synonymous terms. There may be expenditures without real . . . costs (for example, production loss). And there may be social costs without expenditures."[1] Among his conclusions were that "expanding health and medical facilities may frequently increase unit costs" and that "sound economic analysis will have a greater cogency in the original statement, and the subsequent experience will be more likely to earn respect for the acumen of the health official, as predictions prove accurate."

Several years later Dorothy Rice noted: "The value of human life expressed in terms of lifetime earnings is a basic tool of the economist, program planner, government administrator, and others who are interested in measuring the social benefits associated with investments in particular programs. For public programs . . . the valuation of human lives is a basic requirement for the proper calculation of the benefits to be derived." She then attempted "improved, refined, comprehensive, and up-to-date estimates of the present value of lifetime earnings," with consideration of age, sex, color, and educational level. In so doing she examined the approaches taken by others and noted that all were devised for a specific use and were not readily adaptable for other purposes. She then reviewed the basic assumptions and economic concepts employed as respects life expectancy, labor force participation, earnings, housewife services, productivity increases, allowance for consumption, the pattern of lifetime earnings, male and female differentials, and educational differences (by sex and color). Discounting for the value of money changes with time (noting the general agreement among economists on the use of discounting but the lack of agreement on the discount rate), she used 4 percent. Her major findings were that peak discounted lifetime earnings occur in the young adult ages, and that peak lifetime earnings for men are nearly double those for women. She also concluded that education pays off, but the payoff is greater for white males than for others.[2]

Subsequently, Mrs. Rice discussed the direct and indirect costs of illness (the full economic costs resulting from illness, disability, and premature death). Noting the effects on medical expenditures, earnings, employment, housewife services, transfer payments (for example, pensions and relief), taxes, losses to the GNP, life expectancy, labor force participation, productivity increases, allowance for consumption, and discounting, she placed the economic toll of illness, disability, and death in 1963 at $105 billion, made up as follows: $34.3 billion for health care; $21.0 billion for morbidity loss to the economy; and $49.9 billion as the present value of the lost output of those who died. Mrs. Rice estimated the total economic cost for 1967 to be $125 billion.[3]

A year later Mrs. Rice approached the measurement and application of illness costs and reached the same conclusions, noting that the application of cost-benefit analysis to public health programs is still in its infancy, due in part to a divergence of opinion on specific techniques, but that it is nonetheless a useful tool.[4]

It will be noted that one of the problems in these economic approaches is that of evaluating the services of housewives. In 1975 in the United States housewife services were valued on an average of $4,705 a year, with the maximum value placed at $6,417 for women aged 25 to 34, the child-rearing ages. It is stated that whichever cost approach is used, "they all indicate that work performed in the home by housewives accounts for a very large amount of unpaid work."[5]

About this same time Herbert Klarman noted the complexity of attempting to measure cost-benefit in the delivery of health services through health programs, including the calculation of investment versus consumption, the rate of discount, calculating health and medical care expenditures for a disease (direct costs), calculating output loss (indirect costs), earnings, rate of unemployment, consumption, application of cost-benefit calculation, and applications to medical research.[6]

Rashi Fein has noted that while recent years have witnessed much effort and change with respect to the economics of health, the economists do not answer two questions that government budgeteers raise: (1) Are the specific programs for which support is sought "good" investments, and (2) which particular programs should be favored over other programs? While Mr. Fein is addressing himself to the financing of government programs and to the matter of making choices, his comment is certainly applicable to a consideration of financing preventive medicine, either privately or publicly, since essentially the choices made are dependent on the relationship between costs and benefits. Then, as Mr. Fein points out, the basic consideration is the need to specify the objectives of a program.[7] Elsewhere, Mr. Fein has commented: "One might examine a set of questions related to the costs of preventive services and the benefits to be derived from such services. Do expenditures on preventive services represent an investment that yields high economic returns? Are such services an investment worth making?" He goes on to say: "I believe the usefulness of the game is in question because of the weakness of data and that the correctness of the game is in question because of the weakness of analytical techniques." Mr. Fein has also pointed out that "in cost-effective analysis we seek the optimal, economically efficient combination of... inputs so that resources are not wasted and so that they might, therefore, be available for other purposes. The object is to minimize cost per unit of output . . . cost effectiveness does tend, at times, to move closer and closer to benefit-cost analysis."[8]

Robert Grosse believes—perhaps naturally for an economist—that the costs of health services should be a significant consideration in deciding upon government health policies and programs. He makes the point that: "Whether the cost is worthwhile depends on our values and on our ability to estimate and evaluate the benefits from the various alternatives" (so that choices might be made).[9] This is particularly germane when choices have to be made between the various elements of curative medicine on the one hand and the different aspects of preventive medicine on the other. Traditionally, as has been noted, the scales have dipped sharply in favor of curative medicine. Mr. Grosse, in discussing cost-benefit analysis in relation to public programs, has said that "the multiplicity of dimensions of output and their basic incommensurabilities, both with costs and with outputs of other claimants for public expenditure, still require the use of value judgments and political consensus." He points out: "Resources expended on arthritis control may reduce pain and disablement, but will have little effect on mortality rates."

Alan Williams, a British economist, has noted that there are not enough resources to satisfy the desires of a community for things that will improve the quality of life, that this calls for choices and priorities, and that this, in turn, leads to the concept of cost-benefit analysis based on the proposition that services should be provided only if their benefits outweigh their costs. In so doing, he argues, it is assumed that it is possible to separate one service from another, to choose between them, to estimate the outcomes and to value these, to estimate the cost of providing each service, to weigh the costs and benefits against each other, and to decide to end those services whose costs are greater than the benefits. He contends that there are difficulties to sustaining such assumptions. The standard analytical device of holding benefits constant, he argues, converts a cost-benefit study into a cost-effectiveness study, which is acceptable if all alternatives generate the same benefits and if one is not interested in determining whether anything at all should be done if for all alternatives the costs exceed the benefits. Too often, he says, attention is devoted to the actual expenditures for medical services, with relatively little attention to such other costs as loss of output resulting from work disability, additional family expenditures, and the loss of leisure time.

Admitting, then, that cost-benefit analysis is a difficult matter, the author nonetheless recognizes that the underlying issues continue to be present and have to be resolved.[10]

In October 1976 the American Public Health Association considered a resolution, "Cost-Benefit of Prevention," which commented that "heavy dollar investments in treatment of illness have not had a proportionate impact on the quality of health of the total public . . . [that] there are limits to our economic resources for the health of people . . . [and that] the focus of health care should be on prevention and health promotion."[11]

In the expressions of these concepts it is observable that distinction is made between cost-effectiveness as an economic measurement on the one hand and cost-benefit on the other. A decade ago Herbert Klarman said of the difference: "Cost-effectiveness analysis is a special, narrower form of cost-benefit approach that economists have evolved in the past generation. . . . The cost-benefit approach represents an attempt to apply systematic measurement to projects or programs of the public sector where market prices are lacking and external effects in production and consumption loom important. . . . The cost-benefit approach is characterized by (1) the objective of enumerating as completely as possible all costs and all benefits expected and (2) the recognition that costs and benefits tend to accrue over time. In principle, cost-effectiveness analysis partakes of both of these characteristics. . . . Under cost-effectiveness analysis the enumeration of benefits need not be so complete as under the cost-benefit approach. Rather, certain results are specified and all other results are regarded as held constant or perhaps of secondary importance. Cost-effectiveness, rather than cost benefit, is employed when various benefits are difficult to measure . . . it is taken for granted that the results sought can be 'afforded.' "[12]

Klarman and Guzick have made the point that cost-benefit analysis is the close analogue in the public sector to supply-and-demand analysis in the private sector of the economy and that in either case the aim is to arrive at the best possible allocation of society's scarce resources given the diversity of other desires and wants that compete for those resources. The cost component they recognize as the dollar value of manpower, supplies, and facilities needed to mount a prevention campaign; estimating the benefit component is more difficult, particularly since doubts are being raised as to the validity of the more conventional approaches to valuing benefits in the health field, which recognize (1) direct tangible benefits (the use of resources that could be saved), (2) indirect tangible benefits (savings in lost earnings due to morbidity or premature mortality), and (3) intangible benefits (such as the relief of grief, sympathy, and discomfort). They point out that this approach values livelihood rather than human life and that the well-to-do are worth more than the poor, whites more than blacks, men more than women, and young adults more than children or old people. The question posed, they reason, is whether society wishes to accept such value judgments. They conclude that for the time being it may be sensible to retreat from cost-benefit analysis to cost-effectiveness analysis.[13]

Because of their relatedness to industrial health programs it is of value here to review certain findings of the cost-effectiveness of specific prevention programs, and then to review certain findings as respects the cost-effectiveness of programs for physical examinations and multiphasic health screening.

THE EFFECTIVENESS OF SPECIFIC PROGRAMS

There are several examples of the cost saving inherent in preventive programs. Such examinations are most usually made for the purpose of evaluating the cost-effectiveness of a specific type of government program or to attempt to illustrate the preventive potential as respects a specific disease. The following material summarizes the findings of some studies concerned with coronary heart disease, cancer, chronic renal disease, tuberculosis, influenza, hypertension, urinary tract infection, measles, mental retardation, the fluoridation of drinking water, cigarette smoking, the effects of improved diets and nutrition, and the effects of air pollution control.

The study concerned with *coronary heart disease* attempts to provide a basis for a hypothesis on the benefits to be derived from a prevention program. It was based on data of a sample of the population (5,000) of Framingham, Massachusetts. The study was conducted in the two-year period 1948–1950 and represented 12 years of followup. It measured various risk factors associated with coronary heart disease: (1) age, (2) serum cholesterol, (3) blood pressure, (4) weight, (5) hemoglobin concentration, (6) cigarette smoking, and (7) electrocardiogram abnormality, of which the first, the fifth, and the last were considered uncontrollable. The others were considered amenable to control action. The per person cost of the control program was estimated as shown in Table 25-1.

Table 25-2 shows the estimated economic costs of an incident of coronary heart disease. The expected incidence of heart disease by deciles of risk and the observed incidence of heart disease under the program were as shown in Table 25-3. The study concluded:

Table 25-1. Estimated per capita cost of prevention program for 12-year period (1969).

Annual medical examination	@ $50.00 × 12 years	= $600.00
Nutritionist/dietitian consultation	Monthly for 6 visits	
	Quarterly for 2 visits	
	Annually for 11 visits	
	19 @ $3.75 =	71.25
	Total for 12 years =	$671.25

This assumption yields reductions in incidence which are in reasonable agreement with the experiences of the Anti-Coronary Club and the National Diet-Heart Study. When screening is used to identify the high-risk segment of the population, a control program seems to yield dollar return which is significantly in excess of the total program cost, even though the 12-year period for which the hard data exist is considered by itself. A general conclusion supported by this analysis is that a screening and prevention program against coronary heart disease, either through private medical practice or through some public agency, appears to be a profitable undertaking from the viewpoint of saving dollars and lives. . . . Confident prediction of the benefits of preventive programs requires following approximately 100,000 people for about five years. The cost of such an undertaking runs on the order of $100 million.[14]

An examination of cost and cost-effectiveness of a program for the detection of *uterine cancer* states:

A study was made by the Cancer Control Branch to assess the potential direct benefits of a screening program to demonstrate control of

Table 25-2. *Consequences and expected costs of an incident of coronary heart disease (1969).*

Sudden death, 14% of cases	Funeral		$ 1,000
	Income loss*		28,000
		Total	$29,000
Nonsudden death, 4% of cases (2-day survival)	Funeral		$ 1,000
	Physician and hospital		330
	Income loss		28,000
		Total	$29,330
Myocardial infarction, 45% of cases	Physician and hospital		$ 1,906
	Income loss		1,750
		Total	$ 3,656
Acute coronary insufficiency, 10% of cases	Physician and hospital		$ 408
	Income loss		583
		Total	$ 991
Angina pectoris, 27% of cases	Physician		$ 40
	Income loss		135
		Total	$ 175

* Income loss is based on an annual salary of $7,000.

Table 25-3. Expected coronary heart disease compared with observed coronary heart disease (2,187 men, 30 to 62 years old at entry).

Deciles of Risk	Expected Number of CHD Cases	Actual Number of Cases
10	90.5	82
9	47.1	44
8	32.6	31
7	25.0	33
6	19.7	22
5	15.6	20
4	11.5	13
3	8.6	10
2	6.0	3
1	3.4	0

cancer of the uterus. . . . It was found that the life-saving effects of the case-finding activities of the program alone could return, after five years of operation, nine dollars for every dollar invested in the total program. . . . Considered as basic costs in the appraisal of this program were the total anticipated project grant expenditures, including those for training and those applied to improvement of uterine cancer control procedures. Added to these were the costs of hospital and medical care for the women expected to be found with cancer in early stages. Considered as benefits were the savings in estimated earning power of these same women, and the savings in the medical costs that would have accrued had their cancers been allowed to progress to later stages. Although such projects customarily find numbers of cases of cancer that already have reached the later invasive stages, these were eliminated from the computation. Because of the nature of cancer of the uterus, most of these cases would have been discovered soon without the program. Therefore the program could not be credited with the lives that are saved in this group.[15]

Table 25-4 presents some of the planning information that was developed in this study.

Table 25-5 is an estimated cost-benefit analysis approach by HEW to examine the costs and gains from *cancer-control* programs. One result of these findings was that the cost per death averted for the oral and colon-rectum programs was so high that HEW recommended emphasis on research and development rather than on a control program to demonstrate and extend current technology.[16]

Table 25-4(a). Undiscounted grant costs related to examinations, case finding, and deaths averted (1966).

Five-year grant program costs	$73,750,000
Examinations:	
Number performed	6,712,000
Number per case found	80.7
Grant cost, per examination	$10.99
Cancer cases found:	
Number	83,182
Grant cost, per case found	$887
Cancer deaths averted:	
Number	34,206
Grant cost, per death averted	$2,156

Table 25-4(b). Estimated 1968-1972 program costs and benefits, discounted to present values.

Program costs:	
Grant awards	$ 68,086,000
Treatment of cases found at early stages	50,652,000
Total	$ 118,738,000
Program benefits:	
Earnings saved	$ 998,319,000
Treatment costs averted by programs	73,045,000
Total	$1,071,364,000
Benefit-cost ratio	9.0

Source: Cancer Control Branch, Division of Chronic Diseases, Public Health Services, U.S. Department of Health, Education, and Welfare, October 11, 1966.

Table 25-5 HEW cancer control program, 1968-1972.

	Uterine Cervix	Breast	Head and Neck	Colon-Rectum
Grant funds total (in thousands of dollars)	$97,750	$17,750	$13,250	$13,300
Number of examinations (in thousands)	9,363	2,280	609	662
Cost per examination	$10.44	$7.79	$21.76	$20.10
Examinations researched per case found	87.5	167.3	620.2	496.0
Total cases found	107,045	13,628	982	1,334
Cost per case found	$913	$1,302	$13,493	$ 9,970
Total deaths averted	44,084	2,936	303	288
Cost per death averted	$2,217	$6,046	$43,729	$46,181

A program for the detection of *cancer of the cervix* in Manitoba examined the results of over 100,000 Pap smears in 1968. The cost of the test was $470,000, with an additional $500,000 for further examination by a physician. It was estimated that 70 percent of the new cases of cervical cancer were discovered through the screening program. The total cost of the program, including followup treatment, was $1,200,000 in 1968. The benefits of the program were questioned. There was no significant reduction in the incidence of cases of invasive cancer of the cervix over a five-year period, although perhaps eight to ten lives were saved as a result of the program.[17]

Herbert E. Klarman in 1968 examined the cost-effectiveness of the treatment of *chronic renal disease*. Noting that some 6,000 persons whose life span could otherwise have been appreciably prolonged through such treatment die every year of this disease (a large majority of whom are between ages 15 and 54 years), the study recognized that such treatment is expensive. The study estimates that nine life-years would be gained by the use of dialysis at a cost of $11,600 per life-year for treatment in the hospital and $4,200 for treatment at home. Transplantation, on the other hand, would produce an estimated 17 life-years gained at a cost of $2,600 per life-year. The study concludes "that transplantation is economically the most effective way to increase the life expectancy of persons with chronic kidney disease.[18]

The cost-effectiveness of a screening program for *tuberculosis* among lower-economic-level, older, and inner-city people at Grady Memorial Hospital in Atlanta was evaluated recently (recognizing that with the steady decline of tuberculosis the yield from screening programs had dropped steadily, to the point where the Public Health Service in 1972 recommended that X-ray screening for tuberculosis in the general community be discontinued). The program at Grady cost $30,000. Over a three-year period, of more than 48,000 X-rays taken only 22 new cases of active tuberculosis were discovered—at a cost of $1,800 per case. It was concluded that screening of the general population was unjustified on a cost-effective basis.[19] (Here it is of interest that in 1920 one in every 20 cows was infected with bovine tuberculosis, causing 25 percent of the incidence of the disease among children. Today, as a result of preventive measures, only one in 20,000 cows has the disease and its complete eradication is expected by 1995. The relation of the benefits of the effort to the costs is 3.6 to 1.)

A recent examination of the effects of mass screening for tuberculosis in Denmark revealed that the benefits of early diagnosis were an estimated 1 to 2 million kroner, the benefits resulting from savings in such expenses as hospitalization costs and disability pensions. The cost, however, was 23 million kroner for X-ray screening, 42 million kroner in lost working hours, and unknown amounts for such things as administration, followup, and cases that would have healed spontaneously. The study found that the "impact on the public's health is found to be almost nil" and that "when the program's economic costs and benefits are estimated, the balance is negative."[20]

From the standpoint of a nation, the World Bank has noted that ill health reduces the availability of labor, impairs the productivity of workers and capital goods, wastes current resources, and impedes the development of natural resources, animal wealth, and tourism. Noting a study of tuberculosis control in Korea that concluded that an optimal disease program would produce a return of $150 for each dollar spent from increased work life and decreased absenteeism, and noting also the 19 percent increase in productivity in Indonesia that resulted from the treatment of anemic workers (producing a benefit-cost ratio of 280 to 1), the Bank nonetheless cautions that not all endeavors in disease eradication produce such dramatic results. It states that "government programs should be designed on the basis of cost-effectiveness studies."[21]

In 1975 Klarman and Guzick made a cost-benefit analysis of a program for *influenza* vaccination in the United States. In so doing they recognized the difficulty in valuing the intangible benefits of a reduction in the incidence of disease, especially the valuing of gains in life expectancy. They also recognized the problem of "the unique shift and drift of the influenza virus." The analysis further took cognizance of the fact that a communicable-disease immunization program cannot evaluate the good that flows to the community at large from having vaccinated some people —an indirect benefit. The authors found that a vaccination program for people over age 65 costs $7,240 for each death postponed if the effectiveness rate is 70 percent. The net cost per life-year gained is $700 and $410 respectively. If allowance is made for vaccinations already in place, the adjusted figures are $6,280 and $3,240 respectively, and $600 and $310 respectively. The study notes that in the adjusted figures the difference in net cost is almost 50 percent, while the difference in vaccine effectiveness is only 40 percent.[22]

Concerning *hypertension,* the Inter-Society Commission for Heart Disease Resources has stated: "In the long run it should be less expensive to control hypertension than to care for those who become disabled and economically unproductive as a consequence of the disease."[23] Following this line of reasoning Dr. Jeremiah Stamler and others have noted that morbidity, disability, and mortality from hypertensive disease alone costs the nation, in direct and indirect costs, over $1.7 billion a year. In addition, and assuming that hypertension is a contributing factor in about a third of premature heart attacks and strokes, the indirect costs would be another $5.9 billion and direct costs another $2 billion. Thus on the basis of 1967 data the total cost of hypertensive disease would exceed $9 billion a year. Noting that the Veterans Administration experience indicates that effective comprehensive control programs can reduce morbidity, disability, and mortality from hypertensive disease by over 50 percent, the authors conclude that it is reasonable to project a program that would benefit half of the 20 to 25 million hypertensives by 1978 and to further conclude that the effort "will in the long run save money."[24]

It has been reported in England that "the diagnosis of *urinary tract infection* in a latent stage ... is now possible ... at a reasonable cost "and that the cost of antenatal screening for bacteriuria may well be offset by the reduction of expenditures on hospital admissions for acute pyelonephritis," which, it is estimated, would be prevented in 20 percent of the cases, at a yearly saving of £440,000. The value of bacteriuria screening among school children and nonpregnant adult women, it was recognized, is undetermined.[25]

In 1969 a cost-benefit analysis of *measles immunization* was made. Estimating the loss of workdays and school days, the cost of hospital stays and physician services, and the cost of care for the resultant mentally retarded, it was thought that the number of cases avoided by immunization would have cost $531 million. Offsetting the $108 million cost of immunization, the net economic gain was estimated to be $423 million. It was thought that the savings would include 550,000 hospital days, 5,000 physician visits, 3,244 cases of *mental retardation,* 1,600,000 workdays, and 32,000 school days.[26]

A subsequent study of the economic savings from a measles immunization program from 1963 (when live measles virus vaccines were licensed) to 1972 is quite revealing. Prior to 1963 there were an estimated 4 million cases of measles, 4,000 cases of measles encephalitis, and 400 measles-associated deaths each year. Since 1963 approximately 60 million doses of vaccine have been distributed. By 1968 the incidence of measles had dropped to 220,000 cases a year. In 1968, federal funding expired and the incidence increased sharply, to 850,000 cases in 1971, when federal funds were restored. The study estimates that over the ten-year period, 24 million cases of measles were averted by the use of immunization and that 2,400 lives were saved. Physician visits saved are estimated at over 12 million, and hospital days saved at 1.6 million. In addition, it is estimated that 7,900 cases of mental retardation were obviated. It is also estimated that the premature deaths prevented and the mental retardation prevented saved 709,000 years of productive life. The cost of the program for vaccine production, distribution, administration, and promotional activities averaged $3 per dose of vaccine, or a total of $180 million. For 1972 it is estimated that the cost of the immunization program was somewhat in excess of $15 million and that the economic benefits exceeded $285 million (physician services saved, hospital care avoided, long-term institutional care for the mentally retarded avoided, and lost production resulting from morbidity and premature mortality eliminated).[27]

A recent report by HEW on the cost-effectiveness of a measles immunization program estimates a cost of $108 million for the program. The savings are estimated at $970 million in direct costs and $253 million in indirect costs.[28]

The cost benefit of rubella vaccination has also been considered. Here it was noted that the cost of rubella vaccination (first licensed in 1969) has been estimated to be $3 a dose, but that if combined with measles vaccine, which it is assumed would have been given anyway, the cost would be $1. On the basis of a recent study it is concluded that the economic benefits of a rubella vaccination program were 80 percent greater when females age 12 were vaccinated than was the case when children age 2 were vaccinated. Such vaccination at age 12 reduced natural infections by 80 percent and the expectation of congenital rubella by 95 percent.[29]

A study of the cost savings from the *fluoridation of drinking water* found that the cost of dental care was reduced significantly, 42.2 percent and 47.6 percent respectively, in two communities over a ten-year period. This study, of course, would be of interest to insurance programs for dental care.[30]

Dr. Ernst Wynder of the American Health Foundation has maintained that for individuals participating in a *smoking-cessation* program the effects of smoking cost some $14.2 billion in terms of excess hospital and medical services, days lost from work, and premature death. At the conclusion of the smoking-cessation program, which cost $1.3 billion, these costs were reduced to $10.4 billion, or a net saving of $2.5 billion.[31]

A study of the differences in the use of health services by smokers, nonsmokers, and ex-smokers age 20 and older at the Kaiser Foundation Health Plan has concluded that cigarette smoking habits should be taken into account in estimating medical care utilization and costs. The study found that health examinations were used least by smokers and most frequently by ex-smokers. This distinction was most pronounced among lower-income persons. Frequency of physician visits, however, was highest among male smokers and female ex-smokers, and male smokers age 60 and over made the greatest demand by far for hospital services.[32]

The U.S. Department of Health, Education and Welfare has estimated that *improved diets and nutrition* could result in savings of health care costs alone in excess of $30 billion, some one-third of the total of such expenditures. The principal elements in these projected savings would be an estimated 20 to 25 percent reduction in the incidence of heart disease, producing a saving of some $7 billion in health care expenditures; a 10 percent reduction in mental health disabilities; 3 million fewer children born with birth defects; a 50 percent reduction in the incidence and severity of, and expenditures for, dental care; a reduction of $21 billion in medical costs associated with respiratory and infectious diseases; and an improvement in survivability to age 65 from approximately 65 percent to 90 percent.

The Environmental Protection Agency has maintained that more than 100,000 deaths a year could be prevented if *air pollution* were to be adequately reduced. This, it is maintained, implies a 7.7 percent reduction in the total mortality rate from all causes, which is presently 1.9 million a year. The economic benefits of such a reduction are estimated to be $14.5 billion a year. The cost of a program to reduce combustion-related pollution (sulfates and particulates) would be $9.5 billion. Thus the net cost-effectiveness of such a program based on these estimates would be $5 billion a year.[33]

Concerning seat belt installations in automobiles, Morton S. Hilbert, as president of the American Public Health Association, in 1975 told that Association that $1,000 would be saved for every dollar so invested.

While these programs pertain to the population at large, they are nevertheless related to industrial health programs. Heart disease, hypertension, cancer, urinary tract infections, chronic renal disease, and influenza shorten the lives of valued employees. They strike not only employees but their dependents as well, with resultant costs to the health insurance program. Employers, then, should have an active interest in any public or community programs whose purpose is to reduce the incidence or effects of such diseases. Consideration should also be given to including such preventive measures within the scope of the industrial health program. Measles and mental retardation can affect the costs of dependent coverages under health insurance programs. Again, an active interest in community immunization programs is warranted and consideration might be given to the inclusion in the industrial health program of immunization for employees' families. The potential savings from a cessation of heavy cigarette smoking and the correction of improper diets are such that they warrant emphasis in any health education program. The same is true as respects the use of seat belts and other motor vehicle safety precautions. The matter of reducing air pollution, the benefits of which accrue to the entire community, but including employees and their families, are such as to argue for employer interest and concern.

Regardless of critical attitudes toward the methodology employed in such studies, the findings cannot fail to be of use to employers in many ways in relation to their industrial health programs. The potential economic gain is considerable.

EFFECTIVENESS OF PHYSICAL EXAMINATIONS AND SCREENING

Regarding the effects of general physical examinations and multiphasic health screening on medical care utilization and expenditures, disability, and premature death, and the relationship of these preventive measures to industrial health and insurance programs, the following are some thoughts on the subject and some findings about their cost-effectiveness.

Teeling-Smith, a British economist, suggests that costs can be subdivided into four factors: (1) the cost of the screening test in terms of equipment, materials, and manpower; (2) the cost of contacting the desired public and ensuring that they attend for examinations; (3) the cost of the numbers of false positives that will require further investigation if the tests are of low specificity; and (4) the cost of repeated screening determined by the incidence and duration of the disease under scrutiny. He describes the benefits as dependent on several factors: (1) the seriousness of the disease for the individual and the community; (2) the prevalence of the disease, which will determine the number of cases detected; (3) the sensitivity of the tests, which will determine the proportion of the total number of cases brought to treatment; and (4) the extent to which the disease can be affected by treatment after it is detected. He expresses an opinion that, whereas the early detection and treatment of tuberculosis has resulted in financial savings because mainly young adults were affected, if heart disease, which affects mainly the elderly, could be prevented by early diagnosis, it would lead to a drain on the economy rather than a saving.[34]

Harry E. Emlet, Jr., an engineer, has examined the cost benefits of different models of multiphasic screening.[35] Mr. Emlet has also discussed the reduction in physician time that can result from multiphasic screening using paramedical personnel:

> While the results obtained with the limited data readily available cannot be considered final, they do strongly suggest that when physician man-hours are a fixed resource:
>
> 1. Introduction of screening into a population previously not screened or examined annually will generate a physician workload for followup diagnosis and treatment which is greatly in excess of the physician resources available in that population.

2. Introduction of screening into such a population could actually reduce rather than increase benefits realized, because of a greater diversion of physician man-hours from treatment to screening and diagnosis.

3. Introduction of screening into a population previously examined annually can result in a sizable release of physician man-hours.

These conclusions lead to a further conclusion: if multiphasic screening is not to swamp the medical care system and possibly decrease the current rate of benefits, it must be introduced in a planned manner with careful attention to the above conclusions and to other means of increasing the ratio of physicians available to physicians needed. Clearly, while AMHS, introduced in a wisely planned manner, may be able to help increase this ratio significantly, it can by itself offer only a portion of the complete answer to the physician shortage.

Dr. Richard Sparks of the Harvard Medical School has recently written: "Health maintenance has become our national obsession . . . the annual physical has been extensively promoted by physicians and enthusiastically accepted by patients as an effective means of maintaining health. But is it?" Noting the development of multiphasic health screening in the United States and the interest of legislators in the subject, Dr. Sparks says that "unfortunately this preventive zeal is expensive." Reviewing many of the studies that have been made of the subject, Dr. Sparks feels that in many instances the differences produced by such examinations in terms of disability, time lost from work, or mortality are not significant and are unimpressive and says: "The record is clearly dismal." Reasons for this include, in some cases the fact that some elements of health screening are not sensitive enough to detect diseases in their early stages to allow for effective treatment (such as lung cancer), and in others that the patient will not comply with the recommendations of the physician (for example, proper diet or reduced cigarette consumption). In other cases, he comments that while the test has a high diagnostic efficiency among the ill it offers an unacceptably low yield among the healthy (for example, detection of rectal cancer). In yet other cases the testing process fails, Dr. Sparks maintains, because of the unwillingness of the physician to act upon the information derived. Dr. Sparks is of the opinion, however, that testing for hypertension is feasible because it can readily detect cases where symptoms are lacking, because physicians usually initiate prompt and effective treatment, patient compliance is good, it is inexpensive ($2.66 per discovered case), and the beneficial effects are clearly established. He feels that Pap smear is also of proven value and that mammography is a useful test, although he notes the problem of the potential radiation hazard and that it is expensive ($408 to $1,200).

In the main, however, Dr. Sparks concludes that most diseases can only be detected after symptoms become apparent and that there is no convincing evidence that treatment of diseases before the onset of definite symptoms offers any long-term advantage over treatment initiated after symptoms appear. He states that "the annual physical examination has proved to be little more than an elaborate and expensive ritual that has not fulfilled its promise."[36]

Stuart Schweitzer has presented a methodological framework that permits the evaluation of early diagnostic tests in terms of their cost-effectiveness. He notes that the term "preventive medicine" represents "a bundle of services" and that even in the case of illnesses that are amenable to diagnosis, effective intervention, and cure, complications abound in the areas of disease patterns, diagnosis strategy, and treatment procedures. He also notes that only recently has informed thought begun to question the desirability of offering massive amounts of preventive care and that the Kaiser program has abolished annual multiphasic screening examinations in favor of less frequent testing supplemented by special examinations when symptoms appear. He comments that "preventive health care comes at a high cost in terms of resources utilized, and planners must be well aware of the trade-offs and costs inherent in any new health program."

Noting that testing makes errors, that some tests are inherently dangerous, and that some diseases exhibit the same course and outcome regardless of detection, he says: "It seems naive to talk about the efficacy or cost-effectiveness of a diagnostic test outside of a time framework" (with some diseases a repeat test is not necessary; with others an annual test is too infrequent). He argues that with a cost-effective test, "the most efficient allocation of public and private funds might result from public subsidy of the procedure through a national health insurance scheme, with the consumer left to pay the remainder of the cost of the test and the full cost of more frequent tests, if desired. A system of subsidies rather than full payment might well be incorporated into a health delivery and financing system." Noting that choices are extremely difficult, that there are serious gaps in the methodology, and that economic analysis cannot provide a total answer for society's allocation decisions, the author concludes that there is "a real and current need for this sort of analysis."[37]

Dr. Scott I. Allen of Group Health Association has commented that more studies are needed to determine the value of periodic checkups for populations such as members of prepaid plans under intermittent surveillance by physicians for ordinary common illnesses. Noting the amount of physician time at GHA doing routine physical examinations, he comments that "the overall value of annual physicals in an economic sense depends on the units of measure and on costs to the plan, to the member, and to society.... Clearly it would be fallacious to consider only prepayment dollar costs and savings accruing to a group practice plan without analyzing benefits to the individual."[38]

Dr. Sidney Garfield has asserted that prepayment of medical care reduces inhibitions to use services and stimulates patients to flood the medical care system with trivial demands (people he identifies as the "worried-well").[39] However, a recent study of the Kaiser program in Oregon concluded: "The relative impact of the group of patients described as the 'worried-well' is much smaller than has been postulated. Clearly, the majority of contacts in a medical care system seem to be for legitimate sick-care services, even though it is possible that many of these visits may be handled by paramedical personnel. In addition, a large number of new morbidities disclosed by physical examinations further limits the impact of the 'worried-well.' "[40]

As to the cost of examinations in relation to their worth, the Group Health Cooperative of Puget Sound is reported to have developed serious doubts several years ago as to whether the expenditure of $185,000 a year for routine physical examinations for its 37,000 adult members was justifiable. GHC subsequently discontinued the examinations.[41]

While, again, these expressions of the cost-effectiveness of physical examinations and health screening concern the population at large, they are, nonetheless, of interest to employers and labor unions. Certainly not everyone would be in complete agreement with the opinions expressed. These opinions do, however, provide food for thought. In certain instances they furnish some degree of guidance to employers and unions as to the nature or content of screening programs. Obviously more definitive knowledge of the effectiveness of examinations and screening techniques is sorely needed so that an adequate appraisal, including that of their cost-effectiveness, might be forthcoming.

REFERENCES

1. Burton A. Weisbrod, "Does Better Health Pay?" *Public Health Reports,* June 1960.
2. Dorothy P. Rice, "The Economic Value of Human Life," *American Journal of Public Health,* November 1967.
3. _____, "The Direct and Indirect Cost of Illness," Joint Economic Committee, Congress of the United States, 1968.
4. _____, "Measurement and Application of Illness Costs," *Public Health Reports,* February 1969.

5. Wendyce H. Brody, "Economic Value of a Housewife," HEW Research and Statistics Note 9-75, August 8, 1975.
6. Herbert E. Klarman, *The Economics of Health* (New York: Columbia University Press, 1965).
7. Rashi Fein, "On Measuring Economic Benefits of Health Programmes," *Medical History and Medical Care*, 13.
8. ———, paper delivered before the New York Academy of Medicine, April 25, 1974.
9. Robert N. Grosse, "Cost-Benefit Analysis of Health Service," *Annals of the American Academy of Political Science*, January 1972.
10. Alan Williams, "The Cost-Benefit Analysis and Public Expenditures," *British Medical Bulletin*, 1974.
11. *The Nation's Health*, November 1975.
12. H. E. Klarman, "Cost Effectiveness Analysis Applied to the Treatment of Chronic Renal Disease," *Medical Care*, January-February 1968.
13. Klarman and Guzick, "Economics of Influenza," New York University, 1975.
14. John P. Davis, "Cost-Benefit Analysis of a Screening Program Followed by Risk Factor Control for Coronary Heart Disease," paper presented at Analytic Services, Inc., August 18-22, 1969.
15. Dorothy P. Rice, Testimony before the Senate Subcommittee on Health of the Elderly, September 20-22, 1966.
16. Subcommittee on Economy in Government, Joint Economic Committee, U.S. Congress, "The Analysis and Evaluation of Public Expenditures: PPB System," 1969.
17. *Employee Benefit Plan Review*, 326-2, 9-70 revised.
18. Klarman, "Cost Effective Analysis Applied to the Treatment of Chronic Renal Disease," *Medical Care*, January-February 1968.
19. Alan O. Feingold, M.D., "Cost Effectiveness of Screening for Tuberculosis in a General Medical Clinic," *Public Health Reports*, November-December 1975.
20. Horwitz and Darrow, "Principles and Effects of Mass Screening: Danish Experience in Tuberculosis Screening," *Public Health Reports*, March-April 1976.
21. World Bank, *Health*, March 1975.
22. Klarman and Guzick, op. cit.
23. Report of the Inter-Society Commission for Health Disease Resources, New York, 1972.
24. Jeremiah Stamler, M.D., et al., "Hypertension," Northwestern University Medical School, circa 1974.
25. A. W. Asscher, M.D., "The Early Diagnosis of Urinary Tract Infections" (London: Office of Health Economics, 1965).
26. Axnick, Shavell, and Witte, *Public Health Reports*, August 1969.
27. Witte and Axnick, "The Benefits from Ten Years of Measles Immunization in the United States," HEW, Center for Disease Control, undated.

28. HEW, *Forward Plan for Health,* August 1976.
29. Schoenbaum et al., "Benefit-Cost Analysis of Rubella Vaccination Policy," *New England Journal of Medicine,* February 5, 1976.
30. Doherty and Powell, "An Economic Analysis of the Cost-Effectiveness of Fluoridation Programs," University of Connecticut, Storrs, undated.
31. Dr. Ernst Wynder, *The Nation's Health,* June 1976.
32. Friedman et al., *Medical Care,* December 1974.
33. *The Nation's Health,* April 1976.
34. G. Teeling-Smith, "Economic Aspects of Screening Programs," *Proceedings of the Royal Society of Medicine,* August 1968.
35. Harry E. Emlet, Jr., "A Preliminary Exploration of Cost-Benefits of Multiphasic Health Screening," paper presented at Analytic Services, Inc., August 6, 1968. Also: "The Use of Cost-Benefit Analysis in Solutions to National Health Problems, May 1-3, 1968; and "Four Additional Facts Affecting the Cost-Benefits of Multiphasic Health Screening," August 18, 1969.
36. *The New York Times,* July 25, 1976.
37. Stuart O. Schweitzer, "Cost Effectiveness of Early Disease Detection," *Health Services Research,* Spring 1974.
38. Scott I. Allen, paper published by the Group Health Association, Washington, D.C., undated.
39. S. R. Garfield, M.D., "The Delivery of Medical Care," *Scientific American,* April 1970.
40. Jackson and Greenlick, "The Worried-Well Revisited," paper presented before the American Public Health Association, November 5, 1973.
41. *Employee Benefit Plan Review,* 326-1, 9-70 revised.

26

The Cost-Effectiveness of Industrial Health Programs

The pointed interest in this book is, of course, the cost-effectiveness of all aspects of a broadly conceived industrial health program. When the breadth of American industry is considered, in terms of its diversification, its efficiency, the investment and operating expenses involved, and the numbers of people employed, it is astonishing that so relatively little is known of the cost-effectiveness of industrial health programs, even where quite comprehensive programs are in operation and have been for some time. In no other aspect of its operation would a business enterprise (or even government or nonprofit place of employment) tolerate knowing so little about what is being done and what is being accomplished in economic terms.

David V. McCallum, safety engineer at Eastman Kodak, said that "we have never really addressed the cost of human destruction as a result of health and safety hazards." He argues that had the cost of prevention been included in the price of a machine, product, facility, or service in the first place, the far larger cost (to employers, health and welfare funds, insurance programs, and publicly financed programs) arising from accidents and health hazards would have been eliminated. "Such waste," he says, "is inexcusable." He contends:

Once the cost of prevention is recognized as infinitesimal when compared to the high price of restoration, disability, or dependent welfare, then economics will provide incentives to assure for the obviously needed remedial measures which prevent injury in the first place.... The absence or lack of economic analysis has been the major underlying reason for neglect of emphasis on preventive health and safety measures.... We fail to examine total economics and are inclined to consider safety and health in terms of initial expense rather than measuring the cost of safeguards against losses which can be incurred by their omission . . . we are finding out . . . that the cost of incorporating safety and health in design is far less expensive than the losses we are experiencing when these items are overlooked as fundamental priorities in design.[1]

Similarly a study released by The Conference Board in 1974 commented that industry has made little effort to validate the cost-benefit assumptions behind in-house employee health care programs, or even to subject them to traditional management standards of accountability.[2]

To be discussed here are the concept of the cost-effectiveness of industrial health programs, the problems encountered in arriving at measurements of the cost-effectiveness of such programs, some actual findings of such cost-effectiveness, and the need for better information.

THE CONCEPT

The paramount concept of an industrial health program is, or should be, a humane one: that of the health and well-being of employees, of their dependents, and of the community at large. The purpose and the goal should be the reduction of accidents and illnesses, of disabilities, and of premature deaths to the irreducible minimum. From an economic standpoint the concept is one of reducing absenteeism, of decreasing labor turnover and the cost of training new personnel, of reducing workmen's compensation claims and insurance costs, and of relieving the drain on health and welfare funds, all of which are measurable. Beyond this, but certainly of equal importance, are such economically unmeasurable elements related to an industrial health program as increased efficiency and production, improved employee morale and management-labor relations, and bettered public relations through the exemplification of corporate social responsibility in the welfare not only of employees and their families but of the entire community as well.

Evidences of such an economic concept are readily to be found. A quarter of a century ago a report to the president of the United States said that 75 percent of the employers surveyed found in-plant medical programs useful in job placement, in the promotion of safety, and in improved employer-employee relations. The report commented: "Although indirect benefits cannot be measured easily, they constitute an important part of the value of an in-plant medical program."[3]

The National Association of Manufacturers a few years later said that a forward-looking industrial health program can:

> Improve employee health and working conditions.
> Increase efficiency.
> Reduce time lost due to illness.
> Reduce the incidence of accidents.
> Reduce workmen's compensation claims.
> Reduce labor turnover sharply.
> Increase production in relation to payroll dollars.
> Increase customer satisfaction.
> Improve public relations status.
> Improve employee-management relationship and employee morale.[4]

In 1960 the National Health Council said: "Many an occupational health program actually produces savings far greater than its costs. . . . Numerous studies have shown that [industrial] health services costs are largely offset—and often more than made up for—by the measurable monetary savings that health services made possible."[5]

The following year a safety guide of the Manufacturing Chemists Association made the point that off-the-job safety is an essential component of an effective industrial health program and that it is good economy for a business organization or other place of employment since operating costs are lowered by reductions in the need for substitute personnel and personnel training, by reductions in product spoilage and damaged equipment accompanying the training of new employees, by reduced workloads for supervisory personnel, by reduced health and disability-insurance claims, by reduced payments from the salary-continuance program, by reduced life and accident insurance claims, and by improved employee morale and public relations. Additionally, the guide noted, there is less drain on community health resources and facilities, on rehabilitation agencies, and on various types of welfare agencies. The effect on the health and welfare of the employee and family was also noted.[6]

In 1963 Kenneth H. Klipstein, then president of the American Cyanamid Company, noting the progress that has been made in controlling disease, said: "Reduced death rate and reduced work time due to illness have contributed huge economic benefits. One study by Arthur D.

Little, Incorporated indicates that some of these measurable savings amount to more than $7 billion a year."[7]

In obtaining the views of business leaders in 1972 The Conference Board found that several spoke of the cost benefits of a strong preventive medical and health maintenance program, including the view that in-plant health facilities can minimize the work time lost for minor medical attention, can provide prompt care in the case of health problems of a serious nature, and can retain a maximum control over such problems as malingering, unnecessary treatment, absenteeism, and the level of medical fees. There was a view that an expansion of in-house medical services in many situations is a more effective and less costly solution than relying primarily on outside services.[8]

That there is reasonableness underlying such concepts is evident from what was shown in Chapter 6 concerning the reductions among employed persons in relatively recent years in the incidence of premature death, labor turnover, and absenteeism and in the costs of salary-continuance benefits, workmen's compensation claims, and the payment of health and disability insurance benefits.

THE PROBLEM

The problem, as has been indicated earlier, is that of adequately evaluating the economic concept of industrial health programs—in a word, of determining their cost-effectiveness. There are, of course, some underlying conceptual problems that serve as a hindrance to the development of such knowledge. One of these is the now long-established traditional concept that industrial health is principally concerned with industrial accidents. Another is inherent in the view that insurance coverages or other employee benefits are provided for both occupational and nonoccupational accidents, illnesses, disabilities, and premature deaths, that these are an assumed element in the costs of operation (and as such are tax-deductible), and therefore that any obligations on the part of the employer have been disposed of.

Beyond these, the methodological difficulties in evaluating the cost-effectiveness of an industrial health program, the problems of deciding which factors should be included, the values to be placed on each, and retrieval of data for a period of time prior to the inception of the program for comparative purposes are very real problems. Statistical gathering is a costly endeavor and choices must be made as to the relative value of data gathered vis-à-vis such costs. The collection of comparative data after a statistical process has been established is difficult since many factors involved can be subject to change with time: the industrial processes, materials, and equipment used; the plant design and layout; the makeup of the workforce by such factors as age, sex, educational level, and place of residence; the amount of overtime work; and changes in union contracts. And finally there is the problem of comparing the findings in one plant with those in another of similar type, particularly since probably no two places of employment at present maintain statistics, if they have them at all, on a comparable basis.

It should be noted, furthermore, that an industrial health program involves many elements of cost other than those of the medical department. Included are such matters as the time of executives, personnel staff, and supervisors; the costs of engineers, chemists, architects, and specialized consultants; the cost of safety devices and equipment, of restructuring the plant layout, of adequate ventilation and lighting, and of substituted processes, equipment, or materials; the cost of training workers; the cost of health education; the cost of the health and disability insurance and salary-continuance programs; and the cost of workers' time off for medical department consultation. Beyond this, the various costs of improving the health of the adjacent community are to be considered.

Some interesting examples of the recognition of various aspects of these problems are available. In 1960 the National Health Council noted that the costs of an industrial health program should be documented, including expenditures for personnel, supplies, amortization of equipment, the fair rental value of the space assigned to the medical department, and its share of the general overhead. The number of preplacement and periodic examinations, of treatments, of laboratory tests, and of plant inspections was noted as being among the elements that should be included in such cost appraisals. Against these costs, it was felt, should be detailed evidence of the effects produced for the dollars spent, including the effects on absentee rates, injury rates, workmen's compensation claims, and other measurable results; the Council recognized that because an industrial health program is preventive in nature, its value cannot always be demonstrated in dollars and cents.

In 1972 the President's Report on Occupational Safety and Health commented: "In 1972 the National Safety Council, under contract with OSHA, completed a study to determine whether cost-benefit analyses could be made at the establishment level to show the benefit of effective safety and health programs in economic terms. The National Safety Council concluded, after surveying responses from over 4,200 firms, that information on items of direct costs of accidents and the costs of safety programs is generally available at the establishment level, but that the indirect costs of accidents generally are not. Previous accident cost studies indicate that indirect costs are at least as great as direct costs, probably greater, and the lack of this information means that it is not feasible to perform cost-benefit analyses at the establishment level at this time." That report, in discussing the OSHA attempts to measure the cost of compliance with new standards in relation to the safety and health benefits provided, said: "Economic-impact studies are not traditional cost-effectiveness evaluations. They contain inherent limitations—the difficulty of economic prediction; the paucity of reliable injury and illness data; the understandable inability to objectively quantify pain, suffering, and loss of bodily function and loss of life. Impact studies are, nonetheless, valuable safeguards against adopting standards that might otherwise result in unnecessarily high compliance costs to employers."

The Conference Board in 1974 found that three in four employers had made no attempt to evaluate the effectiveness of their nonoccupational care programs. Of those that had, a few mentioned attempts at cost-benefit analysis. Some reported statistical analyses that verified benefits but without relating them to costs. A few reported having a healthier workforce with less loss of time, a reduction in absenteeism, and increased productivity as a result of their programs. A few reported that evaluations were either planned or pending. The inadequacy of evaluation to date was laid to such matters as the difficulty in conducting such research, particularly on a comparative basis with conditions before the program was initiated; the problem of determining short-term gains in comparison with long-term gains and relating these to changing health conditions unrelated to the industrial health program; some ambiguity concerning the aims of the program; and a reaction that an industrial health program includes compassionate goals as well as profitability.

In 1975 Ronald W. Conley, in examining the cost benefit of rehabilitation programs, noted that such programs have an extremely wide range of services and concluded that present methods for their measurement are fraught with serious limitations. He said that "cost-benefit analyses have promised much [but that] in practice they have delivered little but confusing and contradictory benefit-cost ratios, hedged with many doubts and frequently punctuated with the caveat that these ratios are illustrative, preliminary, tentative, speculative (i.e. not to be taken seriously)."[9]

The type of problem that can face industry in coping with occupational health hazards has been evidenced by a discussion at the Harvard School of Public Health in 1976 of vinyl chloride in relation to the health of workers. Noting the relationship to the development of a rare neoplasm, angiosarcoma of the liver, and that 350,000 workers today are exposed to a potential hazard in the production of vinyl chloride and the fabrication of plastic products, the panel recognized that if the polyvinyl chloride products industry were to be shut down prior to substitutes being found, some 2 million jobs and $75 billion worth of production would be lost.[10] Such an example makes evident the potential economic effects of such considerations on the workers within an industry, on the industry itself, on the consumer, and, eventually, on the national economy. The choices presented by such problems are, to say the least, unfortunate.

Yet another aspect of the problem of cost-effectiveness studies was recognized in 1976 by Nicholas A. Ashford. Defining cost-benefit analysis as "a methodology for evaluating and comparing the consequences of different policy alternatives," Mr. Ashford noted that there are a variety of approaches to using such analyses and recognized that the goals must be specified and that the possible alternatives must be considered. He took cognizance of the limitations inherent in cost-benefit analyses, including the nonmeasurable factors, either costs or benefits. He also noted that estimating the costs and benefits of occupational injuries is far simpler than such efforts as respects occupational disease, since the occupational cause of disease is often difficult to determine, particularly if multiple causes are involved, and since its onset is very frequently delayed for many years, or even generations. Mr. Ashford commented:

> Attempts at formal cost-benefit analysis have been far less influential in management decisions to improve job health and safety practices than the threat of fines imposed under the OSH Act for violations found during an inspection, or the pressure of collective bargaining. Cost-benefit analysis is even less likely to be effective in altering the behavior of workers; nor does the OSH Act impose enforceable duties on workers.[11]

Such retardants to the measurement of cost-effectiveness are not isolated instances, as evidenced by the responses to the author's survey of 15 medical directors of large major corporations with quite developed industrial health programs. The responses quoted here are from those corporations that do not at present maintain cost-effectiveness data or whose data the respondents do not consider to be meaningful. These corporations were diversified and included manufacturing, utilities, communications, publishing, and insurance.

> We do not have the very detailed data available . . . in fact, we would not expect to have the type of cost-effectiveness data you seek for some years yet.

> Our company health programs, including our alcoholism activity, lack definitive information on absenteeism, turnover, and so on as certain job classifications (that is, office versus factory employees) come under different record keeping.

Even though we have had a program for some years we have not as yet been able to put together any cost-effectiveness program. We are at present working on methods to obtain this information but it is too early for any worthwhile figures.

While we have had a medical program for 60 years . . . there is no way that we can measure the effectiveness on savings in absenteeism, labor turnover, and so forth since we have no records for periods without a program. . . . While we are, in some cases, able to isolate the cost of a specific phase of our health programs, such as conducting an electrocardiogram examination, we are unable to separate the overall cost of the program from the operation of the business as a whole.

It has been many years since we undertook a study of medical service costs. Our management control is on workload, staffing, and types of services.

Starting in 1970, a retrenchment program was instituted . . . because of financial necessity. Part of this program did away with our computerized retrieval system. As a result we are unable to provide information concerning cost-effectiveness of our industrial health program.

We are engaged in a reorganization of our health program . . . and have found it necessary to review costs and project them for future budgeting. Unfortunately, we have no specific information on the effectiveness of the program regarding absenteeism, health insurance claims, etc.

The cost-effectiveness of occupational health programs has not been studied adequately.

This department has never made any detailed in-depth study of the cost-effectiveness of our industrial health program.

Unfortunately, I have no information on cost-effectiveness of industrial health programs. . . . I personally believe that such figures are impossible to obtain accurately. For example, when analyzing absenteeism due to, say, flu or the effectiveness of flu shots, it's difficult to tell whether a person actually has the flu or just a bad cold, whether he would have gotten the flu even if he hadn't received the inoculation, or whether two people are absent the same amount of time . . . given the fact that they have the same illness. In general, I believe industrial health programs do have a positive effect . . . but I have given up trying to quantify it in dollars.

The very nature of industrial health programs . . . makes it difficult or impossible to obtain convincing evidence as to the cost-effectiveness of these programs . . . we have not extracted any data from our experience which would be useful.

Many attempts have been made to justify the existence of medical, industrial hygiene, and safety programs in industry.... I do not believe these programs lend themselves to cost-effectiveness analysis.... Several attempts have been made to quantify the cost-effectiveness of executive health-maintenance examinations... but what value should be attributed to the diagnosis of obesity, hypertension, or sociopathic personality is more difficult to determine.

It is virtually impossible to provide the information you asked for... it is so extensive that there is no means at our disposal of providing that information... it would mean assigning a number of people... for a great amount of time... and we just, at this time, do not have the manpower or time.

The data are not easily obtained and their value is significantly decreased when compared with data from other organizations unless care is exercised to assure that the basis of the data are truly comparable.... It is my belief that the more effective occupational health programs are based upon the responsibility of management to provide a safe working environment.

The information you seek is not readily available from an insurance carrier and is not accessible in our records... we have no cost figures for our policyholders' industrial health programs, nor do we have any way of quantifying the cost of not preventing ill health.

The question is quite naturally posed: What information and experience do health insurers have that would throw light on the question of the cost-effectiveness of industrial health programs or of preventive medicine? For both insurance companies and the Blue Cross–Blue Shield plans the answer is: nothing. (The findings at the Kaiser Foundation Health Plan are discussed subsequently.)

In an endeavor to gather information for this book, the author addressed queries to 18 of the larger insurance companies, which write over half the health insurance protection provided by insurance companies. The queries were limited to just one aspect of preventive medicine: insurance coverages for the cost of physical examinations. The following was found:

In some companies no cost-effectiveness data are available. The premium charged by one company anticipates an annual utilization of 50 percent. Another anticipates the following utilization rate:

Adults	65%
Children age 0 to 6 months	100%
Children age 6 months to 5 years	75%
Children age 5 years and over	50%

One company experienced a 12 percent utilization during the first six months the coverage was in force. Another company reports that on one insured group of 225 lives the utilization has been heavy but that actual data are not available. Yet another company said that it does not actively encourage the coverage because of the unknown utilization at risk.

With respect to the effect of such examinations, where they are covered in the insurance benefits no data are available on the medical expenses paid by the insurance company. The following are some divergent impressions:

> Medical expenditures would increase as would overall costs.
> Utilization would not be affected.
> Multiphasic screening invites utilization.
> Experience with multiphasic screening produced a slight adverse swing under major medical expense insurance coverages.
> Minor illness claims could be increased.
> A higher level of claims may be established on a continuing basis.
> Utilization would rise, then fall.
> Overall medical costs may eventually be reduced.
> Immunization coverage does not increase utilization.
> Utilization would eventually diminish.

Where such examinations are covered, again no data are available on the amount of disability-insurance claims paid. Four companies, however, expected that physical examinations would reduce such claims; one felt that they would have no effect. (It was noted that for years insurance underwriters have assumed that preemployment examinations have a beneficial effect on disability-insurance loss ratios.)

Invited comments produced the following observations:

- The coverage is inappropriate except as part of an all-inclusive comprehensive medical care package, such as HMOs.
- The use of preventive services can largely be anticipated and budgeted for through the use of insurance.
- In the long run such coverages should keep employees in a better state of health and hopefully reduce insurance losses.
- Such coverages will tend to lower losses as this form of insurance becomes more universal.
- Such coverages are socially desirable and should lessen premature deaths and disability.
- Physical examinations, however financed, should have a positive effect on productivity and absenteeism (although one study of the British NHS showed that 25 years of experience displays no reduction in absenteeism).

The findings of this survey, then, reveal very little, although what is shown is significant and of interest. These findings, it must be noted, have to be qualified in terms of the relative recentness of such insurance coverages, as well as the small amount of such coverages purchased to date. It is unfortunate that it is not yet feasible for insurers to engage in cost-effectiveness explorations based on their own experience. The probability of the reduction of health insurance claims through such preventive measures as physical examinations and other aspects of industrial health programs is such that study of the subject is merited. Required would be a retrospective claims analysis and a study of future claims over a period of time to determine the degree of savings generated, if any.

It is conceivable that one research effort under way may serve to throw some light on the subject. The Metropolitan Life Insurance Company has contributed funds to finance a research project at the Washington University Demonstration Medical Care Program in St. Louis. The project, which commenced in 1973, has been described as the "first controlled research study to examine and measure the effects of ambulatory and preventive care on the rate of hospitalization by a group of persons enrolled in a prepaid service program as compared with a carefully matched control group of families receiving health services under the traditional fee-for-service method."[12]

SOME FINDINGS

Despite what has been said in the preceding section, there are some findings of studies of the cost-effectiveness of industrial health programs, or of specific aspects of such programs, that provide valuable information for all places of employment. While these findings are noncomparable, since the methodology employed differs from case to case, since most examine only one or two aspects of a program, and since some are hardly more than informed impressions, they nonetheless are of significance in the indications they give of the cost-effectiveness that can be inherent in industrial health programs.

Thus a study at the University of Michigan two decades ago found that among machinery manufacturing firms, those companies not having industrial health programs paid $.93 for every $100 of payroll for workmen's compensation insurance, while those having such programs paid only $.56 per $100 of payroll. The savings in such insurance costs alone were more than double the cost of the medical and nursing personnel. The study also found that among general merchandise stores the costs of workmen's compensation was $.54 per $100 payroll for those not having industrial health programs compared to $.37 per $100 payroll for those with such programs.[13]

About that same time a survey conducted by the National Association of Manufacturers found that of 1,625 industrial health programs, all but five considered their programs "paying propositions." Over 90 percent of the employers in the survey experienced a reduction in accident frequency, in the incidence of occupational diseases, in absenteeism, and in insurance premiums. Labor turnover, on the average, dropped 27.3 percent; absenteeism declined 29.7 percent; and workmen's compensation premiums dropped 28.8 percent.[14]

The following are some examples of the cost-effectiveness of specific industrial health programs at larger places of employment.

In 1949 it was reported of *Socony-Vacuum Oil Company's* East River plant, with 1,500 employees, that as a result of an in-plant medical department established in 1946 there had been a substantial decrease in working days lost and in the amount of sickness benefits paid. Disability payments decreased from $115,726 in 1946 to $66,565 in 1949. Mandays lost from nonoccupational illness decreased from 10,466 in 1947 to 5,435 in 1949.[15]

At the *Denver Federal Center* in 1960 a health service for the 3,425 federal employees at the Center cost $38,000 a year. In its first year of operation, sick leaves declined by more than 20,000 hours, producing a saving of $32,800 in that one area of concern alone. For federal employees not covered by the program sick leave at that same time increased 13 percent.[16]

In 1961 the program at the *Olin Mathieson Company* reduced workmen's compensation costs alone by $28,000 a year. It also reduced labor turnover, absenteeism, and time lost from accidents. At the company's North Carolina plant, with 2,400 employees, an industrial health program was established in 1950. By 1959 workmen's compensation claims were half the rate for all other places of employment in North Carolina, health insurance claims were reduced 30 percent, absenteeism was reduced 10 percent, and the rate of labor turnover was half the national average. In addition, lost-time accident cases declined from 60 a year to 9 (one in 1954), days lost from accidents declined from 7,000 a year to 4,800, accident frequency dropped from 18 to 3, and the accident severity rate dropped from 2,000 to 1,400 (none in 1954).[17] (The reason for the especially good record in 1954 was not made evident.)

The following was shown concerning the cost-effectiveness of the program instituted at the end of 1963 at the *Mountain States Telephone Company*, with 27,000 employees in 1967. In the five years preceding establishment of the program sickness-disability benefits paid by the company for nonoccupational illnesses and injuries averaged $20.73 per $1,000 of payroll. In the four years following the inauguration of the program these payments dropped to $17.79. The total days of absence because of sickness per employee per year dropped from 10.03 to 5.92, over a 40 percent decrease. The dollar savings for the year 1967 for nonoccupational disability benefits was $894,999. For job-related injuries, for which the company pays disability benefits usually higher than those required by workmen's compensation, as well as complete medical care, the cost per year dropped from $0.86 per $1,000 of payroll to $0.61. That figure represented a decrease of 25 percent and produced an estimated saving of $169,839 from the period 1964 to 1967. Other estimates of savings brought about by the industrial health program were shown to be:

- $6,560 a year from in-house treatment of occupational injuries, not including the saving in employee time if treatment had occurred elsewhere.
- $2,148 a year from the in-house treatment of nonoccupational injuries (most often occurring at home prior to leaving for work or while en route to work) in terms of the employee time that would have been lost had treatment occurred elsewhere.
- $13,311 a year in work time saved as a result of the in-house treatment of nonoccupational illnesses. Savings in the payment of sickness-disability benefits were not identified, nor was the value of increased employee morale resulting from such services estimated in dollars.
- $2,697 a year in work time saved by providing injections (allergens for hyposensitization, hormones, iron, vitamins) at the request of the private physician. Again, improved employee morale was not evaluated.
- $17,640 a year through the conduct of in-house preemployment examinations rather than having them done by other sources.
- $5,720 a year from the conduct of in-house physical examinations of long-service and executive employees rather than having them done by other sources.

Savings in sickness-absence payments for the years 1964 to 1967 were estimated to be $2,329,811.

The cost-effectiveness of the fitness examinations for employees whose work performance reflected a health condition (poor motivation, poor personal discipline, mental illness, or alcoholism) was not evaluated. Also not evaluated, because there was no reasonable way found to estimate the dollar value, were such medical staff activities as consultation on environmental conditions in work areas, maintenance of knowledge of job requirements, and assistance in the process of rehabilitation. The balancing cost of the program was not shown except for the program in

THE COST-EFFECTIVENESS OF INDUSTRIAL HEALTH PROGRAMS 381

Denver (see Chapter 18), where the costs of the medical department for the years 1964 to 1967 were $321,666 and the estimated gains were $204,784. Costs for medical care and disability due to occupational injuries decreased over a four-year period, producing a saving of $169,839.

It was also recognized that other factors (for example, improved job design, increased attention by supervisors, an improved safety program, or improved legal or administrative handling of occupational injury cases) can have a bearing on the results obtained. Thus the study commented that "such abstract concepts as length of life, morale, and prevention of specific untoward events relating to health have no precise monetary values and if we try to make such measurements, we risk losing . . . compassion for human values.[18]

The *Norton Company* at Worcester, Massachusetts, has found that its industrial health program, introduced in 1911 but considerably expanded since that time, has reduced its disability rate to the point where it is only 40 percent of the national average. The services provided for its 4,500 employees include, beyond the customary, such efforts as physical examinations, noise-level determinations, and dust, chemical, and radiation surveys.[19]

The *Gates Rubber Company*, while having no quantitative evidence, feels that its program for nonoccupational health care enhances the quality of employee health care as a result of the familiarity of the clinic physician with the conditions and stresses of the employee's work, that it discourages unnecessary hospitalization, that it results in a reduction in time lost as a result of illness, that workmen's compensation and sickness-benefit payments are reduced, and that employees are less likely to have prolonged and costly illnesses as a result of periodic physical examinations and the prompt receipt of necessary care.[20]

With respect to the programs of smaller places of employment, the following findings some years ago are of interest.

- One small firm provided part-time nursing services to its employees at a yearly cost of $1,000. The first year, workmen's compensation losses were reduced $4,700.[21]
- A plant with 600 employees saved twice the cost of its industrial health program in its first year of operation.
- A company with 300 employees established an industrial health program. The result was a 61 percent decrease in workmen's compensation costs.[22]
- A heavy foundry with 200 employees that had had a yearly accident-frequency rate of 108 established an industrial health program in 1947. By 1950 the accident rate was zero.

- In absenteeism costs alone, one small plant saved twice the cost of its industrial health program.
- A plant with 115 employees installed a dispensary and thereby reduced labor turnover 25 percent.[23]
- In a plant with 160 employees, workmen's compensation claims were reduced $4,500 during the first year of operation of its industrial health program. The cost of the medical department was $1,200.
- In 1946 the *Allen Manufacturing Company* joined the Hartford Small Plant Medical Service, a cooperative effort for making industrial health services available. In the following six years, labor turnover at the plant was reduced 87 percent, workmen's compensation premiums were reduced 24 percent, and the incidence of occupational accidents dropped to zero.
- A plant with 220 employees reduced workmen's compensation losses $3,300 after installation of an industrial health program, which cost $2,500.
- At a plant with 115 employees direct savings from absenteeism alone were more than double the cost of its industrial health program.
- In one small plant the lost-time rate from occupational accidents was reduced from 39.5 percent to 7 percent in one year as a result of the installation of an industrial health program.
- A plant with 175 employees employed a nurse. In one year there was a 52 percent reduction in lost-time accident frequency and an 82 percent reduction in lost-time accident severity.
- Another small plant, which became associated with the Hartford Small Plant Medical Service, reduced eye-injury cases 50 percent, reduced accident frequency 50 percent, reduced lost time 55 percent, and in one year reduced the cost of physician services $1,900.[24]

While many of the gains reported in the foregoing (reductions in labor turnover and absenteeism, improved employee morale, increased efficiency and production) are not given in dollars, their economic value alone to a place of employment is readily apparent.

There is also available some concrete information concerning the cost-effectiveness of specific aspects of an industrial health program.

Safety Programs

One such specific aspect is that of the establishment of a safety program. The following are certain of the findings.

At *Bethlehem Steel* a safety program was formalized in 1916. Since that time serious injuries to employees have declined 95 percent. In 1974 the company invested some $5 million in protective equipment and other safety appliances.[25]

THE COST-EFFECTIVENESS OF INDUSTRIAL HEALTH PROGRAMS   383

At *Western Electric*, as a result of a safety program, 19,463,000 man-hours of work were performed without a disabling injury by 1960.

At the *Convair* plant in San Diego 21,814,000 man-hours of work were performed without a major injury by 1960 as a result of its safety program.

At the *IBM* plant in Poughkeepsie, New York, there had been no disabling injury in 17,600,000 man-hours of work by 1960 as a consequence of its safety program.

At the *Du Pont* rayon plant at Old Hickory, Tennessee, there had been no disabling injuries in 29 million working hours by 1960 as the result of a safety program.

At the *W. E. Upjohn Company* a safety program was started in 1949, and in 1967 an industrial hygiene program was developed. Disabilities per million hours worked declined from 12.1 in 1950 to 5.4 in 1975 (with a low point of 3.0 in 1971). Days lost per million hours worked were 221 in 1950 and 284 in 1951; in 1974 they were 55 and in 1975 they were 72 (with a low of 43 in 1969). Table 26-1 shows the figures for three specific years. The reason for the adverse experience in very recent years is not known.[26]

Table 26-1. Results of the W. E. Upjohn Company's industrial health program.

	1968	1972	1975
Hours worked (in thousands)	8,250	8,412	11,361
Lost-time disabilities	63	32	61
Disabilities per million hours worked	7.0	3.8	5.4
Days lost	1,343	520	822
Days lost per million hours worked	163	61.8	72.3
Average days lost time per disability	22	16.2	12.4
Outside medical, compensation, and wage cost	$51,750	$47,064	$92,577
Major accidents		103	183
Accident frequency rate per million hours worked		12.2	16.1
Personal injuries		642	1,079
Frequency rate per million hours worked		76.3	94.9

Boise Cascade has 29,000 employees plus those at foreign installations. In 1972 the company operated 21 lumber mills, 19 plywood plants, 19 corrugated container plants, 25 composite can plants, 6 kitchen cabinet plants, 13 pulp and paper mills, 13 manufactured housing plants, 3 particleboard plants, 2 door plants, 2 fiberboard plants, 5 envelope plants, 6 paper distribution centers, 23 office products distribution centers, 20 building service centers, and 26 wholesale distribution centers. In 1972 the company became concerned about safety. Working with its insurer, the Employers of Wausau, the company developed an accident-prevention plan under policy established by management. Safety directors and industrial hygienists were employed, a safety manual was developed, training was provided to all plant managers, and a safety training program instituted, including monitoring of hazards, performance evaluation, the collection and analysis of accident statistics, and the measurement of the effectiveness of the program. By 1975 there were 1,668 fewer accidents than in 1973, a 33 percent reduction, producing a saving to the company of $2.1 million, based on an average cost per accident of $1,255 in direct costs and $5,020 in indirect costs. Fatalities decreased 33 percent in 1974 and 67 percent in 1975. For these same years total accidents decreased 19 and 17 percent respectively, accidents resulting in work loss decreased 20 and 7 percent respectively, and accidents without work loss decreased 19 and 22 percent respectively.[27]

That progress has been made in industrial safety has been shown in a report of the Manufacturing Chemists Association based on a survey of employers between 1946 and 1975. Over that period of time the employee hours of exposure in the firms surveyed increased 250 percent while deaths from injuries increased almost 100 percent (50 deaths among 85 firms in 1975), permanent total disabilities doubled (2 in 1975), permanent partial disabilities declined one-sixth (125 in 1975), temporary total disabilities decreased slightly (3,275 in 1975), the accident-frequency rate dropped from 7.65 in 1946 to 3.36 in 1975, and the severity rate dropped from 689 to 433. The frequency rate for small employers (1 to 250 employees) was 22.61 in 1975, compared to 2.41 for firms with over 5,000 employees. The severity rate for small employers in that year was 605, for the large employers it was 412 (however, it was 219 for firms employing from 250 to 1,000 employees).[28]

Home Health Care

James L. Bischoff of *Upjohn Company* has discussed the benefits that can accrue from including home health care benefits in employee health insurance coverages. He maintains that savings in the cost of health care result, that lengths of hospitalizations are reduced, that there are savings when one considers the time employees might need to take from the job to visit family members who otherwise would be hospitalized, and that an improved climate of employer-employee relations results. Naturally, a prerequisite for such a program is the existence in the community of a qualified home health care agency. Citing the experience in Rochester, New York, Mr. Bischoff says that *Eastman Kodak,* as a result of such action, has saved $160,000 a year in health care costs. At Upjohn, following an essentially similar approach, the company reported an average saving of $325 per case for the 12-month period ending June 30, 1975, based on an average reduction of 4.74 days of hospitalization per hospitalized case. Total cost savings were a minimum of 5 percent for hospitalized cases and from 12 to 16 percent for nursing home cases. Furthermore, the company found that through the use of home health services, employees returned to work faster.[29]

Physical Examinations

Another specific aspect of an industrial health program on which some cost-effectiveness information is available is that of periodic physical examinations or multiphasic screening examinations for employed people.

In 1960, at a medium-size food manufacturing company that provided physical examinations for all its employees, the health insurance claims were studied by the type of medical service used by those employees, separating those that resulted from the examinations, and estimated what the medical costs would have been in the absence of the early detection provided by the examinations. It was found:

- That of 29 health insurance claims in excess of $200, 19 resulted from the examination findings.
- That in 9 of the 19 cases no saving was attributed to the early diagnosis.
- That for the remaining 10 cases there was an estimated saving of $4,791, or $479 per individual case, which otherwise would have produced medical costs of $924.10 each. In addition, it was estimated that 26 weeks of time loss were saved, and perhaps one death.[30]

The following year it was estimated that the physical examination program for the employees of the *Seamen's Bank for Savings* resulted in a reduction in absenteeism from 7 percent to below 3 percent, a reduction in group life insurance premiums to 2.7 percent of payroll, a marked gain in productivity, and improved employee morale, with a decrease in malingering. The cost of the program was 2.7 percent of payroll.[31]

In 1968, according to Dr. Emerson Day, physical examinations performed at the Strang Clinic in New York City resulted in a reduction of nearly 50 percent in workmen's compensation claims, and claims for occupational diseases were reduced 63 percent. The average expense per examinee was $74.20, with an initial examination charge of $70.

Dr. Carl Zenz, medical director of *Allis Chalmers Corporation*, said that prompt preventive steps may reduce the need for extremely high-cost medical care. He said that *Westinghouse Corporation* reports savings from its executive health examination program of at least four times the cost, that a manufacturing plant with a periodic examination program found that medical claims over a 12-year period were 10 percent to 20 percent below the expected rate, and that a Columbia City, Indiana, medical plan reported an adjusted rate of hospitalization about 60 percent of the national average because of its emphasis on preventive medicine and early detection. He said, further, that a survey of more that 20,000 examinations of male executives in several clinics estimated a reduction of the hazard of death of nearly 50 percent, and when subjects had multiple examinations the hazard of death was further reduced.[32]

The survey of insurance companies previously mentioned revealed that those companies do not have cost-effectiveness information available as respects physical examinations. As was shown, their impressions of the effects of such examinations on health insurance claims ran both ways.

With respect to multiphasic screening examinations, Richard Clinchy, president of *Del-Val Financial Planners,* Bethlehem, Pennsylvania, has said that multiphasic screening provides a new benefit for the individual in a participating group unlike any other fringe benefit in that it is preventive in nature. He maintains that testing offers some assistance in the area of controlling health insurance costs and health maintenance expenditures, that hospital utilization and total dollars expended might be significantly reduced, and that ultimately a reduction in morbidity and mortality should result.

As substantiation Clinchy cites a case where, "after accounting for the cost of health screening, the average expense for medical claims alone for the unscreened employees was $237 per year higher than for those employees who had the screening benefit." In another company (the *Century Electric*) it was found that "claims incurred by the unscreened group should total $117,858 versus claims incurred by the screened persons of $35,067. This indicates a raw savings of $243, approximately, per person. When the cost of the screening program, which was $40 per man last year [1973], is included, the total saving amounts to $283 per man per year." Further cited is a pilot program of the *Shop and Mills Health and Welfare Trust* (carpenters) in St. Louis. The projected per capita savings were reported to be $190 as compared to a per capita costs for multiphasic testing of $40. The program also found that the cost savings took place immediately and that claims occurring during a 12-month period of the testing were actually smaller per claim than otherwise. The author states:

THE COST-EFFECTIVENESS OF INDUSTRIAL HEALTH PROGRAMS

There may be a period during which expenditures for the screening will balloon employee benefit costs [but] if the plan is properly used, however, this increased expense should very quickly be overshadowed by a reduction in health and Workmen's Compensation claim levels [and] therefore, both labor organizations and employers should be careful how multiphasic testing expenditures are treated since the start-up expense should be returned many times over in claim dollar savings.[33]

Robert Demsey has analyzed the criteria for decision making with respect to multiphasic health testing, including the functional objectives, the quality of the tests, the physical nature of the program (fixed or mobile), the population group to be served, integration with the medical care system (physician followup, approach to operation, health education, and costs). With respect to costs he notes that the average cost per participant varies widely, but that generally it would range from $30 to $45 for a sound testing program. These figures do not include supplemental physician followup, but do include all computer and clerical costs. Not all prevention programs are so successful. For example, there is a study by the National Association of Manufacturers of one plant employing 595 people and another plant of the same company employing 426 people, the first having an annual multiphasic screening program and the second not having such a program. In this case the resultant sickness and disability claims, including insurance premiums, have been 500 percent higher for the plant without the multiphasic program.[34]

One writer reports that several years ago a number of actuaries (not identified) felt that the short-term effect of screening would be an immediate rise in health costs as a result of the identification of illnesses, but that in the long run it would offer a control device to keep claims stabilized. This, however, did not prove to be the case. Instead, "twenty-nine cases have been documented that either an immediate drop in claims occurred or expenditures for claims had leveled out without regard to inflation. $1,200,000 was saved by one group of 7,100 lives after only two years of screening." The principal area of saving appeared to be a reduction in hospitalization for diagnostic purposes. It was additionally noted that disability-income expenditures also dropped almost in direct proportion to the reduction in medical care costs. Workmen's compensation had less significant reductions but absenteeism showed a significant drop, with an average reduction of lost time of 2.6 days a year per person, a saving that is said "to more than support the modest cost of screening."[35]

The most complete examination of the cost-effectiveness of multiphasic screening is the study under way at the Kaiser Health Foundation Plan at Oakland, California. The study was commenced in 1964 with

10,700 of the plan's members then aged 35 to 54 selected at random, practically all of whom were employed persons or their dependents. From these a study group and a control group were identified, the former being prodded to have checkups annually (the cost of examinations is included in the plan benefits). During the period 1964 to 1971, 54.3 percent of the study group had four or more examinations, compared to 13.3 percent for the control group. During each of the seven years, 60 to 70 percent of the study group were examined, contrasted with 20 to 24 percent for the control group. The most significant finding of the study was the dramatic difference between the two groups in the rate of disability for males born between 1910 and 1919. There was a significant reduction in self-rated disability and reported time lost from work and a lower self-reported utilization of medical services by the sick among those having more frequent examinations. For the younger group of males and for the females there was little difference in the incidence of disability. The study concludes that while multiphasic tests do not reduce the occurrence of chronic conditions they may result in better control of these conditions.

A significant difference was also revealed for the same group of older males with respect to mortality: the study group having 183 deaths, compared to 217 for the control group (the greatest distinction was death from cancer, 40 compared to 62; heart disease being about the same for both groups). No noticeable difference in the mortality rate was evident for females or for men born between 1920 and 1929. However, in three of the four age/sex categories, the rate of "potentially postponable" deaths (judgmental, based on hypertension, hypertensive cardiovascular disease, and certain forms of cancer) was considerably higher for the control group.

There was no statistically significant difference with respect to hospital utilization for the two groups of older men, although the findings are considered inconclusive. Women in the study group, however, had persistently greater hospitalization than those in the control group. The preliminary findings recognize that the hospitalization data have been difficult to interpret since the screening may have generated hospital utilization for therapeutic management of disabilities uncovered, but at once may have reduced hospital use if the screening resulted in prevention or postponement of a serious illness.

An economic evaluation of the findings was made for the males born between 1910 and 1919 (in which any income from the California temporary disability benefits law was ignored). It was recognized as a preliminary cost-benefit analysis. The conclusion was that annual savings to the study group averaged $142.28 per man over the seven-year period, or a total of $996 per man for the period 1964–1971. When allowance was made for the slight difference in laboratory costs and for the costs of the multiphasic screening examination (an adjusted annual cost of $23), the net actual saving for the study group was $117.42 per man per year, or a total of $822 for the seven-year period. When the per man saving between 1965 and 1971 in the study group of $54 on hospitalization expenses is added to this, the net economic gain from reduced disability, premature death, and medical care expenditures was $876 per man for the period 1964 to 1971.[36]

It should be noted with respect to this study that its authors are extremely cautious in deriving any conclusions from the data and that they feel it would be misleading to conclude that the study demonstrates beyond doubt the value of screening examinations. In speaking with others interested in preventive medicine, I have found that many questions and doubts persist with respect to the application of the findings of this study.

Rehabilitation

Workmen's compensation insurers have found that the processes of rehabilitation result in a cost-effectiveness of $5 for every $1 spent. In 1967 it was reported that the Liberty Mutual Insurance Company, which provides rehabilitation services at its own centers, was saving more than $3 million on 130 workmen's compensation cases involving injuries to the spinal cord at a cost of $1,000 per patient, with 40 percent being able to return to some form of employment. As of January 1958, of 5,107 cases treated at those centers, 87 percent were substantially improved and, of these, 81 percent returned to work. Their average age was 44. With very serious cases, such as amputees, savings of from $5,000 to $50,000 per case have been realized. For more routine cases the average estimated saving in workmen's compensation benefits was about $1,000 per case.[37]

Elmer Rasmussen of the Continental Casualty Company has also maintained that through rehabilitation, cost savings can also be achieved on major medical expense insurance coverages and disability insurance but that there must be a well-administered and -planned company program. He cites two examples of major medical coverages where $2,500 and $2,000 respectively were saved by taking appropriate rehabilitation action at insurance company expense.[38]

ALCOHOLISM CONTROL PROGRAMS

Somewhat oddly, more cost-effectiveness information is available for alcoholism control programs at places of employment than for the broader aspects of industrial health programs. One reason for this is probably the fact that to compile data on one aspect of an industrial health program is much simpler and easier (although in itself it can be quite complex) than for such a program in its entirety. Another reason is probably the encouragement and assistance provided by many organizations concerned with alcoholism, principally the National Council on Alcoholism and the National Institute on Alcohol Abuse and Alcoholism. Regardless of the reason, the following is a very brief condensation of the savings found to accrue from alcoholism control programs at places of employment (the details are contained in my book *Alcoholics and Business*).

Scovill Manufacturing Company, with 27,000 employees in 1973, has experienced a 78 percent recovery rate at an annual estimated saving of $186,550 after an average cost of $1,100 per alcoholic employee for treatment and rehabilitation.

Illinois Bell Telephone Company, with 45,000 employees, has experienced a recovery rate of 72 percent of the alcoholic cases. While no evaluation of such matters as savings in overtime pay, employee turnover, premature deaths, increased production, or insured medical expenses saved by the alcoholism control program was made, and although the costs of the program were not shown, it was found that sickness-disability cases of more than seven days dropped from 662 in the five years before the program was started to 356 in the following five years, at a saving of $459,000. In the same time period, off-the-job accidents decreased 66 percent, and job accidents decreased 80 percent.

At *General Motors* the Pontiac Division alcoholism control program saved $9,878 on disability-insurance benefits and 10,850 lost work hours each year on just 25 alcoholic employees. At the Oldsmobile Division, the program experienced a 49 percent decrease in lost man-hours, a 29 percent decrease in disability-insurance benefits, and a 56 percent decrease in leaves of absence among 117 hourly workers at a cost of $11,114.

Allis-Chalmers Corporation has experienced as a result of its alcoholism control program an estimated saving of $80,000 a year.

McDonnell Douglas Corporation, as a result of its program, has saved $3 million over a four-year period in lost productivity alone.

The New York Transit Authority saves $2 million a year; the program costs $130,000 a year.

The program for *Federal Civilian Employees* has projected a saving of $1.25 billion over a five-year period, or a saving of $17 for every $1 spent on the alcoholism control program.

The *U.S. Postal Service,* experiencing a 75 percent recovery rate through its program, estimates a potential net saving of $1,864,000 after the costs of the alcoholism control program.

An unidentified manufacturing company experiences an annual saving of $100,650, against a cost of $11,400 for its program.

These examples unfortunately are not comparable, nor are they complete evaluations of either the costs of such programs or of their economic effectiveness. They leave one wanting to know considerably more than is shown. Nonetheless, they do make it reasonably clear that industrial health programs are cost-effective from solely an economic point of view.

POLLUTION CONTROL

The cost-effectiveness of specific efforts at containing environmental pollution recently reported is also of interest. On January 19, 1977, Lewis W. Lehr, president of *3M,* said that 19 projects in the company's two-year-old "Pollution Prevention Pays" program saved the company $11 million and eliminated 73,000 tons of air pollution, 500 million gallons of polluted waste water, and 2,800 tons of sludge annually. At that same time Dr. Earle B. Barnes of *Dow Chemical Company* reported that the company's $3 million investment in air pollution control in Louisiana was yielding $100,000 a year in recovered chemicals. It has been reported by EPA that the *J. R. Simplot Company* at Caldwell, Idaho, had completely ceased massive discharge of potato-processing wastes into the Boise River by spraying the nutrients on pastureland, thereby doubling its annual forage production and supporting 26,000 cattle.[39]

THE NEED FOR BETTER INFORMATION

It is unfortunate that more abundant and conclusive evidences of the cost-effectiveness of industrial health programs do not exist. There is sore need for such information, for only by this means can unbelievers and doubters be converted, the indifferent activated, and the antagonists to the concept brought to heel. Beyond that, more comprehensive data would serve management interests by presenting them with meaningful facts about their own program, its positive aspects and its failings, and also enabling them to make comparisons with other programs that function under essentially similar circumstances.

The question is then presented: Where are those in industry, organized labor, insurance, or government to turn for information that will provide guidance to assist them in arriving at a sound concept of an industrial health program, its scope and content, and the best way in which to execute its functioning? Under present circumstances anyone who desires to proceed must do so on conjecture, albeit of an informed nature. This means placing at risk sizable sums of money and personnel resources without the customary reasonable assurance of positive economic results, measurable or unmeasurable.

In 1973 Dr. A. Walter Hoover of NIOSH commented that one of the problems facing industrial health programs is that of the development of a practical methodology for cost-benefit analysis. He noted that past studies have not readily established the certainty of a good return for expenditures and that the preventive medicine aspects have not been consistently justified on a cost-benefit basis. Because of this, the Division of Occupational Health Programs at NIOSH is presently conducting a pilot study of cost-benefit analyses.[40]

Dr. Hallet A. Lewis, director of occupational medicine at Kaiser Foundation International, has discussed a study in which he is engaged on behalf of NIOSH, the purpose of which is to develop a conceptual model for the development of cost-benefit information for industrial health programs. Need for the project, which started in July 1974, lies in the concept that cost-benefit information on occupational programs, where it is available at all, is of limited use because no standardized method has been formulated for calculating either the costs or the benefits, that few comprehensive cost analyses are available, and that even less has been done in objectively assessing the values derived from those that have been done. It is recognized that while industrial health programs have less tangible benefits, these benefits have a measurable economic value, however difficult to quantify. Therefore the project is attempting to identify the intangible benefits so that the real effects of an occupational health program may be calculated more precisely. Accordingly a dollar value is being assigned to such matters as the value of a human life and the different elements of the quality of life as these may be affected by illness or injury. The project recognizes that a well-constructed model should provide the opportunity to examine, both statistically and dynamically, responses of the system to changed conditions and should permit the accurate forecasting of the cost of an evolving occupational health program over a planning span of three to seven years.

It is recognized that the model that is developed should be applicable to a variety of enterprises and should be able to accommodate to changes imposed by future OSHA regulations. It should be suitable for automation and computerization. Included in the considerations are the composition of the employee population, the capital investment involved, the operating costs of the program, the services provided, reductions in overall medical charges, and other benefits derived from the program, including favorable employee reaction, improved work performance, reduced absenteeism, and job enrichment. The project is being tested through the occupational health program of the major steel manufacturing facility.[41] The difficulties inherent in the endeavor are well recognized. This effort should eventually be of considerable assistance to places of employment in decision making regarding an industrial health program, changes in the concept and content of such a program, in evaluating the results of a program, and in comparing both the costs and the results of a specific program with others of a similar nature.

Meanwhile, in an attempt to assist the process of evaluating an industrial health program, Dr. James L. Craig, then medical director of TVA and presently corporate medical director at General Mills, Inc., in 1974 discussed the subject of such a cost analysis. He notes that critical to a cost study is the necessity of keeping good records on program activities, including the record of each employee, the type of medical personnel who interviews the employee, the nature of the interview, whether or not the condition is job-related, the type of examination given, the nature of the clinical service, and the disposition of the case. Such data enable the tabulation of program activities and estimates of the costs of the various services, utilizing time-and-motion information since personnel costs represent a large majority of the total expense of the program. On this basis, and assuming a total cost of a program of $1 million, the following is the hypothetical cost by various activities:

Interviews	$ 8,958.60
Immunization	17,064.00
Counseling	25,596.00
Nonoccupational treatment	191,970.00
Special examinations	127,980.00
Occupational treatment	116,461.80
Replacement examinations	255,960.00
Periodic examinations	255,960.00

Dr. Craig then notes that a more difficult analysis is that of determining the benefits from an industrial health program, including, for example, deciding on acceptable units of measurement for such matters as the availability of health services, reduced workmen's compensation costs, decreased health insurance expenditures, a lowered rate of absenteeism, and increased productivity, as well as the evaluation of such "value benefits" as improved employee morale, fringe benefits to help attract new employees, selective placement procedures for the employment of the handicapped, improved community relations, and the motivation of employees to seek early treatment.

Dr. Craig suggests that a general method of determining benefits would be to develop procedures to monitor the health status of employees over a time period and to match this with an adequate control group. As a first step in a practical approach to the problem it is suggested that there be a determination of the goals and objectives of an industrial health program. Next would be a determination of the alternative methods of providing the services (for example, full-time in-plant health staff with a wide variety of specialty services, a full-time in-plant staff associated with a medical superstructure to provide more specialized services, a part-time staff, or an occupational health service provided on a contractual basis). Added to this would be a list of the desired occupational health activities. With this suggested method, Dr. Craig feels that it is possible to develop some fundamental data as to which method is most efficient.

Dr. Craig then discusses some specific examples of approaches that can be taken. Concerning preemployment examinations, the cost of these can be compared to the potential liability of not providing them, recognizing that this approach ignores "value benefits," such as damage to plant equipment or injury to other workers. Another example is tetanus immunization where the cost of an active immunization program within a certain time frame can be compared with other alternatives, the basis of comparison including the possibility of having a case of tetanus result from an industrial injury. A final example concerns the treatment of occupational injuries and disease where methodology can be applied to determine the types of treatment to be provided by the industrial health program and the cost of alternative methods of providing the service, including the actual cost of the service, the work-loss differential, the cost of time and transportation for treatment outside the plant, and the effect of outside treatment on the employee's early return to work. The fact that outside personnel are unfamiliar with the health record of the employee as well as the conditions of the workplace is another factor in favor of establishing in-plant services that can quickly determine the best course of treatment.

Cost-benefit analysis, it is felt, can be utilized to determine whether the industrial health services should be provided to employees in remote areas or whether it is more feasible to bring such employees to a central medical office, or to take the service to them through the use of mobile units, or to contract for these services with a local physician, or to not provide the services at all.

Dr. Craig is convinced that each activity of an industrial health program can be analyzed by some method. He maintains that until health economists and systems engineers develop sophisticated techniques for making such analyses his proposal provides a practical, common-sense approach to evaluating the cost-benefit effectiveness of an occupational health program and its various component activities.[42]

REFERENCES

1. David V. McCallum, "Interdisciplinary Teamwork in the Health/Safety Professions," paper presented before the American Medical Association, September 21, 1976.
2. The Conference Board. "Industry Roles in Health Care," New York, 1974.
3. The President's Commission on the Health Needs of the Nation, *Building America's Health,* Vol. 4, 1952.
4. HEW, "Small Plant Health and Medical Programs," 1958.
5. National Health Council, "The Health of People Who Work," New York, 1960.
6. Manufacturing Chemists Association. "Off-the-Job Safety," Washington, D.C., 1961.
7. Kenneth H. Klipstein, "Your Health and Industry," paper presented before the National Health Forum, March 18, 1963.
8. The Conference Board, "Top Executives View Health Care Issues," New York, 1972.
9. Ronald W. Conley, "Program Evaluation: Selected Readings," Research Utilization Laboratory of ICD Rehabilitation and Research, Washington, D.C., 1975.
10. Public-Health Rounds at the Harvard School of Public Health, March 18, 1976.
11. Nicholas A. Ashford, *Crisis in the Workplace* (Cambridge, Mass.: M.I.T. Press, 1976).
12. George M. Wheatley, M.D. *Bulletin of the New York Academy of Medicine,* January 1973.
13. HEW, op cit.
14. Ibid.
15. Klem and McKiever, "Small Plant Health and Medical Programs," Public Health Service Publication No. 215, 1952.
16. National Health Council, op. cit.
17. Norman Collisson, "Management and an Occupational Health Program," *Archives of Environmental Health,* February 1961.
18. Bond, Buckwalter, and Perkin, "An Occupational Health Program," *Archives of Environmental Health,* September 1968.
19. Karl T. Benedict, M.D., "Industrial Health Protection," *National Safety News,* June 1975.
20. The Conference Board, op. cit.
21. National Health Council, op. cit.
22. National Conference on Labor Health Services, American Labor Association, 1958.
23. HEW, op. cit.
24. National Health Council, op. cit.
25. Bethlehem Steel advertisement.
26. Communication received by the author from Paul F. Woolrich.

27. Doris M. Baldwin, "Cutting Costs Through Safety Management," *Job Safety and Health*, June 1976.
28. Manufacturing Chemists Association. "Industry Injury Report," Washington, D.C., 1976.
29. John L. Bischoff, "Home Health Care and the Insurance Industry," *Best's Review*, 1976.
30. National Health Council. op. cit.
31. Clark and Clarke, "Investment in Man," *Atlanta Economic Review*, July 1961.
32. Carl Zenz, M.D., paper delivered before the National Foundation of Health, Welfare, and Pension Plans, Inc. (now the International Foundation of Employee Benefit Plans), reported in *Employee Benefit Plan Review*, February 1972.
33. Richard A. Clinchy III, "Multiphasic Health Testing: A Two-Edged Sword," *CLU Journal*, April 1974.
34. Robert H. Demsey, "Multiphasic Health Testing," *Prevention and Involvement*, 1972.
35. Branzlued, "Multiphasic Health Testing: A Tool for Employee Benefits and Financial Risk Engineering," unpublished paper, September 2, 1973.
36. Morris Collen, M.D., et al., "Multiphasic Checkup Evaluation Study," *Preventive Medicine*, June 1973.
37. W. Scott Allan, "Rehabilitation as a Practical Insurance Program," *The Journal of Insurance*, undated.
38. Elmer J. Rasmussen, "Control of Claim Cost Through Rehabilitation," paper presented before the Health Insurance Association of America, October 26, 1964.
39. *The New York Times*, January 20, 1977.
40. A. Walter Hoover, M.D., "The Future of Occupational Health Programs," *Journal of Occupational Medicine*, June 1973.
41. Hallet A. Lewis, M.D., "Determining the Cost Effectiveness of Occupational Health Programs," in *Health Care Issues for Industry* (New York: The Conference Board, 1974).
42. James L. Craig, M.D., "A Practical Approach to Cost Analysis of an Occupational Health Program," *Journal of Occupational Medicine*, July 1974.

Part VIII

The Future

To attempt to forecast the future of such a dynamic subject as industrial health in the United States would call for a combination of the wisdom of Solomon and the mystic powers of the Oracle at Delphi. He who would dare the enterprise hazards contradiction at the hand of subsequent events.

It is not impossible, however, to take a look at the more immediate future and to observe certain of the forces that are at work, each of which will have its influence on the course of the future. It also is possible to observe some of the needs that can lead to more rapid progress.

In discussing the teaching of preventive medicine, one doctor said that the industrial health field is wide open for expansion. The increasing interest in relatively recent years on the part of employers, workers, labor unions, and insurers would appear to guarantee such a forecast. So also would the efforts of a great many voluntary organizations, many of which are employer-financed. The increasing role of several segments of government at all levels should hasten the process, particularly those concerned with occupational health and safety, with product safety, and with the creation of a more healthful environment. Certainly the potential inherent in the Occupational Safety and Health Act and the various types of relatively recent legislation concerned with product safety and environmental health should have a very considerable influence on the future. So, too, should the growing interest on the part of journalists, consumer groups, and the public at large.

The questions remain: How can improvement in and expansion of the development of industrial health in its broad aspects best be brought about? Where should the responsibility be placed and who is best qualified to accomplish the task? What is the best way to go about it? These are important questions, and the answers are not readily found. Experimentation and research will be needed.

One thing is certain: Advancement in the future must be based on soundly conceived concepts. Too frequently today concerns over both industrial and environmental health are surrounded and confused by strongly partisan viewpoints, by emotionalism, and by current fads that come and go like the whooping crane, even developing short-lived cults to be recognized for what they are. A book review by Whitney Balliett in the December 9, 1961, issue of *The New Yorker,* while addressed to literary subjects, is nonetheless valid here: "Writers who attract cults are, like lapdogs, a pitiable lot. Their worshippers coddle them, overcelebrate their virtues, shush their faults, and frighten away prospective and perhaps skeptical readers with an apologist's fervor. In short, these cultists establish a pampering sanctuary around the writer, softening him, and making him far more inaccessible than when he was simply ignored."

Thus it must be recognized that in industrial and environmental health matters much is as yet not clearly known about the subject, much more needs to be known, and called for is a hardheaded sorting of facts from fiction. The subject is far too important to people, to the economy, and to the nation to be hampered by sensationalism. Arnold Toynbee has wisely observed that both undue conservatism and undue egalitarianism are to be avoided in all such matters since both forces work against creativity and, in combination, they mount up to a formidable repressive force. In the January 1962 issue of the *Pennsylvania Gazette* he wrote: "There are two present-day adverse forces that are conspicuously deadly to creativity. One of these is a wrongheaded conception of the function of democracy. The other is an excessive anxiety to conserve vested interest, particularly the vested interest in acquired wealth." And well might be heeded the admonition of St. Lactantius: The first point of wisdom is to discern what is false, the second is to know what is truth.

The future of industrial health unquestionably depends upon a combination, a blend, of such practically minded persons as employers and union leaders; upon many types of scientists, engineers, and architects; upon a large corps of dedicated health professionals; and upon dreamers who are capable of permitting their imaginations to take wing and soar with a sense of vision toward practical and useful ends. George Bernard Shaw said in "In the Beginning" that first one must imagine, then one must will that which is desired, and then one creates.

In the arts the great masters are not the radicals, not the violent innovators, despite the fact that in the perspective of history they might appear so. Rather, the giants are those possessed of such bigness of concept that they are capable of entering a period filled with innovation, with unhinging and confusing developments, and of bringing a sense of unity and purpose to the whole. The field of industrial health has been experiencing a convulsive period of innovation and development. The result has been a running off in several directions at once. In the confusion, others tend to stand pat, perhaps more so than is their wont, out of an understandable fear of the consequences of disorder. What is needed, then, are people of unusual breadth and scope who can tie together the many developments, evolving from them a new but sound pattern that will put to use the valids of the innovations, cast aside the colorful foibles, and bring forth a system commensurate with all that has gone before and with the socioeconomic situation in the United States in the 1980s.

While this can appear to be somewhat esoteric for matters having to do with economics, health, business, and government, it is nonetheless valid. The concern is the health and welfare of human beings and of a society.

The future, then, will require prudent executive leadership and boldness of decision. It will call for the best available forces among our health professionals, our scientists, our universities, and our foundations. It will call for initial capital outlay. It will call for soundly conceived approaches by government at all levels.

27

Some Aspects of the Future of Industrial Health

There are many aspects of the future of industrial health, all of which are significant. They are, furthermore, closely interrelated, so much so that one of the current problems is that of establishing adequate intercommunication and cooperation among those concerned with the various aspects of the subject. Particularly is this so if the broad view of industrial health is to be considered, that is, concerns not alone with occupational hazards but also with the health of the community as affected by the products offered for sale and the environment in which employees and their families live. Such special health problems as mental illness and personality disorders, alcoholism, and drug addiction serve to add to the complexity of the interrelatedness of the various aspects of industrial health concerns.

Since industrial health efforts are clearly devoted to preventive medicine it is of interest to note the recent comments of Dr. George Rosen of Yale University on the future of preventive medicine. Dr. Rosen believes that preventive medicine today is in a transitional period and that there is a clear need for a national health policy. He feels that the National Health Policy Planning and Resources Development Act of 1974 can have an important impact on future activities and programs for health maintenance, disease prevention, and health education. He views with hope the change in prestige for preventive and occupational medicine that can come from the establishment of the American Board of Preventive Medicine and the American College of Preventive Medicine; from the relatively recent emphasis on such disciplines as epidemiology, biostatistics, environmental health, occupational health, health education, behavioral science, and demography; and from the doubling since 1950 of the number of schools of public health. While recognizing the many traditional problems of the relatively low status of preventive medicine, Dr. Rosen nonetheless feels that recent changes in the curriculum of the medical schools at least provide a beginning for possible future evolution "if the incentives toward preventive medicine were to outweigh whatever presumed disadvantages exist." He comments, however, that "as long as private clinical practice is considered more rewarding financially and in terms of professional status, preventive medicine will be at the end of the procession."[1]

Here it should not pass unnoted that one of the objectives of the 1976 HEW *Forward Plan for Health* is the prevention of illness, disease, and accidents through improvement in health education, health promotion, and preventive health services, including immunization, control of cigarettes content, fluoridation of water, improved nutritional status, improved child health, the reduction in adverse reactions to certain drugs, the promotion of physical fitness, the advancement of epidemiology, and improved data gathering.[2]

Without doubt the major role in the future of industrial health must be played by employers, whether private or public. Called for will be an increasingly active interest on their part in all the many aspects of industrial health. That this is frequently but by no means universally recognized by employers becomes increasingly evident.

In 1972 The Conference Board reported on the views of corporate executives as to the role of employers in the future with respect to health care issues. Nearly half the executives interviewed believed that in the years ahead employer involvement in health care will be more comprehensive and more expensive. One in six expected an expansion of industrial health programs, some limiting this expansion to occupational risks, others anticipating greater in-house attention to nonoccupational health care, including a full range of primary care on an ambulatory basis, periodic health inventories, counseling, health education, and on-the-job rehabilitation for both nonoccupational and occupational health problems. One in three of the executives forecast changes in employee health benefit programs, including more comprehensive insurance coverages, increasing union demands, and a greater share of the cost being borne by the employer. A few of the executives asserted that business should assign its highest priorities to such problems as water and air pollution, work hazards, and unsafe products. As one executive said: "Business had a hand in producing some of these problems; and it should, in my opinion, take an active role in their solution."[3]

Similar recognition was noted in 1977 by Dr. John Zapp of Du Pont when he stated: "Industry is concerned about its contributions to untimely deaths and untimely illness and recognizes the need for more thorough testing of chemicals than was thought necessary in the past."[4]

One thing is certain: Increasingly, employers have been, and will continue to be, held responsible for the economic consequences of the illnesses, injuries, and premature deaths among their employees—and to a certain extent the dependents of those employees—whether occupationally or nonoccupationally induced. The entire trend since World War II has been toward more comprehensive benefits under health insurance, disability insurance, salary-continuance programs, life insurance, and workmen's compensation. There is every reason to surmise that this trend will continue through either voluntary or compulsory means. More and more of these costs are being borne by employers and, in terms of the proportion of total payroll they represent, are considerable. Employers, then, have every reason to become actively involved in industrial health in its broadest aspects. This means not alone the establishment of industrial health programs either singly or on a cooperative or consortium basis with the paramount purpose of eliminating occupational hazards, but equally the prevention of nonoccupational illness, accidents, and premature deaths to the maximum extent possible. It means a commitment to and an involvement in all aspects of the health of the community where employees and their families work and live. It means an active involvement in the availability and costs of health services. It means an active concern with such matters as water supply, the sewage system, the prevention of communicable diseases, highway safety and the prevention of all types of accidents, and the eradication of hazardous environmental factors. It means the development of health education programs that will make individuals aware of their own contribution to health and safety hazards.

Unquestionably this propels employers into areas of concern to which they are not accustomed, in which they may feel an absence of adequate knowledge, and in which they might feel an unrelatedness to their own operations. Such an extension of interests costs money and this, in turn, can affect their position in the competitive marketplace. Yet, in one way or another, these costs are being borne by places of employment at present, and the competitive factor can be overcome by the achievement, through whatever means, of a degree of uniformity of such efforts among all places of employment by the voluntary establishment of basic standards through employer organizations. In the long view, the entire economy would benefit, and the health and welfare of people would be improved.

Most certainly particular, and at times difficult, problems are presented in the case of smaller employers, marginal businesses, agriculture, and oddly, government employers. But these problems are not without solution. Again, the costs of health and safety hazards, at the place of employment or within the community at large, are significantly borne by such employers in one way or another, and recognition of this fact should serve to spur an interest on the part of these employers.

Workers, individually or collectively, also have a role of significance to play. In a certain way it can appear strange that workers, for the most part to date, have performed a relatively inactive role in industrial health. Concerns can focus considerably on such matters as wages and hours, inflation, layoffs and unemployment, and improved employee benefits including health and disability insurance, paid vacations, and pension programs. Concern is also frequently expressed over working conditions, but health hazards in their broader aspects, and particularly as respects the environment of the community in which they and their families work and live, are frequently ignored. Through the processes of collective bargaining and the operation of health and welfare funds workers can do much, in instances where such actions are needed, to eliminate occupational health and safety hazards, to ensure that insurance benefits will be realistic, to influence the availability and costs of health services, to improve the health aspects of community environment, and to induce workers and their families as individuals to live safer and more healthful lives. The need and value for such increased activity on the part of workers have been noted recently by several sources.

Stellman and Daum have expressed the thought that one approach to the reduction of occupational health and safety hazards is for workers to become fully aware of their rights under OSHA to review surveys and health records and to call for inspections. They also argue that the unions should negotiate for a fund to be used to find new engineering processes to eliminate work hazards and to monitor the work environment. The unions, they say, should develop effective safety committees, should demand regular medical examinations provided by the employer, and should employ a physician to review all medical reports.[5]

In a letter to *The New York Times* dated December 22, 1976, Melvin A. Glasser of the UAW, while noting some disappointments in the administration of OSHA, commented that OSHA has created a climate in which organized workers can cooperate more effectively in industrial health and can negotiate health and safety programs with employers that allow for a greater union involvement in safeguarding the worker's health. He cited one such case where the injury rate dropped 30 percent and "thousands of neglected hazards have been corrected." Union members, he said, "have gained rights that allow them to participate in making the vital decisions necessary to recognize and correct unsafe and health-threatening conditions."

Similarly Morris Joselow at the College of Medicine and Dentistry of New Jersey sees a more active role for the labor force in protecting itself from work hazards, in monitoring the work environment, and in alerting government to potential problems, although he notes that to date labor has "largely lacked the technical knowledge needed to assess the safety of the work environment." He predicts that "major efforts will be undertaken by the labor force . . . to train workers to assess occupational hazards."[6]

Most certainly, all such activities on the part of workers should to the maximum extent possible be performed in a cooperative relationship with employers, community resources, and government. In this way optimal results might reasonably be expected.

Insurers also can play a vital role in the future developments of industrial health. As has been shown, workmen's compensation insurers actively participate in many ways in the reduction of occupational health and safety hazards and in the advancement of the processes of rehabilitation. Health insurers, on the other hand, display an amazing absence of activity that might lead to the prevention of illnesses and accidents, disability, and premature death. While they have indeed improved such insurance coverages tremendously over the past two decades, they give evidence of remarkably little interest in industrial health efforts. Considering that the vast majority of their insureds are employed people and their dependents and that the costs of such insurance continually rise, insurers sorely need an active, broad-based program that would encourage and assist the development of industrial health programs as well as contribute to the development of environmental or community health. The little that is done in such directions today is at best sporadic. Needed is a long-range, purposefully conceived program at an industry level. Such a role by insurers could provide a valuable contribution to the advancement of industrial health.

A key factor in the development of industrial health is, of course, a sufficient supply of professional people. This is the sine qua non of the entire subject. Included here is the broad gamut of health professionals: physicians, nurses, health hygienists, health educators, therapists, technicians, and rehabilitation specialists. That they are in short supply has already been discussed. As long as this inadequate supply exists, the future of industrial health will be impeded, perhaps seriously. Schools of medicine, nursing, and public health have a great responsibility to help overcome this shortage. Much more is needed from these sources than is presently the case. Beyond that, employers and unions must be prepared to offer income levels and benefits that will be at least competitive with the compensation paid these professionals in the medical field if competency, let alone vision and enthusiasm, is to be expected.

About 15 years ago, Dr. William H. Steward of the U.S. Public Health Service noted these needs when he said:

> The very term "community medicine" recognizes the dependence of modern medicine on a host of auxiliary health personnel. And we are short of the number of trained people we need in just about every category of these auxiliary health occupations. . . . But there is also a more fundamental reason. So far, public acceptance and support have not been great enough to attract enough young people to the health field. If it were, the problems—whatever they are—would have been worked out because the solution of problems is the essence of support. . . . It is possible that many of these occupations are so new that there has not been time for their importance to modern medical care to be recognized. Possibly when there is a public awareness of the facts, then the acceptance and support will be forthcoming. Then maybe we'll have more nurses, physical therapists, and medical technicians than we know what to do with, all problems of training and job pay having been worked out by public support or minimized by the prestige given to the work. This is possible, although I don't believe it can be predicted with certainty. But if something of the kind doesn't happen, we shall have to revise our plans for the future of medical care and start cutting our coat to fit the cloth.[7]

That the picture has not changed appreciably and that the responsibilities have increased enormously is evident from a much more recent statement by Dr. David Goldstein when he said that in the future it seems evident that both the industrial hygienist and the occupational physician will increasingly have to become environmentalists. He feels that such professionals will have to be equipped with a knowledge of air pollution, water pollution, radiation hazards, and the problems of lasers and microwaves. This means an enlarged responsibility for such professionals. A retarding influence, he recognizes, is the lack of growth among the occupational health professions.[8]

Equally in short supply are the various types of health services that could assist employers and unions in the establishment of industrial health programs, train on-the-job personnel in the roles they can play, and provide the necessary health and inspection services. Smaller employers in all forms of endeavor, including agriculture, are in sore need of such services. Providing such services could be physician groups, HMOs, hospitals, and rehabilitation agencies. Through such services consortia of employers or unions could be formed without too much effort. Needed is a central source to provide the incentive and to spark the enterprise. This might arise from concerned employers in the community, from labor union leaders, or from the health sources themselves.

One aspect of health care, including industrial health, which will receive more intensive consideration in the future, is the confidentiality of medical records. Confidentiality of health records is of particular significance to employed people, and the importance of confidentiality was noted earlier. In January 1977 a government-sponsored study found that most computerized health data systems have failed to adopt precise standards to maintain the confidentiality of patients' records. While the study found that most harm to individual rights involves the use of manual records, Congress nonetheless was urged to deal with the potential abuse of computerized records.[9]

Beyond the health professions and services is the need for other professional personnel: engineers, chemists, architects, statisticians, researchers, and scientists. All such personnel are necessary, as discussed in preceding chapters, to make the workplace as free as possible of safety and health hazards, to ensure the safety of products offered the consumer, and to remove environmental health and safety hazards, however produced. Research is needed to gain more knowledge of the nature, causes, and effects of many of the modern-day health and safety hazards. Experimentation is needed to find the most effective means for the removal of all such hazards. Standards for the further improvement of safety and health, on the job or in the community, are constantly needed and can be developed through existing agencies or organizations on a voluntary basis, taking care that they are adequate to the situation. In some instances care is required that in all such efforts an interrelationship of the many disciplines and agencies is assured on a cooperative basis if fragmentation and unnecessary duplication of effort are to be avoided.

Very recently the World Health Organization made recommendations to the various nations with respect to the health hazards resulting from new environmental pollutants. Included in the recommendations were:

- The need to develop guidelines concerning the types of information that are needed in order to adequately assess the possible health risks associated with the use of new chemicals, including the development of test systems to that end and the establishment of toxicological data banks.

- The need to encourage research into rapid bioassay systems.
- The need for improving health statistics.
- The need for cooperative efforts between occupational and community health services in order to identify health hazards and encourage cost-risk-benefit studies.[10]

In the preceding chapter the need for data that could produce evidence of the cost-effectiveness of industrial health programs was noted. More data are needed to bring about a better evaluation on a more meaningful basis of the costs of ill health and injuries among employed people and their families and of the nature and severity of health and safety hazards among the workforce. Cost-effectiveness data on a more uniform and comparable basis are sorely needed in order to prove or disprove the value of the many different aspects of industrial health efforts and to convince the dubious of the worthiness of the enterprise. In so doing, every aspect of the economic effects of injuries and illnesses discussed throughout the text requires examination. Unfortunately the means for the adequate fulfillment of this need do not presently exist, at least in identifiable form, although government agencies are presently taking some steps in this direction. Employers, through their various organizations, could perform an extremely valuable service if they were to consider this need and delegate the responsibility for fulfilling it.

Government at all levels will unquestionably have a larger role to play in the future of industrial health. This expanded role will concern not alone the elimination of occupational health and safety hazards but also product safety and environmental health. Public policy on these subjects can be expected to expand its areas of concerns as employers, workers, consumer groups, journalists, and the public at large increasingly voice their interest.

The extent to which government, through legislative or regulatory authority, should continue to enter such areas is, of course, a matter of continuing debate. Among other matters to be considered are the costs to the private sector and the effects on the national economy, on unemployment, and on inflation. Furthermore, the question is posed: Should such actions be taken by the federal government, the state governments, or local governments? Or should it be all three and, if so, where is the dividing line to be drawn so that unnecessary duplication of effort and confusion do not result?

Direct cognizance of such problems was taken by Elliot L. Richardson as secretary of commerce when he said on January 19, 1977:

> One of society's most pressing needs is a more rational process of regulation, especially in the regulation of the environment. There is a def-

inite lack of flexibility in much of the environmental legislation. . . . Economic considerations are grossly underrepresented. . . . Regional differences are generally not taken into account. . . . We must inject more economic reality into the regulatory process. We must consider ways to transfer some of the Federal Government's regulatory responsibility to the states.[11]

Also to be overcome is the duplication of assumed or designated responsibility by legislative committees and regulatory agencies at all levels of government. At present these responsibilities are sorely fragmented, with resultant duplication and confusion. In the interest of more rapid and direct accomplishment, steps should be taken to overcome this situation. To do so would be in the public interest. The subject of overcoming otherwise-avoidable injuries, illnesses, disabilities, and premature death is complex enough without these added, and unnecessary, sources of divisiveness, confusion, and loss of public confidence.

That the overcoming of both occupational and environmental hazards is costly has been shown in the earlier text. In 1975 the cost of pollution control to American industry was shown to be $6.5 billion for equipment and facilities alone. The cost of operating such facilities has been estimated to be $3.5 billion in 1975, which is projected to escalate to $12.7 billion by 1984. At present EPA is subsidizing 35 percent of the cost of pollution-reducing research and development projects in such industries as oil refining, chemicals, textiles, paper, lumbering, timber, and dairy products. Hence a cost to the taxpayer is involved. In all, the nation is estimated to have spent $35 billion in pollution control in 1975, of which $16 billion was borne by industry, $12 billion by government, and $7 billion by consumers' direct expenditures. Such proportions clearly indicate the need for careful, thoughtful, well-designed approaches to the elimination or reduction of health hazards.

It cannot pass unnoted that one of the objectives of the 1976 HEW *Forward Plan for Health* mentioned earlier is a tactical plan for environmental and occupational health. The plan includes the monitoring and modification of the environment, the strengthening of the occupational health programs of NIOSH, the reduction and control of toxic substances, the prevention of the exposure of workers to unhealthy and unsafe working conditions, consumer safety, and the conduct of environmental health research.

The role of government in relation to industrial health, then, can very well be expected to expand and to become more all-encompassing. The extent of the role to be played will be determined in part by the degree to which employers do or do not assume a role of responsibility.

Finally, the role of the individual, the employee, the members of his or her family, and the public at large cannot be overlooked. The deterrents to improved health in the United States have been discussed earlier in this text and their relationship to deleterious health practices on the part of the individual is clear. The conclusion is increasingly reached by many students of public health and health economics that personal attitudes toward health and safety are today the major obstacles to an improved health level in this nation.

For example, the National Safety Council has recently noted that one of the major causes of cancer results from the use of tobacco, particularly cancer of the lung, mouth, and larynx, and that emphysema and cardiovascular diseases are causally related to smoking. The NSC recommends major efforts to develop smoking-withdrawal clinics and the development of less harmful smoking products as preventive measures. The NSC has also taken cognizance of the fact that cancers have been shown to be associated with certain dietary conditions (for example, the association of a high intake of fat with breast, ovary, endometrium, prostate, and colon and rectum cancer) and urges as preventive measures public education of the need to lower fat intake and the development of an awareness of the danger within the agricultural and food industries. Noting that foods containing a high carbohydrate/fat ration, low micronutrients, and sizable amounts of nitrate are associated with gastric cancer, the NSC makes similar recommendations in the interest of prevention and also proposes public education of the need to store foods at low temperatures to prevent the formation of nitrate.[12]

The relationship of such individual health and safety practices to industrial health programs and places of employment is, as has been shown, very direct and consequential. The ignoring of safety precautions and working under stress or fatigue results in work injuries. Carelessness on the highway, in the home, and during recreation has the consequence of an untold number of unnecessary accidents. Ignoring such disease conditions as hypertension, diabetes, and heart conditions, the need for adequate rest and for sufficient exercise, and the effects of obesity, excessive smoking, and the abuse of alcohol and drugs accounts for a sizable proportion of all illnesses, disabilities, and premature death. The costs to individuals, families, employers, and society are so serious as to demand much more attention to such individually produced losses. All such occurrences are certainly of consequence to places of employment. Absenteeism and labor turnover are unnecessarily increased. Insurance costs become needlessly inflated. Skilled, capable, and important members of the workforce are unnecessarily lost.

The question is presented: How can such unnecessary loses be overcome? The ready answer is, of course, through safety training and health education by the employer, the labor union, the host of voluntary agencies, or the community. Unfortunately, too frequently such efforts are fragmented or executed in a haphazard manner, with the result that the efforts can be ineffectual. Too frequently the message, for whatever reason, is ignored by the individual. The problem, it would appear, is essentially one of communication, of finding the most effectual means for communicating the importance of the subject to workers, their families, and the community at large. It is a matter of too great importance to people and to society in all its aspects, and of too large proportions, to be ignored to the extent that is presently the case. The individual must be brought to recognize the need to assume responsibility for many aspects of his own health and safety and to realize that there are many aspects of his personal health and safety for which no one but himself can assume responsibility.

In summary, the long-range benefits of a fuller and more active interest in and concern with industrial health in its broadest aspect are clear beyond any doubt. They mean a healthier, more efficient workforce functioning with a higher level of morale. They mean increased productivity and improved customer and public relations. They mean reduced costs of operations at places of employment in many ways: through reduced absenteeism, a lower rate of labor turnover, the preservation of the health of experienced and valued employees at all levels, and reduced insurance and health and welfare fund costs. For the individual and his or her family they mean improved health, a better sense of well-being, and a longer life. For the nation as a whole they mean a healthier and more productive society. To gain these ends, there must be effort, expenditures of funds, and dedication on the part of a great many of the citizenry. It is an endeavor well worth the cost and effort. It is humane. Recognition of its importance can only argue for a broader future for industrial health. As Jawaharlal Nehru has said: "I am not interested in excuses for delay: I am interested only in a thing done."

REFERENCES

1. George Rosen, M.D., *Preventive Medicine in the U.S. 1900-1975* (Boston: Science History Publications, 1975).
2. HEW, *Forward Plan for Health 1978-82*, August 1976.
3. The Conference Board. "Top Executives View Health Care Issues," New York, 1972.
4. *The New York Times*, January 9, 1977.
5. Jeanne M. Stellman and Susan M. Daum, *Work Is Dangerous to Your Health* (New York: Vintage Books, 1973).

6. Morris M. Joselow, "Occupational Health in the U.S.: The Next Decades," *American Journal of Public Health,* November 1973.
7. William H. Steward, M.D., "Community Medicine and Its Problems," paper presented before the American Public Health Association, May 2, 1962.
8. David H. Goldstein, M.D., "The Changing Scope of Occupational Medicine and Environmental Health," *American Journal of Public Health,* April 1970.
9. U.S. Commerce Department, National Bureau of Standards Computers, "Health Records and Citizens Rights," January 1977.
10. WHO, "Health Hazards from New Environmental Pollutants," Geneva, 1976.
11. *The New York Times,* January 20, 1977.
12. National Safety Council, *Safety Newsletter,* October 1976.

Appendixes

    A. Occupational Safety and Health Act of 1970
        B. Established Industrial Health Programs
C. Sources of Information, Assistance, and Guidance
                D. Annotated Bibliography

A. OCCUPATIONAL SAFETY AND HEALTH ACT OF 1970*

An Act

84 STAT. 1590

To assure safe and healthful working conditions for working men and women; by authorizing enforcement of the standards developed under the Act; by assisting and encouraging the States in their efforts to assure safe and healthful working conditions; by providing for research, information, education, and training in the field of occupational safety and health; and for other purposes.

Be it enacted by the Senate and House of Representatives of the United States of America in Congress assembled, That this Act may be cited as the "Occupational Safety and Health Act of 1970".

Occupational Safety and Health Act of 1970.

CONGRESSIONAL FINDINGS AND PURPOSE

SEC. (2) The Congress finds that personal injuries and illnesses arising out of work situations impose a substantial burden upon, and are a hindrance to, interstate commerce in terms of lost production, wage loss, medical expenses, and disability compensation payments.

(b) The Congress declares it to be its purpose and policy, through the exercise of its powers to regulate commerce among the several States and with foreign nations and to provide for the general welfare, to assure so far as possible every working man and woman in the Nation safe and healthful working conditions and to preserve our human resources—

(1) by encouraging employers and employees in their efforts to reduce the number of occupational safety and health hazards at their places of employment, and to stimulate employers and employees to institute new and to perfect existing programs for providing safe and healthful working conditions;

(2) by providing that employers and employees have separate but dependent responsibilities and rights with respect to achieving safe and healthful working conditions;

(3) by authorizing the Secretary of Labor to set mandatory occupational safety and health standards applicable to businesses affecting interstate commerce, and by creating an Occupational Safety and Health Review Commission for carrying out adjudicatory functions under the Act;

(4) by building upon advances already made through employer and employee initiative for providing safe and healthful working conditions;

(5) by providing for research in the field of occupational safety and health, including the psychological factors involved, and by developing innovative methods, techniques, and approaches for dealing with occupational safety and health problems;

(6) by exploring ways to discover latent diseases, establishing causal connections between diseases and work in environmental conditions, and conducting other research relating to health problems, in recognition of the fact that occupational health standards present problems often different from those involved in occupational safety;

(7) by providing medical criteria which will assure insofar as practicable that no employee will suffer diminished health, functional capacity, or life expectancy as a result of his work experience;

(8) by providing for training programs to increase the number and competence of personnel engaged in the field of occupational safety and health;

---

* Public Law 91-596 passed by the 91st Congress, S. 2193, December 29, 1970.

(9) by providing for the development and promulgation of occupational safety and health standards;

(10) by providing an effective enforcement program which shall include a prohibition against giving advance notice of any inspection and sanctions for any individual violating this prohibition;

(11) by encouraging the States to assume the fullest responsibility for the administration and enforcement of their occupational safety and health laws by providing grants to the States to assist in identifying their needs and responsibilities in the area of occupational safety and health, to develop plans in accordance with the provisions of this Act, to improve the administration and enforcement of State occupational safety and health laws, and to conduct experimental and demonstration projects in connection therewith;

(12) by providing for appropriate reporting procedures with respect to occupational safety and health which procedures will help achieve the objectives of this Act and accurately describe the nature of the occupational safety and health problem;

(13) by encouraging joint labor-management efforts to reduce injuries and disease arising out of employment.

DEFINITIONS

SEC. 3. For the purposes of this Act—
 (1) The term "Secretary" mean the Secretary of Labor.
 (2) The term "Commission" means the Occupational Safety and Health Review Commission established under this Act.
 (3) The term "commerce" means trade, traffic, commerce, transportation, or communication among the several States, or between a State and any place outside thereof, or within the District of Columbia, or a possession of the United States (other than the Trust Territory of the Pacific Islands), or between points in the same State but through a point outside thereof.
 (4) The term "person" means one or more individuals, partnerships, associations, corporations, business trusts, legal representatives, or any organized group of persons.
 (5) The term "employer" means a person engaged in a business affecting commerce who has employees, but does not include the United States or any State or political subdivision of a State.
 (6) The term "employee" means an employee of an employer who is employed in a business of his employer which affects commerce.
 (7) The term "State" includes a State of the United States, the District of Columbia, Puerto Rico, the Virgin Islands, American Samoa, Guam, and the Trust Territory of the Pacific Islands.
 (8) The term "occupational safety and health standard" means a standard which requires conditions, or the adoption or use of one or more practices, means, methods, operations, or processes, reasonably necessary or appropriate to provide safe or healthful employment and places of employment.
 (9) The term "national consensus standard" means any occupational safety and health standard or modification thereof which (1), has been adopted and promulgated by a nationally recognized standards-producing organization under procedures whereby it can be determined by the Secretary that persons interested

and affected by the scope or provisions of the standard have reached substantial agreement on its adoption, (2) was formulated in a manner which afforded an opportunity for diverse views to be considered and (3) has been designated as such a standard by the Secretary, after consultation with other appropriate Federal agencies.

(10) The term "established Federal standard" means any operative occupational safety and health standard established by any agency of the United States and presently in effect, or contained in any Act of Congress in force on the date of enactment of this Act.

(11) The term "Committee" means the National Advisory Committee on Occupational Safety and Health established under this Act.

(12) The term "Director" means the Director of the National Institute for Occupational Safety and Health.

(13) The term "Institute" means the National Institute for Occupational Safety and Health established under this Act.

(14) The term "Workmen's Compensation Commission" means the National Commission on State Workmen's Compensation Laws established under this Act.

APPLICABILITY OF THIS ACT

SEC. 4. (a) This Act shall apply with respect to employment performed in a workplace in a State, the District of Columbia, the Commonwealth of Puerto Rico, the Virgin Islands, American Samoa, Guam, the Trust Territory of the Pacific Islands, Wake Island, Outer Continental Shelf lands defined in the Outer Continental Shelf Lands Act, Johnston Island, and the Canal Zone. The Secretary of the Interior shall, by regulation, provide for judicial enforcement of this Act by the courts established for areas in which there are no United States district courts having jurisdiction. [67 Stat. 462. 43 USC 1331 note.]

(b)(1) Nothing in this Act shall apply to working conditions of employees with respect to which other Federal agencies, and State agencies acting under section 274 of the Atomic Energy Act of 1954, as amended (42 U.S.C. 2021), exercise statutory authority to prescribe or enforce standards or regulations affecting occupational safety or health. [73 Stat. 688.]

(2) The safety and health standards promulgated under the Act of June 30, 1936, commonly known as the Walsh-Healey Act (41 U.S.C. 35 et seq.), the Service Contract Act of 1965 (41 U.S.C. 351 et seq.), Public Law 91-54, Act of August 9, 1969 (40 U.S.C. 333), Public Law 85-742, Act of August 23, 1958 (33 U.S.C. 941), and the National Foundation on Arts and Humanities Act (20 U.S.C. 951 et seq.) are superseded on the effective date of corresponding standards, promulgated under this Act, which are determined by the Secretary to be more effective. Standards issued under the laws listed in this paragraph and in effect on or after the effective date of this Act shall be deemed to be occupational safety and health standards issued under this Act, as well as under such other Acts. [49 Stat. 2036. 79 Stat. 1034. 83 Stat. 96. 72 Stat. 835. 79 Stat. 845; Ante, p. 443.]

(3) The Secretary shall, within three years after the effective date of this Act, report to the Congress his recommendations for legislation to avoid unnecessary duplication and to achieve coordination between this Act and other Federal laws. [Report to Congress.]

(4) Nothing in this Act shall be construed to supersede or in any manner affect any workmen's compensation law or to enlarge or diminish or affect in any other manner the common law or statutory rights, duties, or liabilities of employers and employees under any law with respect to injuries, diseases, or death of employees arising out of, or in the course of, employment.

DUTIES

SEC. 5. (a) Each employer—
  (1) shall furnish to each of his employees employment and a place of employment which are free from recognized hazards that are causing or are likely to cause death or serious physical harm to his employees;
  (2) shall comply with occupational safety and health standards promulgated under this Act.
(b) Each employee shall comply with occupational safety and health standards and all rules, regulations, and orders issued pursuant to this Act which are applicable to his own actions and conduct.

OCCUPATIONAL SAFETY AND HEALTH STANDARDS

80 Stat. 381;
81 Stat. 195.
5 USC 500.

SEC. 6. (a) Without regard to chapter 5 of title 5, United States Code, or to the other subsections of this section, the Secretary shall, as soon as practicable during the period beginning with the effective date of this Act and ending two years after such date, by rule promulgate as an occupational safety or health standard any national consensus standard, and any established Federal standard, unless he determines that the promulgation of such a standard would not result in improved safety or health for specifically designated employees. In the event of conflict among any such standards, the Secretary shall promulgate the standard which assures the greatest protection of the safety or health of the affected employees.

(b) The Secretary may by rule promulgate, modify, or revoke any occupational safety or health standard in the following manner:

Advisory committee, recommendations.

(1) Whenever the Secretary, upon the basis of information submitted to him in writing by an interested person, a representative of any organization of employers or employees, a nationally recognized standards-producing organization, the Secretary of Health, Education, and Welfare, the National Institute for Occupational Safety and Health, or a State or political subdivision, or on the basis of information developed by the Secretary or otherwise available to him, determines that a rule should be promulgated in order to serve the objectives of this Act, the Secretary may request the recommendations of an advisory committee appointed under section 7 of this Act. The Secretary shall provide such an advisory committee with any proposals of his own or of the Secretary of Health, Education, and Welfare, together with all pertinent factual information developed by the Secretary or the Secretary of Health, Education, and Welfare, or otherwise available, including the results of research, demonstrations, and experiments. An advisory committee shall submit to the Secretary its recommendations regarding the rule to be promulgated within ninety days from the date of its appointment or within such longer or shorter period as may be prescribed by the Secretary, but in no event for a period which is longer than two hundred and seventy days.

(2) The Secretary shall publish a proposed rule promulgating, modifying, or revoking an occupational safety or health standard in the Federal Register and shall afford interested persons a period of thirty days after publication to submit written data or comments. Where an advisory committee is appointed and the Secretary determines that a rule should be issued, he shall publish the proposed rule within sixty days after the submission of the advisory committee's recommendations or the expiration of the period prescribed by the Secretary for such submission.

Publication in Federal Register.

(3) On or before the last day of the period provided for the submission of written data or comments under paragraph (2), any interested person may file with the Secretary written objections to the proposed rule, stating the grounds therefor and requesting a public hearing on such objections. Within thirty days after the last day for filing such objections, the Secretary shall publish in the Federal Register a notice specifying the occupational safety or health standard to which objections have been filed and a hearing requested, and specifying a time and place for such hearing.

Hearing, notice.

Publication in Federal Register.

(4) Within sixty days after the expiration of the period provided for the submission of written data or comments under paragraph (2), or within sixty days after the completion of any hearing held under paragraph (3), the Secretary shall issue a rule promulgating, modifying, or revoking an occupational safety or health standard or make a determination that a rule should not be issued. Such a rule may contain a provision delaying its effective date for such period (not in excess of ninety days) as the Secretary determines may be necessary to insure that affected employers and employees will be informed of the existence of the standard and of its terms and that employers affected are given an opportunity to familiarize themselves and their employees with the existence of the requirements of the standard.

(5) The Secretary, in promulgating standards dealing with toxic materials or harmful physical agents under this subsection, shall set the standard which most adequately assures, to the extent feasible, on the basis of the best available evidence, that no employee will suffer material impairment of health or functional capacity even if such employee has regular exposure to the hazard dealt with by such standard for the period of his working life. Development of standards under this subsection shall be based upon research, demonstrations, experiments, and such other information as may be appropriate. In addition to the attainment of the highest degree of health and safety protection for the employee, other considerations shall be the latest available scientific data in the field, the feasibility of the standards, and experience gained under this and other health and safety laws. Whenever practicable, the standard promulgated shall be expressed in terms of objective criteria and of the performance desired.

Toxic materials.

(6)(A) Any employer may apply to the Secretary for a temporary order granting a variance from a standard or any provision thereof promulgated under this section. Such temporary order shall be granted only if the employer files an application which meets the requirements of clause (B) and establishes that (i) he is unable to comply with a standard by its effective date because of unavailability of professional or technical personnel or of materials and equipment needed to come into compliance with the standard or because necessary construction or alteration of facilities cannot be completed by the effective date, (ii) he is taking all available steps to safeguard his employees against the hazards covered by the standard, and (iii) he has an effective program for coming into compliance with the standard as quickly as

Temporary variance order.

practicable. Any temporary order issued under this paragraph shall prescribe the practices, means, methods, operations, and processes which the employer must adopt and use while the order is in effect and state in detail his program for coming into compliance with the standard. Such a temporary order may be granted only after notice to employees and an opportunity for a hearing: *Provided*, That the Secretary may issue one interim order to be effective until a decision is made on the basis of the hearing. No temporary order may be in effect for longer than the period needed by the employer to achieve compliance with the standard or one year, whichever is shorter, except that such an order may be renewed not more than twice (I) so long as the requirements of this paragraph are met and (II) if an application for renewal is filed at least 90 days prior to the expiration date of the order. No interim renewal of an order may remain in effect for longer than 180 days.

(B) An application for a temporary order under this paragraph (6) shall contain:

(i) a specification of the standard or portion thereof from which the employer seeks a variance,

(ii) a representation by the employer, supported by representations from qualified persons having firsthand knowledge of the facts represented, that he is unable to comply with the standard or portion thereof and a detailed statement of the reasons therefor,

(iii) a statement of the steps he has taken and will take (with specific dates) to protect employees against the hazard covered by the standard,

(iv) a statement of when he expects to be able to comply with the standard and what steps he has taken and what steps he will take (with dates specified) to come into compliance with the standard, and

(v) a certification that he has informed his employees of the application by giving a copy thereof to their authorized representative, posting a statement giving a summary of the application and specifying where a copy may be examined at the place or places where notices to employees are normally posted, and by other appropriate means.

A description of how employees have been informed shall be contained in the certification. The information to employees shall also inform them of their right to petition the Secretary for a hearing.

(C) The Secretary is authorized to grant a variance from any standard or portion thereof whenever he determines, or the Secretary of Health, Education, and Welfare certifies, that such variance is necessary to permit an employer to participate in an experiment approved by him or the Secretary of Health, Education, and Welfare designed to demonstrate or validate new and improved techniques to safeguard the health or safety of workers.

(7) Any standard promulgated under this subsection shall prescribe the use of labels or other appropriate forms of warning as are necessary to insure that employees are apprised of all hazards to which they are exposed, relevant symptoms and appropriate emergency treatment, and proper conditions and precautions of safe use or exposure. Where appropriate, such standard shall also prescribe suitable protective equipment and control or technological procedures to be used in connection with such hazards and shall provide for monitoring or measuring employee exposure at such locations and intervals, and in such manner as may be necessary for the protection of employees. In

addition, where appropriate, any such standard shall prescribe the type and frequency of medical examinations or other tests which shall be made available, by the employer or at his cost, to employees exposed to such hazards in order to most effectively determine whether the health of such employees is adversely affected by such exposure. In the event such medical examinations are in the nature of research, as determined by the Secretary of Health, Education, and Welfare, such examinations may be furnished at the expense of the Secretary of Health, Education, and Welfare. The results of such examinations or tests shall be furnished only to the Secretary or the Secretary of Health, Education, and Welfare, and, at the request of the employee, to his physician. The Secretary, in consultation with the Secretary of Health, Education, and Welfare, may by rule promulgated pursuant to section 553 of title 5, United States Code, make appropriate modifications in the foregoing requirements relating to the use of labels or other forms of warning, monitoring or measuring, and medical examinations, as may be warranted by experience, information, or medical or technological developments acquired subsequent to the promulgation of the relevant standard.

Medical examinations.

80 Stat. 383.

(8) Whenever a rule promulgated by the Secretary differs substantially from an existing national consensus standard, the Secretary shall, at the same time, publish in the Federal Register a statement of the reasons why the rule as adopted will better effectuate the purposes of this Act than the national consensus standard.

Publication in Federal Register.

(c)(1) The Secretary shall provide, without regard to the requirements of chapter 5, title 5, United States Code, for an emergency temporary standard to take immediate effect upon publication in the Federal Register if he determines (A) that employees are exposed to grave danger from exposure to substances or agents determined to be toxic or physically harmful or from new hazards, and (B) that such emergency standard is necessary to protect employees from such danger.

Temporary standard. Publication in Federal Register.
80 Stat. 381;
81 Stat. 195.
5 USC 500.

(2) Such standard shall be effective until superseded by a standard promulgated in accordance with the procedures prescribed in paragraph (3) of this subsection.

Time limitation.

(3) Upon publication of such standard in the Federal Register the Secretary shall commence a proceeding in accordance with section 6(b) of this Act, and the standard as published shall also serve as a proposed rule for the proceeding. The Secretary shall promulgate a standard under this paragraph no later than six months after publication of the emergency standard as provided in paragraph (2) of this subsection.

(d) Any affected employer may apply to the Secretary for a rule or order for a variance from a standard promulgated under this section. Affected employees shall be given notice of each such application and an opportunity to participate in a hearing. The Secretary shall issue such rule or order if he determines on the record, after opportunity for an inspection where appropriate and a hearing, that the proponent of the variance has demonstrated by a preponderance of the evidence that the conditions, practices, means, methods, operations, or processes used or proposed to be used by an employer will provide employment and places of employment to his employees which are as safe and healthful as those which would prevail if he complied with the standard. The rule or order so issued shall prescribe the conditions the employer must maintain, and the practices, means, methods, operations, and processes which he must adopt and utilize to the extent they

Variance rule.

differ from the standard in question. Such a rule or order may be modified or revoked upon application by an employer, employees, or by the Secretary on his own motion, in the manner prescribed for its issuance under this subsection at any time after six months from its issuance.

Publication in Federal Register.

(e) Whenever the Secretary promulgates any standard, makes any rule, order, or decision, grants any exemption or extension of time, or compromises, mitigates, or settles any penalty assessed under this Act, he shall include a statement of the reasons for such action, which shall be published in the Federal Register.

Petition for judicial review.

(f) Any person who may be adversely affected by a standard issued under this section may at any time prior to the sixtieth day after such standard is promulgated file a petition challenging the validity of such standard with the United States court of appeals for the circuit wherein such person resides or has his principal place of business, for a judicial review of such standard. A copy of the petition shall be forthwith transmitted by the clerk of the court to the Secretary. The filing of such petition shall not, unless otherwise ordered by the court, operate as a stay of the standard. The determinations of the Secretary shall be conclusive if supported by substantial evidence in the record considered as a whole.

(g) In determining the priority for establishing standards under this section, the Secretary shall give due regard to the urgency of the need for mandatory safety and health standards for particular industries, trades, crafts, occupations, businesses, workplaces or work environments. The Secretary shall also give due regard to the recommendations of the Secretary of Health, Education, and Welfare regarding the need for mandatory standards in determining the priority for establishing such standards.

ADVISORY COMMITTEES; ADMINISTRATION

Establishment; membership.

SEC. 7. (a)(1) There is hereby established a National Advisory Committee on Occupational Safety and Health consisting of twelve members appointed by the Secretary, four of whom are to be designated by the Secretary of Health, Education, and Welfare, without regard to the provisions of title 5, United States Code, governing appointments in the competitive service, and composed of representatives of manangement, labor, occupational safety and occupational health professions, and of the public. The Secretary shall designate one of the public members as Chairman. The members shall be selected upon the basis of their experience and competence in the field of occupational safety and health.

80 Stat. 378. 5 USC 101.

(2) The Committee shall advise, consult with, and make recommendations to the Secretary and the Secretary of Health, Education, and Welfare on matters relating to the administration of the Act. The Committee shall hold no fewer than two meetings during each calendar year. All meetings of the Committee shall be open to the public and a transcript shall be kept and made available for public inspection.

Public transcript.

(3) The members of the Committee shall be compensated in accordance with the provisions of section 3109 of title 5, United States Code.

80 Stat. 416.

(4) The Secretary shall furnish to the Committee an executive secretary and such secretarial, clerical, and other services as are deemed necessary to the conduct of its business.

(b) An advisory committee may be appointed by the Secretary to assist him in his standard-setting functions under section 6 of this Act. Each such committee shall consist of not more than fifteen members and shall include as a member one or more designees of the Secretary of Health, Education, and Welfare, and shall include among its members an equal number of persons qualified by experience and affiliation to present the viewpoint of the employers involved, and of persons similarly qualified to present the viewpoint of the workers involved, as well as one or more representatives of health and safety agencies of the States. An advisory committee may also include such other persons as the Secretary may appoint who are qualified by knowledge and experience to make a useful contribution to the work of such committee, including one or more representatives of professional organizations of technicians or professionals specializing in occupational safety or health, and one or more representatives of nationally recognized standards-producing organizations, but the number of persons so appointed to any such advisory committee shall not exceed the number appointed to such committee as representatives of Federal and State agencies. Persons appointed to advisory committees from private life shall be compensated in the same manner as consultants or experts under section 3109 of title 5, United States Code. The Secretary shall pay to any State which is the employer of a member of such a committee who is a representative of the health or safety agency of that State, reimbursement sufficient to cover the actual cost to the State resulting from such representative's membership on such committee. Any meeting of such committee shall be open to the public and an accurate record shall be kept and made available to the public. No member of such committee (other than representatives of employers and employees) shall have an economic interest in any proposed rule.

80 Stat. 416.

Recordkeeping.

(c) In carrying out his responsibilities under this Act, the Secretary is authorized to—

(1) use, with the consent of any Federal agency, the services, facilities, and personnel of such agency, with or without reimbursement, and with the consent of any State or political subdivision thereof, accept and use the services, facilities, and personnel of any agency of such State or subdivision with reimbursement; and

(2) employ experts and consultants or organizations thereof as authorized by section 3109 of title 5, United States Code, except that contracts for such employment may be renewed annually; compensate individuals so employed at rates not in excess of the rate specified at the time of service for grade GS-18 under section 5332 of title 5, United States Code, including traveltime, and allow them while away from their homes or regular places of business, travel expenses (including per diem in lieu of subsistence) as authorized by section 5703 of title 5, United States Code, for persons in the Government service employed intermittently, while so employed.

Ante, p. 198-1.

80 Stat. 499;
83 Stat. 190.

INSPECTIONS, INVESTIGATIONS, AND RECORDKEEPING

SEC. 8. (a) In order to carry out the purposes of this Act, the Secretary, upon presenting appropriate credentials to the owner, operator, or agent in charge, is authorized—

(1) to enter without delay and at reasonable times any factory, plant, establishment, construction site, or other area, workplace or environment where work is performed by an employee of an employer; and

423

(2) to inspect and investigate during regular working hours and at other reasonable times, and within reasonable limits and in a reasonable manner, any such place of employment and all pertinent conditions, structures, machines, apparatus, devices, equipment, and materials therein, and to question privately any such employer, owner, operator, agent or employee.

Subpoena power.

(b) In making his inspections and investigations under this Act the Secretary may require the attendance and testimony of witnesses and the production of evidence under oath. Witnesses shall be paid the same fees and mileage that are paid witnesses in the courts of the United States. In case of a contumacy, failure, or refusal of any person to obey such an order, any district court of the United States or the United States courts of any territory or possession, within the jurisdiction of which such person is found, or resides or transacts business, upon the application by the Secretary, shall have jurisdiction to issue to such person an order requiring such person to appear to produce evidence if, as, and when so ordered, and to give testimony relating to the matter under investigation or in question, and any failure to obey such order of the court may be punished by said court as a contempt thereof.

Recordkeeping.

(c)(1) Each employer shall make, keep and preserve, and make available to the Secretary or the Secretary of Health, Education, and Welfare, such records regarding his activities relating to this Act as the Secretary, in cooperation with the Secretary of Health, Education, and Welfare, may prescribe by regulation as necessary or appropriate for the enforcement of this Act or for developing information regarding the causes and prevention of occupational accidents and illnesses. In order to carry out the provisions of this paragraph such regulations may include provisions requiring employers to conduct periodic inspections. The Secretary shall also issue regulations requiring that employers, through posting of notices or other appropriate means, keep their employees informed of their protections and obligations under this Act, including the provisions of applicable standards.

Work-related deaths, etc.; reports.

(2) The Secretary, in cooperation with the Secretary of Health, Education, and Welfare, shall prescribe regulations requiring employers to maintain accurate records of, and to make periodic reports on, work-related deaths, injuries and illnesses other than minor injuries requiring only first aid treatment and which do not involve medical treatment, loss of consciousness, restriction of work or motion, or transfer to another job.

(3) The Secretary, in cooperation with the Secretary of Health, Education, and Welfare, shall issue regulations requiring employers to maintain accurate records of employee exposures to potentially toxic materials or harmful physical agents which are required to be monitored or measured under section 6. Such regulations shall provide employees or their representatives with an opportunity to observe such monitoring or measuring, and to have access to the records thereof. Such regulations shall also make appropriate provision for each employee or former employee to have access to such records as will indicate his own exposure to toxic materials or harmful physical agents. Each employer shall promptly notify any employee who has been or is being exposed to toxic materials or harmful physical agents in concentrations or at levels which exceed those prescribed by an applicable occupational safety and health standard promulgated under section 6, and shall inform any employee who is being thus exposed of the corrective action being taken.

(d) Any information obtained by the Secretary, the Secretary of Health, Education, and Welfare, or a State agency under this Act shall be obtained with a minimum burden upon employers, especially those operating small businesses. Unnecessary duplication of efforts in obtaining information shall be reduced to the maximum extent feasible.

(e) Subject to regulations issued by the Secretary, a representative of the employer and a representative authorized by his employees shall be given an opportunity to accompany the Secretary or his authorized representative during the physical inspection of any workplace under subsection (a) for the purpose of aiding such inspection. Where there is no authorized employee representative, the Secretary or his authorized representative shall consult with a reasonable number of employees concerning matters of health and safety in the workplace.

(f)(1) Any employees or representative of employees who believe that a violation of a safety or health standard exists that threatens physical harm, or that an imminent danger exists, may request an inspection by giving notice to the Secretary or his authorized representative of such violation or danger. Any such notice shall be reduced to writing, shall set forth with reasonable particularity the grounds for the notice, and shall be signed by the employees or representative of employees, and a copy shall be provided the employer or his agent no later than at the time of inspection, except that, upon the request of the person giving such notice, his name and the names of individual employees referred to therein shall not appear in such copy or on any record published, released, or made available pursuant to subsection (g) of this section. If upon receipt of such notification the Secretary determines there are reasonable grounds to believe that such violation or danger exists, he shall make a special inspection in accordance with the provisions of this section as soon as practicable, to determine if such violation or danger exists. If the Secretary determines there are no reasonable grounds to believe that a violation or danger exists he shall notify the employees or representative of the employees in writing of such determination.

(2) Prior to or during any inspection of a workplace, any employees or representative of employees employed in such workplace may notify the Secretary or any representative of the Secretary responsible for conducting the inspection, in writing, of any violation of this Act which they have reason to believe exists in such workplace. The Secretary shall, by regulation, establish procedures for informal review of any refusal by a representative of the Secretary to issue a citation with respect to any such alleged violation and shall furnish the employees or representative of employees requesting such review a written statement of the reasons for the Secretary's final disposition of the case.

(g)(1) The Secretary and Secretary of Health, Education, and Welfare are authorized to compile, analyze, and publish, either in summary or detailed form, all reports or information obtained under this section. *Reports, publication.*

(2) The Secretary and the Secretary of Health, Education, and Welfare shall each prescribe such rules and regulations as he may deem necessary to carry out their responsibilities under this Act, including rules and regulations dealing with the inspection of an employer's establishment. *Rules and regulations.*

CITATIONS

Sec. 9. (a) If, upon inspection or investigation, the Secretary or his authorized representative believes that an employer has violated a requirement of section 5 of this Act, of any standard, rule or order promulgated pursuant to section 6 of this Act, or of any regulations prescribed pursuant to this Act, he shall with reasonable promptness issue a citation to the employer. Each citation shall be in writing and shall describe with particularity the nature of the violation, including a reference to the provision of the Act, standard, rule, regulation, or order alleged to have been violated. In addition, the citation shall fix a reasonable time for the abatement of the violation. The Secretary may prescribe procedures for the issuance of a notice in lieu of a citation with respect to de minimis violations which have no direct or immediate relationship to safety or health.

(b) Each citation issued under this section, or a copy or copies thereof, shall be prominently posted, as prescribed in regulations issued by the Secretary, at or near each place a violation referred to in the citation occurred.

Limitation.

(c) No citation may be issued under this section after the expiration of six months following the occurrence of any violation.

PROCEDURE FOR ENFORCEMENT

Sec. 10. (a) If, after an inspection or investigation, the Secretary issues a citation under section 9(a), he shall, within a reasonable time after the termination of such inspection or investigation, notify the employer by certified mail of the penalty, if any, proposed to be assessed under section 17 and that the employer has fifteen working days within which to notify the Secretary that he wishes to contest the citation or proposed assessment of penalty. If, within fifteen working days from the receipt of the notice issued by the Secretary the employer fails to notify the Secretary that he intends to contest the citation or proposed assessment of penalty, and no notice is filed by any employee or representative of employees under subsection (c) within such time, the citation and the assessment, as proposed, shall be deemed a final order of the Commission and not subject to review by any court or agency.

(b) If the Secretary has reason to believe that an employer has failed to correct a violation for which a citation has been issued within the period permitted for its correction (which period shall not begin to run until the entry of a final order by the Commission in the case of any review proceedings under this section initiated by the employer in good faith and not solely for delay or avoidance of penalties), the Secretary shall notify the employer by certified mail of such failure and of the penalty proposed to be assessed under section 17 by reason of such failure, and that the employer has fifteen working days within which to notify the Secretary that he wishes to contest the Secretary's notification or the proposed assessment of penalty. If, within fifteen working days from the receipt of notification issued by the Secretary, the employer fails to notify the Secretary that he intends to contest the notification or proposed assessment of penalty, the notification and assessment, as proposed, shall be deemed a final order of the Commission and not subject to review by any court or agency.

(c) If an employer notifies the Secretary that he intends to contest a citation issued under section 9(a) or notification issued under subsection (a) or (b) of this section, or if, within fifteen working days of the issuance of a citation under section 9(a), any employee or representative of employees files a notice with the Secretary alleging that the period of time fixed in the citation for the abatement of the violation is unreasonable, the Secretary shall immediately advise the Commission of such notification, and the Commission shall afford an opportunity for a hearing (in accordance with section 554 of title 5, United States Code, but without regard to subsection (a)(3) of such section). The Commission shall thereafter issue an order, based on findings of fact, affirming, modifying, or vacating the Secretary's citation or proposed penalty, or directing other appropriate relief, and such order shall become final thirty days after its issuance. Upon a showing by an employer of a good faith effort to comply with the abatement requirements of a citation, and that abatement has not been completed because of factors beyond his reasonable control, the Secretary, after an opportunity for a hearing as provided in this subsection, shall issue an order affirming or modifying the abatement requirements in such citation. The rules of procedure prescribed by the Commission shall provide affected employees or representatives of affected employees an opportunity to participate as parties to hearings under this subsection.

80 Stat. 384.

JUDICIAL REVIEW

SEC. 11. (a) Any person adversely affected or aggrieved by an order of the Commission issued under subsection (c) of section 10 may obtain a review of such order in any United States court of appeals for the circuit in which the violation is alleged to have occurred or where the employer has its principal office, or in the Court of Appeals for the District of Columbia Circuit, by filing in such court within sixty days following the issuance of such order a written petition praying that the order be modified or set aside. A copy of such petition shall be forthwith transmitted by the clerk of the court to the Commission and to the other parties, and thereupon the Commission shall file in the court the record in the proceeding as provided in section 2112 of title 28, United States Code. Upon such filing, the court shall have jurisdiction of the proceeding and of the question determined therein, and shall have power to grant such temporary relief or restraining order as it deems just and proper, and to make and enter upon the pleadings, testimony, and proceedings set forth in such record a decree affirming, modifying, or setting aside in whole or in part, the order of the Commission and enforcing the same to the extent that such order is affirmed or modified. The commencement of proceedings under this subsection shall not, unless ordered by the court, operate as a stay of the order of the Commission. No objection that has not been urged before the Commission shall be considered by the court, unless the failure or neglect to urge such objection shall be excused because of extraordinary circumstances. The findings of the Commission with respect to questions of fact, if supported by substantial evidence on the record considered as a whole, shall be conclusive. If any party shall apply to the court for leave to adduce additional evidence and shall show to the satisfaction of the court that such additional evidence is material and that there were reasonable grounds for the failure to adduce such evidence in the hearing before the Commission, the court may order such additional evidence to be taken before the Commission and to be made a part of the record. The Commission may modify its findings as to the facts, or make new findings, by reason of additional evidence so taken and filed, and it shall file such modified or new findings, which findings with respect to questions of fact, if supported by substantial evi-

72 Stat. 941;
80 Stat. 1323.

dence on the record considered as a whole, shall be conclusive, and its recommendations, if any, for the modification or setting aside of its original order. Upon the filing of the record with it, the jurisdiction of the court shall be exclusive and its judgment and decree shall be final, except that the same shall be subject to review by the Supreme Court of the United States, as provided in section 1254 of title 28, United States Code. Petitions filed under this subsection shall be heard expeditiously.

(b) The Secretary may also obtain review or enforcement of any final order of the Commission by filing a petition for such relief in the United States court of appeals for the circuit in which the alleged violation occurred or in which the employer has its principal office, and the provisions of subsection (a) shall govern such proceedings to the extent applicable. If no petition for review, as provided in subsection (a), is filed within sixty days after service of the Commission's order, the Commission's findings of fact and order shall be conclusive in connection with any petition for enforcement which is filed by the Secretary after the expiration of such sixty-day period. In any such case, as well as in the case of a noncontested citation or notification by the Secretary which has become a final order of the Commission under subsection (a) or (b) of section 10, the clerk of the court, unless otherwise ordered by the court, shall forthwith enter a decree enforcing the order and shall transmit a copy of such decree to the Secretary and the employer named in the petition. In any contempt proceeding brought to enforce a decree of a court of appeals entered pursuant to this subsection or subsection (a), the court of appeals may assess the penalties provided in section 17, in addition to invoking any other available remedies.

(c) (1) No person shall discharge or in any manner discriminate against any employee because such employee has filed any complaint or instituted or caused to be instituted any proceeding under or related to this Act or has testified or is about to testify in any such proceeding or because of the exercise by such employee on behalf of himself or others of any right afforded by this Act.

(2) Any employee who believes that he has been discharged or otherwise discriminated against by any person in violation of this subsection may, within thirty days after such violation occurs, file a complaint with the Secretary alleging such discrimination. Upon receipt of such complaint, the Secretary shall cause such investigation to be made as he deems appropriate. If upon such investigation, the Secretary determines that the provisions of this subsection have been violated, he shall bring an action in any appropriate United States district court against such person. In any such action the United States district courts shall have jurisdiction, for cause shown to restrain violations of paragraph (1) of this subsection and order all appropriate relief including rehiring or reinstatement of the employee to his former position with back pay.

(3) Within 90 days of the receipt of a complaint filed under 'the subsection the Secretary shall notify the complainant of his determination under paragraph 2 of this subsection.

THE OCCUPATIONAL SAFETY AND HEALTH REVIEW COMMISSION

SEC. 12. (a) The Occupational Safety and Health Review Commission is hereby established. The Commission shall be composed of three members who shall be appointed by the President, by and with the advice and consent of the Senate, from among persons who by reason of training, education, or experience are qualified to carry out the functions of the Commission under this Act. The President shall designate one of the members of the Commission to serve as Chairman.

(b) The terms of members of the Commission shall be six years except that (1) the members of the Commission first taking office shall serve, as designated by the President at the time of appointment, one for a term of two years, one for a term of four years, and one for a term of six years, and (2) a vacancy caused by the death, resignation, or removal of a member prior to the expiration of the term for which he was appointed shall be filled only for the remainder of such unexpired term. A member of the Commission may be removed by the President for inefficiency, neglect of duty, or malfeasance in office. *Terms.*

(c) (1) Section 5314 of title 5, United States Code, is amended by adding at the end thereof the following new paragraph: *80 Stat. 460.*

"(57) Chairman, Occupational Safety and Health Review Commission."

(2) Section 5315 of title 5, United States Code, is amended by adding at the end thereof the following new paragraph: *Ante, p. 776.*

"(94) Members, Occupational Safety and Health Review Commission."

(d) The principal office of the Commission shall be in the District of Columbia. Whenever the Commission deems that the convenience of the public or of the parties may be promoted, or delay or expense may be minimized, it may hold hearings or conduct other proceedings at any other place. *Location.*

(e) The Chairman shall be responsible on behalf of the Commission for the administrative operations of the Commission and shall appoint such hearing examiners and other employees as he deems necessary to assist in the performance of the Commission's functions and to fix their compensation in accordance with the provisions of chapter 51 and subchapter III of chapter 53 of title 5, United States Code, relating to classification and General Schedule pay rates: *Provided,* That assignment, removal and compensation of hearing examiners shall be in accordance with sections 3105, 3344, 5362, and 7521 of title 5, United States Code. *5 USC 5101, 5331. Ante, p. 198-1.*

(f) For the purpose of carrying out its functions under this Act, two members of the Commission shall constitute a quorum and official action can be taken only on the affirmative vote of at least two members. *Quorum.*

(g) Every official act of the Commission shall be entered of record, and its hearings and records shall be open to the public. The Commission is authorized to make such rules as are necessary for the orderly transaction of its proceedings. Unless the Commission has adopted a different rule, its proceedings shall be in accordance with the Federal Rules of Civil Procedure. *Public records.*

28 USC app.

(h) The Commission may order testimony to be taken by deposition in any proceedings pending before it at any state of such proceeding. Any person may be compelled to appear and depose, and to produce books, papers, or documents, in the same manner as witnesses may be compelled to appear and testify and produce like documentary evidence before the Commission. Witnesses whose depositions are taken under this subsection, and the persons taking such depositions, shall be entitled to the same fees as are paid for like services in the courts of the United States.

(i) For the purpose of any proceeding before the Commission, the provisions of section 11 of the National Labor Relations Act (29 U.S.C. 161) are hereby made applicable to the jurisdiction and powers of the Commission. *61 Stat. 150; Ante, p. 930.*

84 STAT. 1605

Report.

(j) A hearing examiner appointed by the Commission shall hear, and make a determination upon, any proceeding instituted before the Commission and any motion in connection therewith, assigned to such hearing examiner by the Chairman of the Commission, and shall make a report of any such determination which constitutes his final disposition of the proceedings. The report of the hearing examiner shall become the final order of the Commission within thirty days after such report by the hearing examiner, unless within such period any Commission member has directed that such report shall be reviewed by the Commission.

(k) Except as otherwise provided in this Act, the hearing examiners shall be subject to the laws governing employees in the classified civil service, except that appointments shall be made without regard to section 5108 of title 5, United States Code. Each hearing examiner shall receive compensation at a rate not less than that prescribed for GS-16 under section 5332 of title 5, United States Code.

80 Stat. 453.

Ante, p. 198-1.

PROCEDURES TO COUNTERACT IMMINENT DANGERS

SEC. 13. (a) The United States district courts shall have jurisdiction, upon petition of the Secretary, to restrain any conditions or practices in any place of employment which are such that a danger exists which could reasonably be expected to cause death or serious physical harm immediately or before the imminence of such danger can be eliminated through the enforcement procedures otherwise provided by this Act. Any order issued under this section may require such steps to be taken as may be necessary to avoid, correct, or remove such imminent danger and prohibit the employment or presence of any individual in locations or under conditions where such imminent danger exists, except individuals whose presence is necessary to avoid, correct, or remove such imminent danger or to maintain the capacity of a continuous process operation to resume normal operations without a complete cessation of operations, or where a cessation of operations is necessary, to permit such to be accomplished in a safe and orderly manner.

(b) Upon the filing of any such petition the district court shall have jurisdiction to grant such injunctive relief or temporary restraining order pending the outcome of an enforcement proceeding pursuant to this Act. The proceeding shall be as provided by Rule 65 of the Federal Rules, Civil Procedure, except that no temporary restraining order issued without notice shall be effective for a period longer than five days.

28 USC app.

(c) Whenever and as soon as an inspector concludes that conditions or practices described in subsection (a) exist in any place of employment, he shall inform the affected employees and employers of the danger and that he is recommending to the Secretary that relief be sought.

(d) If the Secretary arbitrarily or capriciously fails to seek relief under this section, any employee who may be injured by reason of such failure, or the representative of such employees, might bring an action against the Secretary in the United States district court for the district in which the imminent danger is alleged to exist or the employer has its principal office, or for the District of Columbia, for a writ of mandamus to compel the Secretary to seek such an order and for such further relief as may be appropriate.

REPRESENTATION IN CIVIL LITIGATION

SEC. 14. Except as provided in section 518(a) of title 28, United States Code, relating to litigation before the Supreme Court, the Solicitor of Labor may appear for and represent the Secretary in any civil litigation brought under this Act but all such litigation shall be subject to the direction and control of the Attorney General.

80 Stat. 613.

CONFIDENTIALITY OF TRADE SECRETS

SEC. 15. All information reported to or otherwise obtained by the Secretary or his representative in connection with any inspection or proceeding under this Act which contains or which might reveal a trade secret referred to in section 1905 of title 18 of the United States Code shall be considered confidential for the purpose of that section, except that such information may be disclosed to other officers or employees concerned with carrying out this Act or when relevant in any proceeding under this Act. In any such proceeding the Secretary, the Commission, or the court shall issue such orders as may be appropriate to protect the confidentiality of trade secrets.

62 Stat. 791.

VARIATIONS, TOLERANCES, AND EXEMPTIONS

SEC. 16. The Secretary, on the record, after notice and opportunity for a hearing may provide such reasonable limitations and may make such rules and regulations allowing reasonable variations, tolerances, and exemptions to and from any or all provisions of this Act as he may find necessary and proper to avoid serious impairment of the national defense. Such action shall not be in effect for more than six months without notification to affected employees and an opportunity being afforded for a hearing.

PENALTIES

SEC. 17. (a) Any employer who willfully or repeatedly violates the requirements of section 5 of this Act, any standard, rule, or order promulgated pursuant to section 6 of this Act, or regulations prescribed pursuant to this Act, may be assessed a civil penalty of not more than $10,000 for each violation.

(b) Any employer who has received a citation for a serious violation of the requirements of section 5 of this Act, of any standard, rule, or order promulgated pursuant to section 6 of this Act, or of any regulations prescribed pursuant to this Act, shall be assessed a civil penalty of up to $1,000 for each such violation.

(c) Any employer who has received a citation for a violation of the requirements of section 5 of this Act, of any standard, rule, or order promulgated pursuant to section 6 of this Act, or of regulations prescribed pursuant to this Act, and such violation is specifically determined not to be of a serious nature, may be assessed a civil penalty of up to $1,000 for each such violation.

(d) Any employer who fails to correct a violation for which a citation has been issued under section 9(a) within the period permitted for its correction (which period shall not begin to run until the date of the final order of the Commission in the case of any review proceeding under section 10 initiated by the employer in good faith and not solely for delay or avoidance of penalties), may be assessed a civil penalty of not more than $1,000 for each day during which such failure or violation continues.

(e) Any employer who willfully violates any standard, rule, or order promulgated pursuant to section 6 of this Act, or of any regulations prescribed pursuant to this Act, and that violation caused death to any employee, shall, upon conviction, be punished by a fine of not more than $10,000 or by imprisonment for not more than six months, or by both; except that if the conviction is for a violation committed after a first conviction of such person, punishment shall be by a fine of not more than $20,000 or by imprisonment for not more than one year, or by both.

(f) Any person who gives advance notice of any inspection to be conducted under this Act, without authority from the Secretary or his designees, shall, upon conviction, be punished by a fine of not more than $1,000 or by imprisonment for not more than six months, or by both.

(g) Whoever knowingly makes any false statement, representation, or certification in any application, record, report, plan, or other document filed or required to be maintained pursuant to this Act shall, upon conviction, be punished by a fine of not more than $10,000, or by imprisonment for not more than six months, or by both.

65 Stat. 721;
79 Stat. 234.

(h)(1) Section 1114 of title 18, United States Code, is hereby amended by striking out "designated by the Secretary of Health, Education, and Welfare to conduct investigations, or inspections under the Federal Food, Drug, and Cosmetic Act" and inserting in lieu thereof "or of the Department of Labor assigned to perform investigative, inspection, or law enforcement functions".

62 Stat. 756.

(2) Notwithstanding the provisions of sections 1111 and 1114 of title 18, United States Code, whoever, in violation of the provisions of section 1114 of such title, kills a person while engaged in or on account of the performance of investigative, inspection, or law enforcement functions added to such section 1114 by paragraph (1) of this subsection, and who would otherwise be subject to the penalty provisions of such section 1111, shall be punished by imprisonment for any term of years or for life.

(i) Any employer who violates any of the posting requirements, as prescribed under the provisions of this Act, shall be assessed a civil penalty of up to $1,000 for each violation.

(j) The Commission shall have authority to assess all civil penalties provided in this section, giving due consideration to the appropriateness of the penalty with respect to the size of the business of the employer being charged, the gravity of the violation, the good faith of the employer, and the history of previous violations.

(k) For purposes of this section, a serious violation shall be deemed to exist in a place of employment if there is a substantial probability that death or serious physical harm could result from a condition which exists, or from one or more practices, means, methods, operations, or processes which have been adopted or are in use, in such place of employment unless the employer did not, and could not with the exercise of reasonable diligence, know of the presence of the violation.

(l) Civil penalties owed under this Act shall be paid to the Secretary for deposit into the Treasury of the United States and shall accrue to the United States and may be recovered in a civil action in the name of the United States brought in the United States district court for the district where the violation is alleged to have occurred or where the employer has its principal office.

STATE JURISDICTION AND STATE PLANS

SEC. 18. (a) Nothing in this Act shall prevent any State agency or court from asserting jurisdiction under State law over any occupational safety or health issue with respect to which no standard is in effect under section 6.

(b) Any State which, at any time, desires to assume responsibility for development and enforcement therein of occupational safety and health standards relating to any occupational safety or health issue with respect to which a Federal standard has been promulgated under section 6 shall submit a State plan for the development of such standards and their enforcement.

(c) The Secretary shall approve the plan submitted by a State under subsection (b), or any modification thereof, if such plan in his judgment—

(1) designates a State agency or agencies as the agency or agencies responsible for administering the plan throughout the State,

(2) provides for the development and enforcement of safety and health standards relating to one or more safety or health issues, which standards (and the enforcement of which standards) are or will be at least as effective in providing safe and healthful employment and places of employment as the standards promulgated under section 6 which relate to the same issues, and which standards, when applicable to products which are distributed or used in interstate commerce, are required by compelling local conditions and do not unduly burden interstate commerce,

(3) provides for a right of entry and inspection of all workplaces subject to the Act which is at least as effective as that provided in section 8, and includes a prohibition on advance notice of inspections,

(4) contains satisfactory assurances that such agency or agencies have or will have the legal authority and qualified personnel necessary for the enforcement of such standards,

(5) gives satisfactory assurances that such State will devote adequate funds to the administration and enforcement of such standards,

(6) contains satisfactory assurances that such State will, to the extent permitted by its law, establish and maintain an effective and comprehensive occupational safety and health program applicable to all employees of public agencies of the State and its political subdivisions, which program is as effective as the standards contained in an approved plan,

(7) requires employers in the State to make reports to the Secretary in the same manner and to the same extent as if the plan were not in effect, and

(8) provides that the State agency will make such reports to the Secretary in such form and containing such information, as the Secretary shall from time to time require.

(d) If the Secretary rejects a plan submitted under subsection (b), he shall afford the State submitting the plan due notice and opportunity for a hearing before so doing. *Notice of hearing.*

(e) After the Secretary approves a State plan submitted under subsection (b), he may, but shall not be required to, exercise his authority under sections 8, 9, 10, 13, and 17 with respect to comparable standards promulgated under section 6, for the period specified in the next sentence. The Secretary may exercise the authority referred to above until he determines, on the basis of actual operations under the

State plan, that the criteria set forth in subsection (c) are being applied, but he shall not make such determination for at least three years after the plan's approval under subsection (c). Upon making the determination referred to in the preceding sentence, the provisions of sections 5(a)(2), 8 (except for the purpose of carrying out subsection (f) of this section), 9, 10, 13, and 17, and standards promulgated under section 6 of this Act, shall not apply with respect to any occupational safety or health issues covered under the plan, but the Secretary may retain jurisdiction under the above provisions in any proceeding commenced under section 9 or 10 before the date of determination.

Continuing evaluation.

(f) The Secretary shall, on the basis of reports submitted by the State agency and his own inspections make a continuing evaluation of the manner in which each State having a plan approved under this section is carrying out such plan. Whenever the Secretary finds, after affording due notice and opportunity for a hearing, that in the administration of the State plan there is a failure to comply substantially with any provision of the State plan (or any assurance contained therein), he shall notify the State agency of his withdrawal of approval of such plan and upon receipt of such notice such plan shall cease to be in effect, but the State may retain jurisdiction in any case commenced before the withdrawal of the plan in order to enforce standards under the plan whenever the issues involved do not relate to the reasons for the withdrawal of the plan.

Plan rejection, review.

(g) The State may obtain a review of a decision of the Secretary withdrawing approval of or rejecting its plan by the United States court of appeals for the circuit in which the State is located by filing in such court within thirty days following receipt of notice of such decision a petition to modify or set aside in whole or in part the action of the Secretary. A copy of such petition shall forthwith be served upon the Secretary, and thereupon the Secretary shall certify and file in the court the record upon which the decision complained of was issued as provided in section 2112 of title 28, United States Code. Unless the court finds that the Secretary's decision in rejecting a proposed State plan or withdrawing his approval of such a plan is not supported by substantial evidence the court shall affirm the Secretary's decision. The judgment of the court shall be subject to review by the Supreme Court of the United States upon certiorari or certification as provided in section 1254 of title 28, United States Code.

72 Stat. 941; 80 Stat. 1323.

62 Stat. 928.

(h) The Secretary may enter into an agreement with a State under which the State will be permitted to continue to enforce one or more occupational health and safety standards in effect in such State until final action is taken by the Secretary with respect to a plan submitted by a State under subsection (b) of this section, or two years from the date of enactment of this Act, whichever is earlier.

FEDERAL AGENCY SAFETY PROGRAMS AND RESPONSIBILITIES

SEC. 19. (a) It shall be the responsibility of the head of each Federal agency to establish and maintain an effective and comprehensive occupational safety and health program which is consistent with the standards promulgated under section 6. The head of each agency shall (after consultation with representatives of the employees thereof)—

(1) provide safe and healthful places and conditions of employment, consistent with the standards set under section 6;

(2) acquire, maintain, and require the use of safety equipment, personal protective equipment, and devices reasonably necessary to protect employees;

(3) keep adequate records of all occupational accidents and illnesses for proper evaluation and necessary corrective action; — Recordkeeping.

(4) consult with the Secretary with regard to the adequacy as to form and content of records kept pursuant to subsection (a)(3) of this section; and

(5) make an annual report to the Secretary with respect to occupational accidents and injuries and the agency's program under this section. Such report shall include any report submitted under section 7902(e)(2) of title 5, United States Code. — Annual report. 80 Stat. 530.

(b) The Secretary shall report to the President a summary or digest of reports submitted to him under subsection (a)(5) of this section, together with his evaluations of and recommendations derived from such reports. The President shall transmit annually to the Senate and the House of Representatives a report of the activities of Federal agencies under this section. — Report to President. Report to Congress.

(c) Section 7902(c)(1) of title 5, United States Code, is amended by inserting after "agencies" the following: "and of labor organizations representing employees".

(d) The Secretary shall have access to records and reports kept and filed by Federal agencies pursuant to subsections (a) (3) and (5) of this section unless those records and reports are specifically required by Executive order to be kept secret in the interest of the national defense or foreign policy, in which case the Secretary shall have access to such information as will not jeopardize national defense or foreign policy. — Records, etc.; availability.

RESEARCH AND RELATED ACTIVITIES

SEC. 20. (a)(1) The Secretary of Health, Education, and Welfare, after consultation with the Secretary and with other appropriate Federal departments or agencies, shall conduct (directly or by grants or contracts) research, experiments, and demonstrations relating to occupational safety and health, including studies of psychological factors involved, and relating to innovative methods, techniques, and approaches for dealing with occupational safety and health problems.

(2) The Secretary of Health, Education, and Welfare shall from time to time consult with the Secretary in order to develop specific plans for such research, demonstrations, and experiments as are necessary to produce criteria, including criteria identifying toxic substances, enabling the Secretary to meet his responsibility for the formulation of safety and health standards under this Act; and the Secretary of Health, Education, and Welfare, on the basis of such research, demonstrations, and experiments and any other information available to him, shall develop and publish at least annually such criteria as will effectuate the purposes of this Act.

(3) The Secretary of Health, Education, and Welfare, on the basis of such research, demonstrations, and experiments, and any other information available to him, shall develop criteria dealing with toxic materials and harmful physical agents and substances which will describe exposure levels that are safe for various periods of employment, including but not limited to the exposure levels at which no employee will suffer impaired health or functional capacities or diminished life expectancy as a result of his work experience.

(4) The Secretary of Health, Education, and Welfare shall also conduct special research, experiments, and demonstrations relating to occupational safety and health as are necessary to explore new problems, including those created by new technology in occupational safety and health, which may require ameliorative action beyond that

which is otherwise provided for in the operating provisions of this Act. The Secretary of Health, Education, and Welfare shall also conduct research into the motivational and behavioral factors relating to the field of occupational safety and health.

Toxic substances, records.

(5) The Secretary of Health, Education, and Welfare, in order to comply with his responsibilities under paragraph (2), and in order to develop needed information regarding potentially toxic substances or harmful physical agents, may prescribe regulations requiring employers to measure, record, and make reports on the exposure of employees to substances or physical agents which the Secretary of Health, Education, and Welfare reasonably believes may endanger the health or safety of employees. The Secretary of Health, Education, and Welfare also is authorized to establish such programs of medical examinations and tests as may be necessary for determining the incidence of occupational illnesses and the susceptibility of employees to such illnesses. Nothing in this or any other provision of this Act shall be deemed to authorize or require medical examination, immunization, or treatment for those who object thereto on religious grounds, except where such is necessary for the protection of the health or safety of others. Upon the request of any employer who is required to measure and record exposure of employees to substances or physical agents as provided under this subsection, the Secretary of Health, Education, and Welfare shall furnish full financial or other assistance to such employer for the purpose of defraying any additional expense incurred by him in carrying out the measuring and recording as provided in this subsection.

Medical examinations.

Toxic substances, publication.

(6) The Secretary of Health, Education, and Welfare shall publish within six months of enactment of this Act and thereafter as needed but at least annually a list of all known toxic substances by generic family or other useful grouping, and the concentrations at which such toxicity is known to occur. He shall determine following a written request by any employer or authorized representative of employees, specifying with reasonable particularity the grounds on which the request is made, whether any substance normally found in the place of employment has potentially toxic effects in such concentrations as used or found; and shall submit such determination both to employers and affected employees as soon as possible. If the Secretary of Health, Education, and Welfare determines that any substance is potentially toxic at the concentrations in which it is used or found in a place of employment, and such substance is not covered by an occupational safety or health standard promulgated under section 6, the Secretary of Health, Education, and Welfare shall immediately submit such determination to the Secretary, together with all pertinent criteria.

Annual studies.

(7) Within two years of enactment of this Act, and annually thereafter the Secretary of Health, Education, and Welfare shall conduct and publish industrywide studies of the effect of chronic or low-level exposure to industrial materials, processes, and stresses on the potential for illness, disease, or loss of functional capacity in aging adults.

Inspections.

(b) The Secretary of Health, Education, and Welfare is authorized to make inspections and question employers and employees as provided in section 8 of this Act in order to carry out his functions and responsibilities under this section.

Contract authority.

(c) The Secretary is authorized to enter into contracts, agreements, or other arrangements with appropriate public agencies or private organizations for the purpose of conducting studies relating to his responsibilities under this Act. In carrying out his responsibilities under this subsection, the Secretary shall cooperate with the Secretary of Health, Education, and Welfare in order to avoid any duplication of efforts under this section.

(d) Information obtained by the Secretary and the Secretary of Health, Education, and Welfare under this section shall be disseminated by the Secretary to employers and employees and organizations thereof.

(e) The functions of the Secretary of Health, Education, and Welfare under this Act shall, to the extent feasible, be delegated to the Director of the National Institute for Occupational Safety and Health established by section 22 of this Act.

<small>Delegation of functions.</small>

TRAINING AND EMPLOYEE EDUCATION

SEC. 21. (a) The Secretary of Health, Education, and Welfare, after consultation with the Secretary and with other appropriate Federal departments and agencies, shall conduct, directly or by grants or contracts (1) education programs to provide an adequate supply of qualified personnel to carry out the purposes of this Act, and (2) informational programs on the importance of and proper use of adequate safety and health equipment.

(b) The Secretary is also authorized to conduct, directly or by grants or contracts, short-term training of personnel engaged in work related to his responsibilities under this Act.

(c) The Secretary, in consultation with the Secretary of Health, Education, and Welfare, shall (1) provide for the establishment and supervision of programs for the education and training of employers and employees in the recognition, avoidance, and prevention of unsafe or unhealthful working conditions in employments covered by this Act, and (2) consult with and advise employers and employees, and organizations representing employers and employees as to effective means of preventing occupational injuries and illnesses.

NATIONAL INSTITUTE FOR OCCUPATIONAL SAFETY AND HEALTH

SEC. 22. (a) It is the purpose of this section to establish a National Institute for Occupational Safety and Health in the Department of Health, Education, and Welfare in order to carry out the policy set forth in section 2 of this Act and to perform the functions of the Secretary of Health, Education, and Welfare under sections 20 and 21 of this Act.

<small>Establishment.</small>

(b) There is hereby established in the Department of Health, Education, and Welfare a National Institute for Occupational Safety and Health. The Institute shall be headed by a Director who shall be appointed by the Secretary of Health, Education, and Welfare, and who shall serve for a term of six years unless previously removed by the Secretary of Health, Education, and Welfare.

<small>Director, appointment, term.</small>

(c) The Institute is authorized to—
(1) develop and establish recommended occupational safety and health standards; and
(2) perform all functions of the Secretary of Health, Education, and Welfare under sections 20 and 21 of this Act.

(d) Upon his own initiative, or upon the request of the Secretary or the Secretary of Health, Education, and Welfare, the Director is authorized (1) to conduct such research and experimental programs as he determines are necessary for the development of criteria for new and improved occupational safety and health standards, and (2) after consideration of the results of such research and experimental programs make recommendations concerning new or improved occupational safety and health standards. Any occupational safety and health standard recommended pursuant to this section shall immediately be forwarded to the Secretary of Labor, and to the Secretary of Health, Education, and Welfare.

(e) In addition to any authority vested in the Institute by other provisions of this section, the Director, in carrying out the functions of the Institute, is authorized to—

(1) prescribe such regulations as he deems necessary governing the manner in which its functions shall be carried out;

(2) receive money and other property donated, bequeathed, or devised, without condition or restriction other than that it be used for the purposes of the Institute and to use, sell, or otherwise dispose of such property for the purpose of carrying out its functions;

(3) receive (and use, sell, or otherwise dispose of, in accordance with paragraph (2)), money and other property donated, bequeathed, or devised to the Institute with a condition or restriction, including a condition that the Institute use other funds of the Institute for the purposes of the gift;

(4) in accordance with the civil service laws, appoint and fix the compensation of such personnel as may be necessary to carry out the provisions of this section;

(5) obtain the services of experts and consultants in accordance with the provisions of section 3109 of title 5, United States Code;

80 Stat. 416.

(6) accept and utilize the services of voluntary and noncompensated personnel and reimburse them for travel expenses, including per diem, as authorized by section 5703 of title 5, United States Code;

83 Stat. 190.

(7) enter into contracts, grants or other arrangements, or modifications thereof to carry out the provisions of this section, and such contracts or modifications thereof may be entered into without performance or other bonds, and without regard to section 3709 of the Revised Statutes, as amended (41 U.S.C. 5), or any other provision of law relating to competitive bidding;

(8) make advance, progress, and other payments which the Director deems necessary under this title without regard to the provisions of section 3648 of the Revised Statutes, as amended (31 U.S.C. 529); and

(9) make other necessary expenditures.

Annual report to HEW, President, and Congress.

(f) The Director shall submit to the Secretary of Health, Education, and Welfare, to the President, and to the Congress an annual report of the operations of the Institute under this Act, which shall include a detailed statement of all private and public funds received and expended by it, and such recommendations as he deems appropriate.

GRANTS TO THE STATES

SEC. 23. (a) The Secretary is authorized, during the fiscal year ending June 30, 1971, and the two succeeding fiscal years, to make grants to the States which have designated a State agency under section 18 to assist them—

(1) in identifying their needs and responsibilities in the area of occupational safety and health,

(2) in developing State plans under section 18, or

(3) in developing plans for—
(A) establishing systems for the collection of information concerning the nature and frequency of occupational injuries and diseases;
(B) increasing the expertise and enforcement capabilities of their personnel engaged in occupational safety and health programs; or
(C) otherwise improving the administration and enforcement of State occupational safety and health laws, including standards thereunder, consistent with the objectives of this Act.

(b) The Secretary is authorized, during the fiscal year ending June 30, 1971, and the two succeeding fiscal years, to make grants to the States for experimental and demonstration projects consistent with the objectives set forth in subsection (a) of this section.

(c) The Governor of the State shall designate the appropriate State agency for receipt of any grant made by the Secretary under this section.

(d) Any State agency designated by the Governor of the State desiring a grant under this section shall submit an application therefor to the Secretary.

(e) The Secretary shall review the application, and shall, after consultation with the Secretary of Health, Education, and Welfare, approve or reject such application.

(f) The Federal share for each State grant under subsection (a) or (b) of this section may not exceed 90 per centum of the total cost of the application. In the event the Federal share for all States under either such subsection is not the same, the differences among the States shall be established on the basis of objective criteria.

(g) The Secretary is authorized to make grants to the States to assist them in administering and enforcing programs for occupational safety and health contained in State plans approved by the Secretary pursuant to section 18 of this Act. The Federal share for each State grant under this subsection may not exceed 50 per centum of the total cost to the State of such a program. The last sentence of subsection (f) shall be applicable in determining the Federal share under this subsection.

(h) Prior to June 30, 1973, the Secretary shall, after consultation with the Secretary of Health, Education, and Welfare, transmit a report to the President and to the Congress, describing the experience under the grant programs authorized by this section and making any recommendations he may deem appropriate.

Report to President and Congress.

STATISTICS

SEC. 24. (a) In order to further the purposes of this Act, the Secretary, in consultation with the Secretary of Health, Education, and Welfare, shall develop and maintain an effective program of collection, compilation, and analysis of occupational safety and health statistics. Such program may cover all employments whether or not subject to any other provisions of this Act but shall not cover employments excluded by section 4 of the Act. The Secretary shall compile accurate statistics on work injuries and illnesses which shall include all disabling, serious, or significant injuries and illnesses, whether or not involving loss of time from work, other than minor injuries requiring only first aid treatment and which do not involve medical treatment, loss of consciousness, restriction of work or motion, or transfer to another job.

439

(b) To carry out his duties under subsection (a) of this section, the Secretary may—

(1) promote, encourage, or directly engage in programs of studies, information and communication concerning occupational safety and health statistics;

(2) make grants to States or political subdivisions thereof in order to assist them in developing and administering programs dealing with occupational safety and health statistics; and

(3) arrange, through grants or contracts, for the conduct of such research and investigations as give promise of furthering the objectives of this section.

(c) The Federal share for each grant under subsection (b) of this section may be up to 50 per centum of the State's total cost.

(d) The Secretary may, with the consent of any State or political subdivision thereof, accept and use the services, facilities, and employees of the agencies of such State or political subdivision, with or without reimbursement, in order to assist him in carrying out his functions under this section.

Reports.
(e) On the basis of the records made and kept pursuant to section 8(c) of this Act, employers shall file such reports with the Secretary as he shall prescribe by regulation, as necessary to carry out his functions under this Act.

(f) Agreements between the Department of Labor and States pertaining to the collection of occupational safety and health statistics already in effect on the effective date of this Act shall remain in effect until superseded by grants or contracts made under this Act.

AUDITS

SEC. 25. (a) Each recipient of a grant under this Act shall keep such records as the Secretary or the Secretary of Health, Education, and Welfare shall prescribe, including records which fully disclose the amount and disposition by such recipient of the proceeds of such grant, the total cost of the project or undertaking in connection with which such grant is made or used, and the amount of that portion of the cost of the project or undertaking supplied by other sources, and such other records as will facilitate an effective audit.

(b) The Secretary or the Secretary of Health, Education, and Welfare, and the Comptroller General of the United States, or any of their duly authorized representatives, shall have access for the purpose of audit and examination to any books, documents, papers, and records of the recipients of any grant under this Act that are pertinent to any such grant.

ANNUAL REPORT

SEC. 26. Within one hundred and twenty days following the convening of each regular session of each Congress, the Secretary and the Secretary of Health, Education, and Welfare shall each prepare and submit to the President for transmittal to the Congress a report upon the subject matter of this Act, the progress toward achievement of the purpose of this Act, the needs and requirements in the field of occupational safety and health, and any other relevant information. Such reports shall include information regarding occupational safety and health standards, and criteria for such standards, developed during the preceding year; evaluation of standards and criteria previously developed under this Act, defining areas of emphasis for new criteria and standards; an evaluation of the degree of observance of applicable occupational safety and health standards, and a summary of inspection and enforcement activity undertaken; analysis and evaluation of research activities for which results have been obtained under governmental and nongovernmental sponsorship; an analysis of major occupational diseases; evaluation of available control and measurement technology for hazards for which standards or criteria have been developed during the preceding year; description of cooperative efforts undertaken between Government agencies and other interested parties in the implementation of this Act during the preceding year; a progress report on the development of an adequate supply of trained manpower in the field of occupational safety and health, including estimates of future needs and the efforts being made by Government and others to meet those needs; listing of all toxic substances in industrial usage for which labeling requirements, criteria, or standards have not yet been established; and such recommendations for additional legislation as are deemed necessary to protect the safety and health of the worker and improve the administration of this Act.

NATIONAL COMMISSION ON STATE WORKMEN'S COMPENSATION LAWS

SEC. 27. (a) (1) The Congress hereby finds and declares that—
(A) the vast majority of American workers, and their families, are dependent on workmen's compensation for their basic economic security in the event such workers suffer disabling injury or death in the course of their employment; and that the full protection of American workers from job-related injury or death requires an adequate, prompt, and equitable system of workmen's compensation as well as an effective program of occupational health and safety regulation; and
(B) in recent years serious questions have been raised concerning the fairness and adequacy of present workmen's compensation laws in the light of the growth of the economy, the changing nature of the labor force, increases in medical knowledge, changes in the hazards associated with various types of employment, new technology creating new risks to health and safety, and increases in the general level of wages and the cost of living.
(2) The purpose of this section is to authorize an effective study and objective evaluation of State workmen's compensation laws in order to determine if such laws provide an adequate, prompt, and equitable system of compensation for injury or death arising out of or in the course of employment.
(b) There is hereby established a National Commission on State Workmen's Compensation Laws. **Establishment.**
(c) (1) The Workmen's Compensation Commission shall be composed of fifteen members to be appointed by the President from among members of State workmen's compensation boards, representatives of insurance carriers, business, labor, members of the medical profession having experience in industrial medicine or in workmen's compensation cases, educators having special expertise in the field of workmen's compensation, and representatives of the general public. The Secretary, the Secretary of Commerce, and the Secretary of Health, Education, and Welfare shall be ex officio members of the Workmen's Compensation Commission: **Membership.**
(2) Any vacancy in the Workmen's Compensation Commission shall not affect its powers.
(3) The President shall designate one of the members to serve as Chairman and one to serve as Vice Chairman of the Workmen's Compensation Commission.

Quorum.

(4) Eight members of the Workmen's Compensation Commission shall constitute a quorum.

Study.

(d)(1) The Workmen's Compensation Commission shall undertake a comprehensive study and evaluation of State workmen's compensation laws in order to determine if such laws provide an adequate, prompt, and equitable system of compensation. Such study and evaluation shall include, without being limited to, the following subjects: (A) the amount and duration of permanent and temporary disability benefits and the criteria for determining the maximum limitations thereon, (B) the amount and duration of medical benefits and provisions insuring adequate medical care and free choice of physician, (C) the extent of coverage of workers, including exemptions based on numbers or type of employment, (D) standards for determining which injuries or diseases should be deemed compensable, (E) rehabilitation, (F) coverage under second or subsequent injury funds, (G) time limits on filing claims, (H) waiting periods, (I) compulsory or elective coverage, (J) administration, (K) legal expenses, (L) the feasibility and desirability of a uniform system of reporting information concerning job-related injuries and diseases and the operation of workmen's compensation laws, (M) the resolution of conflict of laws, extraterritoriality and similar problems arising from claims with multistate aspects, (N) the extent to which private insurance carriers are excluded from supplying workmen's compensation coverage and the desirability of such exclusionary practices, to the extent they are found to exist, (O) the relationship between workmen's compensation on the one hand, and old-age, disability, and survivors insurance and other types of insurance, public or private, on the other hand, (P) methods of implementing the recommendations of the Commission.

Report to President and Congress.

(2) The Workmen's Compensation Commission shall transmit to the President and to the Congress not later than July 31, 1972, a final report containing a detailed statement of the findings and conclusions of the Commission, together with such recommendations as it deems advisable.

Hearings.

(e)(1) The Workmen's Compensation Commission or, on the authorization of the Workmen's Compensation Commission, any subcommittee or members thereof, may, for the purpose of carrying out the provisions of this title, hold such hearings, take such testimony, and sit and act at such times and places as the Workmen's Compensation Commission deems advisable. Any member authorized by the Workmen's Compensation Commission may administer oaths or affirmations to witnesses appearing before the Workmen's Compensation Commission or any subcommittee or members thereof.

(2) Each department, agency, and instrumentality of the executive branch of the Government, including independent agencies, is authorized and directed to furnish to the Workmen's Compensation Commission, upon request made by the Chairman or Vice Chairman, such information as the Workmen's Compensation Commission deems necessary to carry out its functions under this section.

(f) Subject to such rules and regulations as may be adopted by the Workmen's Compensation Commission, the Chairman shall have the power to—

80 Stat. 378.
5 USC 101.

(1) appoint and fix the compensation of an executive director, and such additional staff personnel as he deems necessary, without regard to the provisions of title 5, United States Code, governing appointments in the competitive service, and without regard to the provisions of chapter 51 and subchapter III of chapter 53 of such title relating to classification and General Schedule

5 USC 5101, 5331.

pay rates, but at rates not in excess of the maximum rate for GS-18 of the General Schedule under section 5332 of such title, and

(2) procure temporary and intermittent services to the same extent as is authorized by section 3109 of title 5, United States Code.

(g) The Workmen's Compensation Commission is authorized to enter into contracts with Federal or State agencies, private firms, institutions, and individuals for the conduct of research or surveys, the preparation of reports, and other activities necessary to the discharge of its duties.

(h) Members of the Workmen's Compensation Commission shall receive compensation for each day they are engaged in the performance of their duties as members of the Workmen's Compensation Commission at the daily rate prescribed for GS-18 under section 5332 of title 5, United States Code, and shall be entitled to reimbursement for travel, subsistence, and other necessary expenses incurred by them in the performance of their duties as members of the Workmen's Compensation Commission.

(i) There are hereby authorized to be appropriated such sums as may be necessary to carry out the provisions of this section.

(j) On the ninetieth day after the date of submission of its final report to the President, the Workmen's Compensation Commission shall cease to exist.

ECONOMIC ASSISTANCE TO SMALL BUSINESSES

SEC. 28. (a) Section 7(b) of the Small Business Act, as amended, is amended—

(1) by striking out the period at the end of "paragraph (5)" and inserting in lieu thereof "; and"; and

(2) by adding after paragraph (5) a new paragraph as follows:

"(6) to make such loans (either directly or in cooperation with banks or other lending institutions through agreements to participate on an immediate or deferred basis) as the Administration may determine to be necessary or appropriate to assist any small business concern in effecting additions to or alterations in the equipment, facilities, or methods of operation of such business in order to comply with the applicable standards promulgated pursuant to section 6 of the Occupational Safety and Health Act of 1970 or standards adopted by a State pursuant to a plan approved under section 18 of the Occupational Safety and Health Act of 1970, if the Administration determines that such concern is likely to suffer substantial economic injury without assistance under this paragraph."

(b) The third sentence of section 7(b) of he Small Business Act, as amended, is amended by striking out "or (5)" after "paragraph (3)" and inserting a comma followed by "(5) or (6)".

(c) Section 4(c)(1) of the Small Business Act, as amended, is amended by inserting "7(b)(6)," after "7(b)(5),".

(d) Loans may also be made or guaranteed for the purposes set forth in section 7(b)(6) of the Small Business Act, as amended, pursuant to the provisions of section 202 of the Public Works and Economic Development Act of 1965, as amended.

ADDITIONAL ASSISTANT SECRETARY OF LABOR

SEC. 29. (a) Section 2 of the Act of April 17, 1946 (60 Stat. 91) as amended (29 U.S.C. 553) is amended by—

84 STAT. 1619

(1) striking out "four" in the first sentence of such section and inserting in lieu thereof "five"; and

(2) adding at the end thereof the following new sentence, "One of such Assistant Secretaries shall be an Assistant Secretary of Labor for Occupational Safety and Health.".

80 Stat. 462.

(b) Paragraph (20) of section 5315 of title 5, United States Code, is amended by striking out "(4)" and inserting in lieu thereof "(5)".

ADDITIONAL POSITIONS

SEC. 30. Section 5108(c) of title 5, United States Code, is amended by—

(1) striking out the word "and" at the end of paragraph (8);
(2) striking out the period at the end of paragraph (9) and inserting in lieu thereof a semicolon and the word "and"; and
(3) by adding immediately after paragraph (9) the following new paragraph:

"(10) (A) the Secretary of Labor, subject to the standards and procedures prescribed by this chapter, may place an additional twenty-five positions in the Department of Labor in GS-16, 17, and 18 for the purposes of carrying out his responsibilities under the Occupational Safety and Health Act of 1970;

"(B) the Occupational Safety and Health Review Commission, subject to the standards and procedures prescribed by this chapter, may place ten positions in GS-16, 17, and 18 in carrying out its functions under the Occupational Safety and Health Act of 1970."

EMERGENCY LOCATOR BEACONS

72 Stat. 775.
49 USC 1421.

SEC. 31. Section 601 of the Federal Aviation Act of 1958 is amended by inserting at the end thereof a new subsection as follows:

"EMERGENCY LOCATOR BEACONS

"(d)(1) Except with respect to aircraft described in paragraph (2) of this subsection, minimum standards pursuant to this section shall include a requirement that emergency locator beacons shall be installed—

"(A) on any fixed-wing, powered aircraft for use in air commerce the manufacture of which is completed, or which is imported into the United States, after one year following the date of enactment of this subsection; and

"(B) on any fixed-wing, powered aircraft used in air commerce after three years following such date.

"(2) The provisions of this subsection shall not apply to jet-powered aircraft; aircraft used in air transportation (other than air taxis and charter aircraft); military aircraft; aircraft used solely for training purposes not involving flights more than twenty miles from its base; and aircraft used for the aerial application of chemicals."

SEPARABILITY

SEC. 32. If any provision of this Act, or the application of such provision to any person or circumstance, shall be held invalid, the remainder of this Act, or the application of such provision to persons or circumstances other than those as to which it is held invalid, shall not be affected thereby.

APPROPRIATIONS

Sec. 33. There are authorized to be appropriated to carry out this Act for each fiscal year such sums as the Congress shall deem necessary.

EFFECTIVE DATE

Sec. 34. This Act shall take effect one hundred and twenty days after the date of its enactment.

Approved December 29, 1970.

LEGISLATIVE HISTORY:

HOUSE REPORTS: No. 91-1291 accompanying H.R. 16785 (Comm. on Education and Labor) and No. 91-1765 (Comm. of Conference).
SENATE REPORT No. 91-1282 (Comm. on Labor and Public Welfare).
CONGRESSIONAL RECORD, Vol. 116 (1970):
    Oct. 13, Nov. 16, 17, considered and passed Senate.
    Nov. 23, 24, considered and passed House, amended, in lieu of H.R. 16785.
    Dec. 16, Senate agreed to conference report.
    Dec. 17, House agreed to conference report.

B. ESTABLISHED INDUSTRIAL HEALTH PROGRAMS

The following is a listing of certain established industrial health programs. The listing is not complete, but it is indicative of what has been done to date. The concept and scope of such programs can differ markedly, as has been shown in the main text.

MANUFACTURING

Abcor, Inc.
A. C. Horn Co.
Air Products and Chemicals, Inc.
Alabama Dry Dock Co.
Allen Manufacturing Co.
Allied Chemical Corporation
Allis-Chalmers Corp.
Aluminum Company of America
AMAX Inc.
American Can Company
American Cyanamid Company
American Hoechst Corporation
American Motors Corporation
American Smelting and Refining Co.
American Sterilizer Co.
AMETEK Inc.
Anaconda Co.
Armco Steel Corporation
Armstrong Cork Co.
Atlantic Companies
Atlantic Richfield Co.
Aviation Corporation
Avondale Shipyards, Inc.
L. S. Ayres & Company, Inc.

Ball Corporation
Babcock & Wilcox Co.
Bayuk Cigars Inc.
Bechtel Corporation
Bell and Howell Co.
Bendix Corporation
Bethlehem Steel Co.
Boeing Co.
Borden, Inc.
Bristol-Myers Co.
Brunswick Corporation
Brush Wellman Inc.
Budd Co.
E. D. Bullard Co.
Burlington Industries, Inc.

Campbell Soup Co.
Carson Pirie Scott & Co.
Caterpillar Tractor Co.

Chrysler Corporation
Ciba-Geigy Corp.
Colgate-Palmolive Co.
Colt Industries Inc.
Cone Mills Corp.
Consolidated Papers, Inc.
Continental Can Co., Inc.
Convair
Corning Glass Works
Creole Petroleum Corp.
Cummins Engine Co., Inc.
Cutler-Hammer, Inc.

Darling Valve & Manufacturing Co.
Deere and Company
Detrex Chemical Industries, Inc.
Diamond Shamrock Corp.
R. R. Donnelley & Sons Co.
Dow Corning Corp.
Dow Jones Chemical Co.
E. I. Du Pont deNemours & Co., Inc.

Eastman Kodak Co.
Eckel Industries, Inc.
Eli Lilly and Company
Endicott Johnson Corp.
Engelhard Minerals & Chemicals Corp.
Exxon Corporation

Farmland Industries, Inc.
Fieldcrest Mills, Inc.
Fingerhut Corp.
Firestone Tire & Rubber Co.
Foote, Cone, & Belding Communications, Inc.
Ford Motor Co.

Gates Rubber Co.
General Dynamics Corp.
General Electric Co.
General Mills, Inc.
General Motors Corp.

General Telephone & Electronics Corporation
General Tire & Rubber Co.
GENTEX Corp.
Georgia Marble Co.
B. F. Goodrich Co.
Goodyear Tire & Rubber Co.
W. R. Grace & Co.
GTE Automatic Electric Incorporated
GTE Sylvania Inc.
Gulf Oil Corp.
Gulf Oil of Canada Limited

Harsco Corp.
Heil Co.
H. J. Heinz Co.
Hercules, Incorporated
Hitchcock Industries, Inc.
Hoerner Waldorf Corp.
Homestake Mining Co.
Honeywell Inc.
Hooker Chemicals and Plastics Corp.
Hudson Bay Mining & Smelting Co., Ltd.
Hughes Aircraft Co.
Hughes Tool Co.

Ingersoll-Rand Co.
Inland Steel Co.
International Business Machines Corp.
International Harvester Co.
International Minerals and Chemical Corp.

Johns Manville Corp.
Jones & Laughlin Steel Corp.

Kaiser Aluminum & Chemical Corp.
Kaiser Industries Corp.
Kawecki Berylco Industries, Inc.
Kennecott Copper Corp.
Kimberly-Clark Corp.
Kimble-Glass Co.
Koehler Mfg. Co.

Lehigh Portland Cement Co.
Lever Brothers Company
Libby, McNeill & Libby
Lone Star Steel Co.

Marathon Oil Co.
Martin Marietta Corp.
McDonnell Douglas Corp.
McNeil Corp.
Mesta Machine Co.

Metpath Inc.
Millipore Corp.
3M Company
Mobay Chemical Co.
Mobile Oil Corp.
Monsanto Company
Motorola, Inc.

Newport News Industrial Corp.
New York Naval Shipyard
Norris Industries, Inc.
Norton Co.

Ohmart Corp.
Oscar Mayer & Co., Inc.
Otter Tail Power Co.
Owens-Corning Fiberglas Corp.
Owens-Illinois, Inc.

Philco-Ford Corp.
Phillips Petroleum Co.
Pickands Mather & Co.
Pillsbury Co.
Pitney-Bowes, Inc.
Potlatch Corp.
Prestolite Co.
Procter & Gamble Co.

Raybestos-Manhattan, Inc.
Raytheon Co.
Reichhold Chemicals, Inc.
Remington Arms Co., Inc.
Republic Steel Corp.
Reynolds Metals Co.
Rockwell International Corp.
Rohm & Haas Co.

St. Regis Paper Co.
Schenley Industries, Inc.
Scott Aviation Division (A-T-O Inc.)
Scott Paper Co.
Scovill Mfg. Co.
G. D. Searle & Co.
Shell Canada, Ltd.
Shell Oil Company
Sherwin-Williams Co.
Socony-Vacuum Oil Co.
Sperry Rand Corp.
Standard Oil Co. of California
Standard Oil Co. (Indiana)
Stauffer Chemical Co.
Sterling Drug Inc.
Stromberg-Carlson Corp.
Sun Oil Co.
Swift & Co.
Sybron Corp.

448   APPENDIXES

Tenneco Inc.
Thompson Hayward Chemical Co.
TRW Inc.

Union Carbide Corp.
Uniroyal Inc.
United States Steel Corp.
Upjohn Co.

Virginia Chemicals Inc.

Westinghouse Electric Corp.
Weyerhaeuser Co.
Wilson & Co., Inc.
Witco Chemical Corporation

Xerox Corp.

Insurance

Aetna Life & Casualty
American Mutual Liability Insurance Co.

Connecticut General Life Insurance Co.
Continental Insurance Co.

Equitable Life Assurance Society of the United States

Fireman's Fund Insurance Company

Hartford Fire Insurance

Insurance Co. of North America

John Hancock Mutual Life Insurance Co.

Kemper Corporation

Liberty Mutual Insurance Co.

Metropolitan Life Insurance Co.
Michigan Mutual Liability Co.
Mutual of Omaha Insurance Co.

New York Life Insurance Co.

Pacific Mutual Life Insurance Co.
Prudential Insurance Co. of America

The Travelers Corp.

Transportation

Akron Canton & Youngstown RR.
Alabama Dry Dock & Shipbuilding Co.
American Airlines, Inc.

Baltimore & Ohio RR. Co.

Chicago, Milwaukee, St. Paul and Pacific RR.
Chicago Transit Authority

Grand Trunk Western Railroad Co.

Massachusetts Bay Transportation Authority
Missouri Pacific RR.

New York City Transit Authority

Pan American World Airways, Inc.
Penn Central Co.

Roadway Express, Inc.

Santa Fe Industries, Inc.
Southern Pacific Co.

Trans World Airlines, Inc.

Union Pacific RR.
United Air Lines, Inc.

Utilities

American Telephone & Telegraph Co.
Arizona Public Service Co.

Baltimore Gas & Electric Co.
Bell Telephone Co. of Canada
Bell Telephone Laboratories, Incorporated
Boston Edison Co.

Cities Service Co.
Commonwealth Edison Co.
Consolidated Edison Co. of New York, Inc.

Detroit Edison Co.

Georgia Power Co.

Illinois Bell Telephone Co.
International Telephone and
  Telegraph Corp.

Long Island Lighting Co.

Michigan Bell Telephone Co.
Mountain States Telephone &
  Telegraph Co.

New England Electric System
New England Telephone & Telegraph
  Co.
New Jersey Bell Telephone Co.
New Orleans Public Service Inc.
New York Telephone Co.
Niagara Mohawk Power Corp.
Northern States Power Co.

Ohio Edison Co.

Pacific Telephone & Telegraph Co.
Peoples Gas Light & Coke Co.
Philadelphia Electric Co.

Southern California Gas Co.
Southern Natural Gas Co.
Southwestern Bell Telephone Co.

Texas Power & Light Co.

Western Electric Co., Inc.
Western Union Telegraph Co.

Service

Control Data Corp.

Finance

Chase Manhattan Bank, N. A.
Chemical Bank
Federal Reserve Bank
First National Bank of Minneapolis
Morgan Guaranty Trust Co. of New
  York
Northwestern National Bank
Seamen's Bank for Savings
United California Bank

Communications

American Broadcasting Co.
Boston Globe
CBS Inc.
The Curtis Publishing Company
McGraw-Hill, Inc.
The New York Times Company
RCA Corp.
Time Inc.
Tribune Co. (Chicago)

Retailing

Carson Pirie Scott & Co.
Emporium Capwell Co.
Gimbel Bros., Inc.
Goldblatt's
R. H. Macy & Co., Inc.
Sears, Roebuck and Co.

C. SOURCES OF INFORMATION, ASSISTANCE, AND GUIDANCE

This Appendix includes the major sources of information, assistance, and guidance for anyone who is interested in establishing or improving industrial health or safety programs, or who is concerned with the surveillance of product safety or aspects of environmental health. Although space does not permit listing the address of every organization included, the reader may refer to state and local medical societies, state and city health departments, and local information services, such as public libraries, municipal offices, health agencies, union locals, directories of consultants, or telephone directories, for convenient sources of assistance.

HEALTH PROFESSIONAL ORGANIZATIONS

American Academy of Compensation
  Medicine
221 W. 57th St.
New York, N.Y. 10019

American Academy of Occupational
  Medicine
801 Old Lancaster Rd.
Bryn Mawr, Pa. 19010

American Association of Industrial
  Nurses, Inc.
170 E. 61st St.
New York, N.Y. 10021

American Association of Industrial
  Physicians and Surgeons
28 E. Jackson Blvd.
Chicago, Ill.

American Conference of Government
  Industrial Hygienists
1014 Broadway
Cincinnati, O. 45202

American Industrial Hygiene
  Association
66 Miller Rd.
Akron, O. 44313

American Medical Association
  Council on Industrial Health
535 N. Dearborn St.
Chicago, Ill. 60610

American Nurses Association
  Industrial Nurses Section
10 Columbus Circle
New York, N.Y. 10019

American Occupational Medical
  Association
150 N. Wacker Dr.
Chicago, Ill. 60606

American Occupational Therapy
  Association
250 W. 50th St.
New York, N.Y. 10019

American Physical Therapy
  Association
1790 Broadway
New York, N.Y. 10019

American Psychiatric Association
1700 18th St., N.W.
Washington, D.C. 20009

American Psychological Association
1200 17th St., N.W.
Washington, D.C. 20036

American Public Health Association
  Occupational Health Section
1015 18th St., N.W.
Washington, D.C. 20036

SOURCES OF INFORMATION, ASSISTANCE, AND GUIDANCE

Canadian Nurses Association
75 Stanley Ave.
Ottawa 2, Ontario, Canada

Drug Abuse Council, Inc.
1828 L St., N.W.
Washington, D.C. 20036

Industrial Health Foundation
5231 Center Ave.
Pittsburgh, Pa. 15213

Industrial Hygiene Association
4400 Fifth Ave.
Pittsburgh, Pa. 15213

National Center for Disease Control
1600 Clifton Rd., N.E.
Atlanta, Ga. 30333

National Council on Alcoholism
733 Third Ave.
New York, N.Y. 10017

National Institutes of Health
Washington, D.C. 20203

National Institute on Alcohol Abuse
 and Alcoholism
5600 Fishers Lane
Rockville, Md. 20852

National League for Nursing
10 Columbus Circle
New York, N.Y. 10019

Occupational Health Institute
55 E. Washington St.
Chicago, Ill. 60602

Occupational Safety and Health
 Administration
Washington, D.C.

Occupational Safety/Health
 Accreditation Program Commission
 (Contact the National Institute of
 Occupational Safety and Health,
 U.S. Post Office, Cincinnati, O.
 45202)

Labor Organizations

AFL-CIO
815 16th St., N.W.
Washington, D.C. 20006

International Labour Organisation
Geneva, Switzerland

Unions with an active interest in
 industrial health can be of
 assistance. Among them are:

International Association of
 Machinists
909 Machinists Building
Washington, D.C. 20036

Oil, Chemical, and Atomic Workers
 International Union
1636 Champa St.
P.O. Box 2812
Denver, Colo. 80201

United Automobile Workers
Solidarity House
8000 E. Jefferson St.
Detroit, Mich.

United Mine Workers of America
 International Union
907 15th St., N.W.
Washington, D.C. 20005

United Paperworkers International
 Union
163-03 Horace Harding Expressway
Flushing, N.Y. 11365

United Rubber, Cork, Linoleum and
 Plastic Workers of America
87 S. High St.
Akron, O. 44308

United Steelworkers of America
Five Gateway Center
Pittsburgh, Pa. 15222

Technical Professional Organizations

Air Pollution Control Association
4400 Fifth Avenue
Pittsburgh, Pa. 15213

American Chemical Society
1155 15th St., N.W.
Washington, D.C. 20036

American Foundrymen's Society
Golf and Wolf Roads
Des Plaines, Ill. 60016

American Institute of Chemical
  Engineers
345 E. 47th St.
New York, N.Y. 10017

American Petroleum Institute
1271 Avenue of the Americas
New York, N.Y. 10020

American Society for Safety
  Engineers
850 Busse Highway
Park Ridge, Ill. 60068

American Society for Testing and
  Materials
1916 Race Street
Philadelphia, Pa. 19103

American Society of Heating and
  Ventilation Engineers
7218 Euclid Avenue
Cleveland, O.

American Society of Heating,
  Refrigeration, and Air Conditioning
  Engineers, Inc.
345 E. 47th St.
New York, N.Y. 10017

American Society of Mechanical
  Engineers, Inc.
345 E. 47th St.
New York, N.Y. 10017

American Textile Manufacturers
  Institute, Inc.
400 S. Tryon Street
Charlotte, N.C. 28285

American Welding Society
2510 N.W. 7th St.
Miami, Fla. 33125

Cadmium Council
292 Madison Avenue
New York, N.Y. 10017

Canadian Safety Council,
  Occupational Section
1765 S. Laurent Blvd.
Ottawa, Ontario, Canada

Combustion Engineers, Inc.
277 Park Avenue
New York, N.Y. 10017

Environmental Protection Agency
Washington, D.C.

Federal Water Pollution Control
  Administration
U.S. Department of the Interior
Washington, D.C. 20242

Food and Drug Administration
Washington, D.C.

Illuminating Engineering Society
51 Madison Ave.
New York, N.Y.

Institute of Electrical and
  Electronics Engineers
345 East 47th Street
New York, N.Y. 10017

Manufacturing Chemists Association
1825 Connecticut Ave., N.W.
Washington, D.C. 20009

National Air Pollution Control
  Administration
Arlington, Va. 22203

National Audio Video Center
GSA
Washington, D.C. 20409

National Center for Urban and
  Industrial Health
Washington, D.C.

SOURCES OF INFORMATION, ASSISTANCE, AND GUIDANCE

National Fire Protection Association
475 Atlantic Ave.
Boston, Mass. 02210

National Institute of Occupational
  Safety and Health
Room 530
U. S. Post Office
Cincinnati, O. 45202

National Safety Council
425 N. Michigan Ave.
Chicago, Ill. 60611

Occupational Safety and Health
  Administration
Washington, D.C.
Regional Offices

U.S. Department of Commerce
Washington, D.C.

U.S. Department of Labor
Washington, D.C.

State departments of labor

Sources of Safety and Health Standards

(Other than those developed by organizations previously shown)

American National Standards
  Institute
1430 Broadway
New York, N.Y. 10018

Atomic Energy Commission
Washington, D.C.

Food and Drug Administration
Washington, D.C.

National Bureau of Standards
Washington, D.C.

National Council on Radiation
  Protection and Measurements
7910 Woodmont Avenue
Washington, D.C. 20014

National Safety Council
425 N. Michigan Ave.
Chicago, Ill. 60611

OSHA
Washington, D.C.

Business Organizations

(Other than those previously noted)

American Iron and Steel Institute
1000 16th St., N.W.
Washington, D.C. 20036

American Management Associations
135 W. 50th St.
New York, N.Y. 10020

Automobile Manufacturers
  Association
320 New Center Building
Detroit, Mich. 48202

Chamber of Commerce of the United
  States
1615 H St., N.W.
Washington, D.C. 20006

The Conference Board
845 Third Ave.
New York, N.Y. 10022

Electronics Industries Association
2001 Eye St., N.W.
Washington, D.C. 20006

Lead Industries Association
292 Madison Avenue
New York, N.Y. 10017

National Association of
  Manufacturers
1776 F St., N.W.
Washington, D.C. 20006

National Electrical Manufacturers
  Association
155 E. 44th St.
New York, N.Y. 10017

In addition, companies with established industrial health programs can be of inestimable value as sources. Many of these have been noted in Appendix B.

Cooperative Programs and Occupational Health Clinics

Alcoholics Anonymous (local)

American Health Foundation
1370 Avenue of the Americas
New York, N.Y.

Beth Israel Medical Center
307 Second Ave.
New York, N.Y.

Better Health Examiners
853 Broadway
New York, N.Y. 10011

Beverly Hills Medical Clinic
Beverly Hills, Calif.

Brookdale Hospital Center
Brooklyn, N.Y.

Columbus Occupational Association
Columbus, Ind.

Duke University Medical Center

Executive Health Examiners
777 Third Ave.
New York, N.Y. 10017

Gates Medical Clinic
Denver, Colo.

Geisinger Medical Clinic
Danville, Pa.

Hartford Plan
Hartford, Conn.

Health Association of Rochester
 and Monroe County
Rochester, N.Y.

Health-Care-A-Van
Morris Plains, N.J.

Health Testing Services
Berkeley, Calif.

HQ Systems, Inc.
Hinsdale, Ill.

Industrial Health Council
Birmingham, Ala.

Institute of Pennsylvania Hospital
Philadelphia, Pa.

International Health Systems, Inc.
Des Plaines, Ill.

Jones Clinic
Hammond, Ind.

Lahey Clinic
Boston, Mass.

Life Extension Institute
1185 Avenue of the Americas
New York, N.Y. 10036

Manufacturers Health Clinic
Winder, Ga.

Mayo Clinic
Rochester, Minn.

Medical Diagnostic Centers, Inc.
Norristown, Pa.

Morristown Memorial Hospital
Morristown, N.J.

Mt. Sinai Hospital
Fifth Avenue and 100th St.
New York, N.Y. 10029

Northwest Industrial Clinic
Minneapolis, Minn.

Petrie Clinic
Atlanta, Ga.

Portland Industrial Clinic
Portland, Ore.

Pratt Diagnostic Clinic
Boston, Mass.

Strang Clinic
55 E. 34th St.
New York, N.Y. 10016

Sutter Clinics, Inc.
St. Louis, Mo.

SOURCES OF INFORMATION, ASSISTANCE, AND GUIDANCE 455

INSURANCE ORGANIZATIONS

American Insurance Association
85 John St.
New York, N.Y. 10038

American Mutual Insurance Alliance
20 N. Wacker Drive
Chicago, Ill. 60606

Blue Cross Association
840 N. Lake Shore Drive
Chicago, Ill. 60611

Health Insurance Association
  of America
919 Third Ave.
New York, N.Y. 10022

National Association of Blue
  Shield Plans
211 E. Chicago Ave.
Chicago, Ill. 60611

In addition, many insurance companies and Blue Cross-Blue Shield plans have consultative services, literature, and films available.

CONSULTANTS

Many different kinds of consultants are available to administrators of industrial health programs. Certain of these are listed.

Aero Pulse, Inc.
Cornwell Heights, Pa.

Aetna Technical Services
Aetna Life & Casualty
Farmington Avenue
Hartford, Conn.

Alnor Instrument Co.
Niles, Ill.

American Air Filter Co., Inc.
Louisville, Ky.

American Health Services
5530 Wisconsin Ave., N.W.
Washington, D.C. 20016

Analytical Instruments Development, Inc.
Avondale, Pa.

Analytical Research Labs
Monrovia, Calif.

Applied Health Research Corp.
Los Altos, Calif.

Argonne National Laboratory
Argonne, Ill.

Asbestos Information Association
Washington, D.C.

Association of Labor-Management
  Administrators and Consultants on
  Alcoholism
300 Wendell Court
Atlanta, Ga. 30336

Autosonic, Inc.
Norristown, Pa.

Bio Industrial Laboratories
Gadsden, Ala.

James H. Botsford (noise control)
Bethlehem, Pa.

Boyce Thompson Institute for Plant
  Research, Inc.
Yonkers, N.Y.

Cesco Safety Products
2727 W. Roscoe St.
Chicago, Ill. 60018

Clayton Environmental Consultants
Southfield, Mich.

Computa-Lab
Rolling Meadows, Ill.

Conam Inspection, Inc.
Tulsa, Okla.

Environmental Compliance Corp.
Venetia, Pa.

Environmental Health Labs, Inc.
Farmington Hills, Mich.

Environmental Research Group Inc.
Ann Arbor, Mich.

General Monitor, Inc.
Costa Mesa, Calif.

Globe Safety Products, Inc.
125 Sunrise Place
Dayton, O. 45407

Health Physics Associates
Highland Park, Ill.

Industrial Bio-Test Laboratories, Inc.
Northbrook, Ill.

Industrial Noise Services, Inc.
Palo Alto, Calif.

Landis, Murray, Sutherland
  (alcoholism)
420 E. 64th St.
New York, N.Y. 10021

National Environmental Instruments
  Co.
Warwick, R.I.

Preventi-Med Consultants, Inc.
North Hollywood, Calif.

Pulsosan Safety Equipment Co.
Flushing, N.Y.

Safety Line Products
P.O. Box 550
Putnam, Conn. 06260

Service Research Associates
  (guide for supervisors)
Chicago, Ill.

Glen Slaughter
2001 Franklin St.
Oakland, Calif. 94612

Systems Control, Inc.
Palo Alto, Calif.

Twin City Testing and Engineering
  Laboratory
St. Paul, Minn.

Westinghouse Health Systems
Box 866
American City Building
Columbia, Md. 21044

Consultation can also be provided by the many manufacturers of safety devices, machinery, and testing instruments. Through them, techniques, systems, services, and hardware are also available for the control of air, water, and soil pollution. Among such sources are the Aluminum Company of America, the Coca-Cola Company, and the Monsanto Company. A list of accredited laboratories is available from the American Industrial Hygiene Association.

LITERATURE

There is a considerable amount of literature available to assist in the formation of an industrial health program and also to implement health and safety education. Sources of such literature not otherwise noted in the references are shown here. Many of the other sources of assistance shown previously also offer valuable literature.

American Cancer Society
52 W. 57th St.
New York, N.Y.

American Diabetes Association
1 E. 45th St.
New York, N.Y. 10017

American Heart Association
1775 Broadway
New York, N.Y. 10019

American Lung Association
1790 Broadway
New York, N.Y. 10019

SOURCES OF INFORMATION, ASSISTANCE, AND GUIDANCE

American National Council
  for Health Education of the Public
800 Second Ave.
New York, N.Y. 10017

American Social Health Association
1740 Broadway
New York, N.Y. 10019

Arthritis and Rheumatism Foundation
10 Columbus Circle
New York, N.Y. 10019

Channing L. Bete Co., Inc.
45 Federal St.
Greenfield, Mass. 01301

Drug Abuse Council, Inc.
1828 L St., N.W.
Washington, D.C. 20036

International Association of
  Machinists
909 Machinists Building
Washington, D.C. 20036

Metropolitan Life Insurance
  Company
One Madison Ave.
New York, N.Y. 10010

McGraw-Hill, Inc.
1221 Avenue of the Americas
New York, N.Y. 10020

National Council on Alcoholism
733 Third Ave.
New York, N.Y. 10017

National Institute on Alcohol Abuse
  and Alcoholism
5600 Fishers Lane
Rockville, Md. 20852

National Kidney Disease Foundation
143 E. 35th St.
New York, N.Y. 10016

National Rehabilitation Association
1025 Vermont Ave., N.W.
Washington, D.C.

National Safety Council
425 N. Michigan Ave.
Chicago, Ill. 60611

Small Business Administration
Washington, D.C.

Spenco Medical Corp.
P.O. Box 8113
Waco, Tex. 76710

PERIODICALS

Annals of Occupational Hygiene
Archives of Environmental Health
Chemical & Engineering News
Chemical Week
Environmental Health and Safety
  News
Environmental Health Perspectives
Environmental News
Industrial Hygiene Digest
International Journal of Occupational
  Health and Safety
Journal of Occupational Medicine
Journal of the Air Pollution
  Control Association
Journal of the American Industrial
  Hygiene Association
Journal of the American Public
  Health Association
National Safety News
Noise Control Engineering
Occupational Hazards
Occupational Health
Occupational Health Nursing
Occupational Mental Health
Occupational Psychology
Occupational Safety and Health
OSHA Action News
OSHA Report

Films

(The following list includes some sources of films on the conduct of an industrial health program or for health educational purposes.)

Aetna Life & Casualty
Allis-Chalmers Corporation
Allstate Insurance Company
American Cancer Society
American Film Productions, Inc.
American Medical Association
American Mutual Insurance Alliance
American Oil Company
American Petroleum Institute
AT&T
Associated Films, Inc.
Association of Instructional Materials
Audio Productions
Austin Productions, Inc.
Automobile Manufacturers Association, Inc.
Bay State Film Productions, Inc.
Boeing Company
Brandon Films
Bray Studios, Inc.
Brentwood Productions
Byron Motion Pictures, Inc.
Canyon Films of Arizona
Caterpillar Tractor Company
Coca-Cola Company
Continental Casualty Company
Coronet Instructional Films
Cune Associates
Walt Disney Educational Materials Company
Douglas Film Industries
Dow Chemical Company
E. I. du Pont de Nemours & Co.
Edison Electric Institute
Employers Insurance of Wausau
Exxon Corporation
Factory Mutual Insurance Company
Ford Motor Company
General Dynamics Corporation
General Electric Company
General Mills, Inc.
General Motors Corporation
Goodyear Tire & Rubber Company
Hartford Fire Insurance Company
Harvest Films
Johnson & Johnson
Kemper Insurance
Lockheed Aircraft Corp.
Manufacturing Chemists Association
McGraw-Hill, Inc.
Metropolitan Life Insurance Company
National Institute of Occupational Health and Safety
National Machine Tool Builders Association
National Safety Council
Nationwide Insurance Companies
New York Telephone Company
Norton Company
Occupational Safety and Health Administration (OSHA)
Phillips Petroleum Company
Portland Cement Association
Sentry Insurance
Standard Oil Company of California
Sun Life Assurance Company of Canada
Travelers Insurance Company
Union Carbide Corp.
U.S. Jaycees
U.S. Steel Corp.
Upjohn Company
Xerox Corp.

COLLEGES AND UNIVERSITIES

Many colleges and universities have an active interest in industrial health. In some instances they train industrial physicians, nurses, and hygienists. In others they train engineers. In others they train public health officers. Many conduct directly related research. In a few instances courses are made available to representatives of places of employment or labor unions. The following are among the colleges and universities with active programs in industrial health training or research.

Auburn University
University of California at Los Angeles
Central Missouri State University
Chicago University
Cincinnati University
Colorado State University
Columbia University
Cornell University, New York State School of Industrial and Labor Relations
Drexel Institute of Technology
Georgia Institute of Technology
Georgia State University
Harvard University
University of Illinois
University of Iowa
Jefferson Medical College
Johns Hopkins University
Louisiana State University
University of Maryland
Massachusetts Institute of Technology
University of Michigan
Michigan State University
University of Minnesota
New York University
North Carolina State University
Northern Illinois University
Ohio State University
University of Pennsylvania
Pennsylvania State University
University of Pittsburgh
University of Rochester
Rutgers—The State University
Texas A & M University
Texas State University
Tulane University
Utah State University
Wayne State University
Yale University
York University at Toronto

Courses of instruction on aspects of occupational health and safety are also made available by such organizations as

Aetna Life & Casualty
Hartford, Conn.

Center for Disease Control
Atlanta, Ga.

Hearing Conservation Noise Control, Inc.
1721 Pine St.
Philadelphia, Pa. 19103

Industrial Health Foundation
5231 Centre Ave.
Pittsburgh, Pa. 15232

Industrial Hygiene Foundation
4400 Fifth Ave.
Pittsburgh, Pa. 15213

Liberty Mutual Insurance Company
Boston, Mass.

National Environmental Research Center
Research Triangle Park, N.C. 27711

National Institute of Occupational Safety and Health
U.S. Post Office
Cincinnati, O. 45202

National Safety Council
425 N. Michigan Ave.
Chicago, Ill. 60611

Saranac Laboratory of the
   Edward L. Trudeau Foundation

Society for Occupational and
   Environmental Health
1714 Massachusetts Ave., N.W.
Washington, D.C. 20036

Travelers Insurance Company
Hartford, Conn.

D. ANNOTATED BIBLIOGRAPHY

The Conference Board. *Industry Roles in Health Care.* New York, 1974. A research report on occupational health programs, their staffing, and their costs. Discussed are such aspects of occupational health as employee placement, periodic health examinations, health education, coping with nonoccupational diseases and accidents, and the problems of alcoholism and drug abuse.

The Economics of Health and Medical Care. Ann Arbor: University of Michigan Press, 1974. A compilation of the proceedings of a conference on the economics of health care. Covers the organization and financing of health services, the effects of supply and demand, the microeconomics of health care, and investments in health.

Follmann, J. F., Jr. *Alcoholics and Business.* New York: AMACOM, 1976. Shown are the scope and costs of alcoholism in places of employment, the identification of the alcoholic employee, what can be done about the problem, the accomplishments to date, and where help and guidance can be obtained. Appendixes include a documentation of employment-centered alcoholism programs and of sources of help and guidance.

Klarman, Herbert E. *The Economics of Health.* New York: Columbia University Press, 1965. Explored are all aspects of the economics of health, including expenditures by business and government, the effects of supply and demand, and the costs and benefits of health programs.

Levinson, Harry. *Emotional Health: In the World of Work.* New York: Harper and Row, 1964. A handbook for the everyday use of the executive in dealing with people in distress. Discussed are the various reactions of employees to mental, emotional, and personality disorders; the problem of identity; the role of the problems of the family and aging; and the role of management.

Maisel, Albert Q., ed. *The Health of People Who Work.* New York: The National Health Council, 1960. Based on the reports to the 1959 National Health Forum of more than 200 industrial medical directors, physicians, nurses, management officials, and other experts in various areas of occupational health. Discussed are the goals of occupational health programs, control of the working environment, the various aspects of occupational health, staffing problems, and the problems of smaller employers.

McLean, Alan, M.D. *Occupational Stress.* Springfield, Ill.: Charles C Thomas, 1974. The views of several authors on the clinical aspects and the psychoanalytical framework of occupational stress. Examines the several elements of job stress as well as the relationship to a place of employment of off-the-job stress, the relationship of occupational stress to work performance, role responsibility, and the importance of health-status assessment.

Rosen, George. *Preventive Medicine in the United States 1900-1975.* New York: Science History Publications, 1975. The author traces the development of preventive medicine in the United States since 1900. Shown are the advancements made over that period, as well as factors that inhibit progress. Included in the discussions are aspects of occupational health programs.

Stellman, Jeanne M., and Susan M. Daum. *Work Is Dangerous to Your Health.* New York: Vintage Books, 1973. Discussed are occupational diseases, the problem of controlling pollution in the workplace, the importance of health records, the matter of measuring and monitoring occupational hazards, chemical and welding hazards, and the effects of stress, noise and vibration, extreme temperatures, and radiation.

U.S. Department of Labor. *OSHA Safety and Health Standards: General Industry Digest.* OSHA No. 2201. Washington, D.C., 1975. A government publication directly related to conformance with the Occupational Safety and Health Act of 1970. Discussed are the safety and health standards developed by the Occupational Safety and Health Administration of the Department of Labor.

Willard, Helen S., and Clare S. Spackman, eds. *Occupational Therapy.* Philadelphia: Lippincott, 1974. Sections by various authors include the treatment approaches in industrial health, the importance of medical records, and occupational therapy for cerebrovascular conditions, sight defects, amputations, rehabilitation cases, and mental health conditions.

Index

Abrams, William, 241
absenteeism
 industrial health programs reducing, 382
 physical examinations and, 186
accidental death and dismemberment insurance, 333, 334
accident insurance, 91
accident prevention, 110–111, 146–147
 at Baltimore Gas and Electric, 277
 insurance companies in, 21
 savings from, 382–384
accidents, 42–44
 alcoholism and, 37
 costs of, 55–56, 87
 deaths from, 29, 81–82
 disability caused by, 79
 among employed people, 71–74
 by handicapped workers, 308
 in industrial health programs, 162–163
 nonoccupational, in industrial health programs, 166–169
 research on, 118
 safety programs reducing, 382–384
 in small workplaces, 280–281
 see also injuries; injuries, occupational
A.C. Horn Company, 284
acute diseases
 disability resulting from, 78
 incidence of, 68
afterlife, 4
Agricola, 10, 12
agricultural wastes, 104
air pollution, 173–175
 costs of, 260–261
 deaths from, 361
 in history, 6, 13
 legislation on, 223
 in public health, 102–104
 work absence correlated with, 79
alcohol abuse, 7
Alcoholics Anonymous, 38, 300
alcoholism, 37–38
 costs of, 54–55, 87
 disability and, 81
 employer responsibility for, 164
 government actions on, 228
 insurance for treatment for, 324, 336
alcoholism control programs
 cost-effectiveness of, 389–391
 costs of, 259
 employer's role in, 197
 in preventive medicine, 146
 problems in, 206, 299–300
 of small businesses, 285
Allen Manufacturing Company, 284, 382
Allen, Scott I., 364–365
allergies, respiratory, 33
Allis-Chalmers Corporation, 390
American Association for Labor Legislation, 16, 17
American Association of Industrial Physicians and Surgeons, 19
American Association of Railway Surgeons, 17
American Board of Preventive Medicine, 18, 241
American Conference of Government Industrial Hygienists, 18
American Industrial Hygiene Association, 174
American Insurance Association, 119–120
American Labor Health Association, 205
American Medical Association
 on confidentiality, 244
 founding of, 17

American Medical Association (*Cont.*)
  industrial health defined by, 156, 158
  on physicians in industrial health programs, 236–237, 241
  survey of 1939 of, 266
  on water pollution, 171–172
American National Standards Institute (ANSI; American Standards Institute)
  accident research of, 118
  founding of, 19
  standards set by, 293, 294
American Public Health Association
  on cost-benefit in prevention, 351–352
  on costs of occupational disability and death, 87
  founding of, 17
  on personnel for industrial health programs, 237
American Society of Heating and Ventilating Engineers, 17
American Telephone & Telegraph Company
  drug abuse program of, 301
  Illinois Bell Telephone Company of, 298, 301–302, 390
  Mountain States Telephone Company of, 256–257, 379–381
  New York Telephone Company of, 184, 185
  Western Electric Company of, 255–256, 269–272, 383
amines, cancer caused by, 14
amyotropic lateral sclerosis, 34–35
Anderson, Odin W.
  on costs of national health services, 140
  on morbidity statistics, 28
  on preventive medicine, 98, 115, 117
anesthesia, history of, 8
angiosarcoma of liver, 71, 373
anthracosilicosis, 144
anthrax bacillus, 13
Arabia, 8

archaeology, 4
Arledge, J.T., 14
aromatic amines, cancer caused by, 14
arsenic
  in air pollution, 173–174
  cancer and, 71
  poisoning by, 12
arthritis, 35
  costs of, 54
  detection of, 145
asbestos
  fibrosis caused by, 14
  union action on, 207
asbestosis, 144
  deaths from, 70
Ascelepius, 8
Ashford, Nicholas A., 209, 374
Assistance to Permanently and Totally Disabled, 314
asthma, 33
Atomic Energy Act (U.S., 1954), 212, 214
Atomic Energy Commission (U.S.), 228
Australopithecus, 5
automobile industry, 321
automobile seat belts, 361
Avicenna, 8
Avnet, Helen, 130

Babylon, ancient, 8
back injuries, 73
Balliett, Whitney, 398
Baltimore Gas and Electric Company, 277
Barnes, Earle B., 391
bathing, history of, 10
Benefit Association of the John Wanamaker Company, 24
benefits
  analyses of, 351–352, 392–394
  under private insurance programs, 321–343
  under social security, 311–312
  total cost of, 91–92

INDEX 465

under workmen's compensation, 306–307
berylliosis, 144
Bethlehem Steel, 382
Bevan, Aneurin, 140
Beveridge Report, 139–140
Bischoff, James L., 384–385
Bismarck, Otto von, 24
Blackburn, Henry, Jr., 23
black lung, *see* pneumoconiosis
blacks
 causes of death in, 31, 33
 drug abuse in, 38
 see also nonwhites
bladder, cancer of
 aromatic amines causing, 14
 occupational causes of, 142
blood, Harvey's work with, 8
blood pressure
 insurance companies pooling information on, 22
 see also hypertension
Blue Cross and Blue Shield, 322–324, 376
 origins of, 25
 preventive medicine and, 334, 336
 rehabilitation under, 343
body weight, studies on, 22
Boeing Company, 183
Boise Cascade, 383–384
breast cancer, 143
Britain
 history of medical care in, 24
 history of public health in, 11–12
 lifespan in, 5
 National Health Service in, 139–140, 346
 working hours regulated in, 16
bronchial asthma, *see* asthma
bronchiectasis, 143–144
bronchitis, chronic
 deaths from, 33
 smoking and, 40, 69, 143, 144
Bronowski, J., 5
brown lung (byssinosis), 70

Buchan, Donald F., 279
buildings, safety features in, 163
Bureau of Labor Statistics (U.S.), 217–218
Bureau of Mines (U.S.), 228
Bureau of Radiological Health (U.S.), 226
burial
 history of, 9
 workmen's compensation benefits for, 307
byssinosis (brown lung), 70

Caesar, Julius, 10
California Cannery Workers, 276–277
California Physicians Service, 25
California Processors, 276–277
cancer
 as cause of death, 29–32
 chemical industry and rates of, 173–174
 cost-effectiveness analysis of prevention of, 355–357
 costs of, 53, 86
 early detection of, 336–337
 as occupational disease, 70–71
 preventive medicine for, 142–143
 research on, 117–118
 smoking and, 40, 410
 vinyl chloride causing, 14, 373
cancer, bladder
 aromatic amines causing, 14
 occupational causes of, 142
cancer, breast, 143
cancer, cervix, 357
cancer, colon-rectum, 142
cancer, liver, 143
 vinyl chloride causing, 14, 71, 373
cancer, lung
 asbestos and, 70
 radioactivity causing, 14, 174
 smoking and, 39–40, 142, 143
cancer, oral, 143
cancer, scrotal, 12, 14

cancer, skin
  industrial causes of, 14
  sunlight causing, 142
cancer, stomach, 143
cancer, uterine
  cost-effectiveness analysis of prevention of, 355
  Pap tests for, 142
capital investment, 250–251
carbon disulfide, 70
carbon in lungs, historical evidence of, 6
carbon monoxide poisoning, 12
carcinogens
  government efforts on, 224
  testing for, 199
cardiovascular diseases, *see* heart diseases
Carey, Mathew, 13
Carnegie Foundation, 17
Carter, Jimmy, 317
Cassuto, Jerry, 194–195
Century Electric, 386
cerebral palsy, 35, 54
cervix, cancer of, 357
Chadwick, Sir Edwin, 13
Chamber of Commerce, U.S.
  on environmental pollution, 171, 174
  industrial health assistance from, 294
CHAMPUS program, 313
Chase, Edward W., 133
chemical industry, cancer rates and, 173–174
Chemical Industry Institute of Toxicology, 198
chemicals
  harmful, 105
  testing of, 198–199
child labor laws, 15, 16
China, ancient, 11, 24
cholera, 13, 36
chromates, 70–71
chromium, 70
chronic diseases
  disability resulting from, 78
  incidence of, 68
  prevention of, 128

cigarettes, *see* smoking
cirrhosis of liver
  alcoholism and, 37
  varying rates of, 129–130
civil liberties issues in public health, 101
Clarke, Robert J., 192
Clean Air Act (U.S., 1970), 103
Clinchy, Richard, 386–387
clinics, in small businesses' health programs, 286–287
Close, Frank M., 25
coinsurance, 323
collective bargaining
  health insurance in, 26, 88
  occupational safety and health in, 110, 205, 207–208, 404
colleges, *see* universities and colleges
Collings, G.H., Jr., 158, 184, 189
Collisson, N.H., 180, 207–208
Colman, J. Douglas, 128
colon, cancer of, 142
Columbus Occupational Health Association, 288
Colwell, Miles O., 189
Committee for Economic Development, 319
Committee on National Health Insurance, 317
communicable diseases
  control of, 106–107, 145
  rates of, 35–36
community
  industrial health physicians in, 239
  in industrial health programs, 169–176, 197–201
  preventive medicine and, 127–131
  in small businesses' health programs, 284–285
Community Health Association of Detroit, 343
community organization, 112
Comprehensive Alcohol Abuse and Alcoholism Prevention, Treatment, and Rehabilitation Act (U.S., 1971), 228
computerized records, 407

Conference Board
  on costs of industrial health, 254–255, 339, 369
  on costs of preventive medicine, 371, 373
  on disability insurance, 327–330
  on employer-provided health care, 167, 402–403
  on health insurance, 324
  industrial health surveys of, 180, 182, 185–191, 194, 199–201
  on life insurance, 333
  on personnel for industrial health programs, 236, 243, 266–267
  on salary-continuance insurance, 332
Conference Board of Physicians in Industry, 19
confidentiality
  in alcoholism control units, 299
  of periodic employee examinations, 192, 196, 206, 270, 272
  of records, 244, 407
  of trade secrets, 215
Conley, Ronald W., 373
Consumer Product Safety Act (U.S., 1972), 225, 226
Consumer Product Safety Commission (U.S.), 226
Continuous Air Monitoring Program (U.S.), 223
Convair, 383
Cooke, Thomas, 11
Cornell Medical School, 17
coronary heart disease, *see* heart diseases
cost-benefit analyses, 351–352, 392–394
cost-effectiveness
  of industrial health programs, 368–394
  need for data on, 408
  of preventive medicine, 347–365
costs
  of air pollution, 103, 174
  in first workmen's compensation law, 15
  of industrial health, 1–2, 156
  of industrial health programs, 248–261
  of preventive medicine, 131–138
  see also financing of health care
costs of health care, 48–62
  under British National Health Service, 139–140, 346
  cost-effectiveness analyses of, 347–365, 368–394
  of employed people, 84–93
  employer responsibilities in, 199
  history of, 8–9
  in history of insurance, 24–27
  preventive medicine in, 133–138
  see also financing of health care
cotton dust, 221
Council on Environmental Quality (U.S.), 224
counseling, in health education, 112
Craig, James L., 393–394
Cummins Engine Company, 288

Daum, Susan M., 209, 404
Davy, Sir Humphry, 12
Day, Emerson, 385
DDT, 105
death(s)
  accidents causing, 42–44
  causes of, historical, 5–7
  causes of, in U.S., 29–30
  of children, Marx on, 14
  consciousness of, 3–4
  on and off job, 72
  overweight and, 42
  premature, among working people, 81–82
  smoking and, 40
Demsey, Robert, 387
Denmark, 357–358
dental care
  fluoridation and costs of, 360
  insurance programs for, 18, 324
Denver Federal Center, 379
diabetes
  detection of, 144

468 INDEX

diabetes (*Cont.*)
  rates of 32
dialysis, 34, 357
diet
  cancer and, 410
  health costs reduced by, 360–361
  in preventive medicine, 147
digestive system diseases, 53
DiMichael, Salvatore G., 121
disability
  costs of, 52
  incidence and duration of, 75–79
  incidence of, 44–46
  life insurance coverage for, 333
disability insurance
  costs of, 89, 259
  history of, 25
  private plans for, 326–331
  public programs for, 314
  salary-continuance insurance and, 331–332
  under Social Security, 311–312
  under workmen's compensation, 306
disabled persons, *see* handicapped persons
diseases
  costs of, 52–55, 349–350
  among employed people, 68–71
  history of, 5–7
  nonoccupational, in industrial health programs, 166–169, 269
  preventive medicine and, 97, 101
  rates of, 30–39
  see also occupational diseases
disposal
  of sewage, 11–12
  of waste, 104
doctors, *see* physicians
Dolinsky, Edward M., 288
Down's syndrome, 145
drownings, 43
drug abuse and addiction, 38–39
  industrial health programs on, 300–302
  insurance for treatment for, 324
Dublin, Louis I., 21, 22

Du Pont, 383

Eastman Kodak, 273, 384–385
economic-impact studies, 373
economics, *see* cost-effectiveness; costs; costs of health care; financing of health care
education
  level of, 130
  see also health education
Edison, Thomas A., 11
Egypt, ancient, 6, 8
electrocardiograms, 23
Ellwood, Paul, 318
Emlet, Harry E., Jr., 362
emotional disturbances, *see* mental illness
Empedocles, 10
emphysema
  deaths from, 33
  smoking and, 69, 143, 144
employees
  confidentiality of records on, 244
  in future of industrial health 404–405, 410
  number of, 67
  physical examinations of, 19, 21, 182–286, 190–193, 206
  private health care programs for, 321–343
  public health care programs for, 305–319
  rights and responsibilities of, under OSHA, 208, 214–215
  in setting up of industrial health programs, 234–235
  of small businesses, 281–282
employers
  cost-effectiveness data of, 374–376
  in future of industrial health, 402–403
  health insurance payments by, 88
  in history of industrial health, 18–20
  in industrial health programs, 178–202, 233–247, 263–278

industrial health responsibilities of, 156–157, 214–215, 218
preventive medicine and, 125
small, problems of, 279–289
employment policies
of handicapped, 176, 201
health in, 182–185
encephalitis, 34, 107
energy conservation, 175
Energy Supply and Environmental Coordination Act (U.S., 1974), 224
England, *see* Britain
environmental controls
government's role in, 222–225
physical examinations less efficient than, 183
environmental health hazards
employer's responsibilities for, 198–199
unions' relation to, 208–209
Environmental Protection Agency (U.S.)
on air pollution, 102, 103
creation of, 224
on environmentally caused disease, 171
epidemic diseases, 6, 32–33
Equal Employment Opportunity Act (U.S.), 157
eugenics, 108
Eulenburg, Albert, 14
Europe
history of housing in, 11
history of medicine in, 8
Ewing, Oscar R., 315, 316
exercise, 147
Exxon, 191–192, 273–275

facilities for industrial health programs, 237–238, 240
famines, 6
Federal Coal Mine Health and Safety Act (U.S., 1969), 86, 228
Federal Energy Administration (U.S.), 224
Fein, Rashi, 137, 319, 350

females
disability rates for, 75, 80
history of laws regulating employment of, 16
mortality data for, 28–29, 31–33
in workforce, 67
fibrosis of lungs, 14
financing of health care
private programs for, 321–343
public programs for, 305–319
see also cost-effectiveness; costs; costs of health care
fines under OSHA, 219–220
firearms, 44
fire insurance, 21
fire prevention
in buildings, 163
off-job, employer actions on, 168
at The New York Times, 259
fires, injuries and deaths from, 42
Flexner Report (Carnegie Foundation), 17
flu, *see* influenza
fluoridation of drinking water, 360
Food and Drug Administration (FDA) (U.S.), 225–226
food and drug laws, 16
food poisoning, 7
Foulger, John, 158
Foundation for Health Care Evaluation in the Twin Cities, 60
foundations
employers in creation of, 200
medical, 60
in preventive medicine, 124
Framingham experiment, 23
Frankel, Lee K., 21
Frank H. Fleer Corporation, 284
Franklin, Benjamin, 12
Freeman, J. Addison, 14
Friendly Societies (Britain, 1750), 24
Fuchs, Victor, 3
on national health insurance 318–319
on preventive medicine, 114, 129, 130, 132
fuel conservation, 175

470 INDEX

Galen, 8
Gandhi, Mahatma, 10
Garfield, Sidney, 365
gas masks, 12
Gates Rubber Company, 288, 381
Gee, Edwin A., 198
General Electric, 260
General Motors
 alcoholism program of, 390
 health care costs to, 321
Germany, 15
Glasser, Melvin A., 205, 404–405
Goldberg, Theodore, 204
Goldstein, David, 406
Goldwater, S.S., 18
gonorrhea, 36
government
 alcoholism program for employees of, 390
 assistance from, 296
 in future of industrial health, 408–409
 group insurance encouraged by, 25–26
 health care financed by, 305–319
 health care planning by, 60
 in history of industrial health, 15–17
 in industrial health programs, 211–228
 jurisdictional conflicts in public health measures by, 102
 preventive medicine and, 125–126
 see also United States
grants, under OSHA, 216
Greece, ancient, 5, 6, 8, 10
Grosse, Robert, 350–351
Group Health Association, 25, 62, 205, 338–339
Group Health Cooperative of Puget Sound, 26, 62, 339, 365
group health insurance, 25–26
 see also health insurance
group life insurance, 333
Guzick, 352, 358

Hadrian (emperor, Rome), 7, 10

Hales, Stephen, 12
Hall, John, 221
Haly Abbas, 8
Hamilton, Alice, 243
Hammurabi's Code, 8
handicapped persons
 rehabilitation of, 118–121, 176, 201
 Veterans Administration programs for, 313
 workmen's compensation and, 307–309
Hansen's disease (leprosy), 36
Harrington, Sir John, 7
Harris, Seymour, 131
Hartford Plan, 285–286
Hartford Small Plant Medical Service, 382
Harvey, William, 8
Haskell Laboratory of Toxicology and Industrial Medicine, 198
Havighurst, Clark C., 60
Hawkes, Thrift G., 183
Hayes, A., 133
hay fever, 33
health
 history of concern for, 5–27
 nonoccupational, employers' role in, 188–197
 nonoccupational, unions on, 204–205
 in U.S., status of, 28–46
 see also industrial health; occupational health
Health Assessment Program, 338–339
health care
 CHAMPUS program for, 313
 history of, 7–9
 home, 384–385
 private insurance plans for, 321–326
 provided by unions, 205
 public programs for, 314
 workmen's compensation benefits for, 306–307
 see also costs of health care
health education
 by employers, 168, 188, 193–197
 insurance companies in, 23

physicians in, 239
in preventive medicine, 111–113
health insurance
  employer costs of, 88, 259–260
  history of, 24–27
  national, 315–319
  preventive medicine in, 134–138
  private plans for, 322–326
Health Insurance Association of America, 329
Health Insurance Plan of Greater New York, 26, 62, 322, 343
  preventive medicine practiced by, 133, 337
Health Maintenance Organization Act (U.S., 1973), 288
health maintenance organizations (HMOs), 61–62, 322–325
  employer assistance in creation of, 200
  periodic examinations by, 193
  preventive medicine and, 337–339
  rehabilitation under, 343
  for small businesses, 288–289
  unions in development of, 205
hearing-conservation programs, 276
hearing impairments, 39, 207
hearing loss, occupational, 70
heart diseases
  as cause of death, 29
  cost-effectiveness analysis of prevention of, 353–355
  costs of, 52–53, 86
  preventive medicine for, 140–141
  rates of, 30
  smoking and, 40
heart size, tables for, 23
Heidelberg Man, 5–6
Hemingway, Gregory, 243
hemodialysis, 34, 357
Henry VIII (king, England), 24
hepatitis, 34
Herlong, A. Sydney, Jr., 309
high blood pressure, *see* hypertension
highway safety, 21–22
Hilbert, Morton S., 227–228, 345, 361

Hilker, Robert, 168–169
Hinkle, Laurence E., Jr., 129
Hippocrates, 8, 10
Hogarth, William, 7
Holcomb, F.W., 192
home, accidents around, 146
home health care, 384–385
homicides, 29, 30
Hoover, A. Walter, 392
Horace, 10
hospital insurance
  history of, 25
  see also health insurance
hospitals
  in health care costs, 51–52, 58, 59
  history of, 8, 26
  NIOSH guidelines for, 219
housewives, economic value of, 350
housing, 11, 13
Hugh of Lucca, 8
humans
  consciousness of health in, 3–4
  value of life for, 347–349
hygiene, personal, 7, 9–12
hypertension
  actuarial study of, 22
  costs of prevention of, 358–359, 363
  on-job testing for, 195
  preventive medicine for, 141
  rates of, 30–31
hypochondriacs, 57–58

IBM
  health-screening and medical data system of, 276
  periodic examinations in, 192, 259
  safety program of, 383
Illinois Bell Telephone Company
  alcoholism program of, 298, 390
  drug abuse program of, 301–302
illnesses, *see* diseases
Illuminating Engineering Society, 19
Im-hotep, 8
immunization, 106–107
  costs of, 359–361
  history of, 8

impairments
  physical, 39
  see also handicapped persons; rehabilitation
incentives for preventive medicine, 133–138
India, ancient, 8
Indonesia, 358
industrial expansion, costs of pollution control and, 260
industrial health
  future of, 401–411
  history of, 5–27
  see also occupational diseases; occupational health
Industrial Health Council of Birmingham, 286
industrial health programs, 155–159, 160–176
  assistance for, 291–296, 450–460
  companies having, 446–449
  cost-effectiveness of, 368–394
  costs of, 248–261
  employers' role in, 178–202
  future of, 401–411
  goals of, 1
  government's role in, 211–228
  private insurance programs and, 321–343
  setting up, 233–247
  for small employers, 279–289
  special problems in, 297–302
  unions' role in, 204–209
  varieties of, 263–278
  see also costs; costs of health care; financing of health care; safety
Industrial Health Service Bureau, 18
industrial hygiene directors, 187
Industrial Hygiene Foundation, 18
industrial hygienists, 187
industrialization, health hazards and improvements from, 2
Industrial Medical Association, 241
  on health education, 194–195
Industrial Medical Association of Preventive Medicine, 18

industrial medicine, 110–111
  see also industrial health; industrial health programs; occupational health
Industrial Revolution, 13–14
industrial wastes, 104
infant mortality, 23, 30, 108
infectious diseases
  control of, 106–107
  rates of, 35–36
inflation
  in disability benefits, 306
  of medical costs, 57
influenza
  cost-benefit analysis of vaccination against, 358
  deaths from, 32–33
  outbreaks of, 107
  prevention of, 144
  work lost because of, 75
injuries
  accidents causing, 42
  nonoccupational, employers' role in, 188–189
  from products, 170
  see also accidents
injuries, occupational, 72–74
  in industrial health programs, 161–163
  safety programs reducing, 382–384
  see also accidents; occupational diseases
inoculations
  during World War I, 16
  see also immunization
insecticides, 104–105
inspections under OSHA, 219
Institute for Occupational Safety and Health, 68
insurance
  benefit patterns of, 61
  costs of, 259–260
  fire, 21
  history of, 24–27
  industrial health programs and, 246
  preventive medicine in, 134–138

private programs for, 321–343
rehabilitation in, 119–121
see also health insurance; life insurance
insurance companies
cost-effectiveness data of, 376, 386
in future of industrial health, 405
health care costs and, 59, 61
in history of industrial health, 20–27
industrial health assistance from, 294–295
programs of, 321–343
workmen's compensation and, 307
Insurance Information Institute, 22
Insurance Institute for Highway Safety, 22
International Labour Organisation, 208
International Ladies' Garment Workers' Union, 26
ischemic heart disease, 30
Israel, ancient, 6

James, George, 179
jaundice, infectious, 107
Jend, William, Jr.,
on employer responsibility, 158, 181–182
on industrial health programs, 152
on nonoccupational health, 191, 193
Jesus of Nazareth, 9–10
Johnson Foundation, 18
Joselow, Morris, 405
Journal of Industrial Hygiene, 18
J.R. Simplot Company, 391

Kahn, Robert L., 181, 255
Kaiser Foundation Health Plan
on cost-effectiveness of multiphasic screening, 364, 365, 387–389
history of, 25
preventive medicine of, 337–338
size of, 61, 322
tests offered by, 132–133
Kay, Sir John William, 13
Kennedy, Edward, 226
Kennedy, John F., 103

Kennicott, Hiram L., Jr., 309
Kepone, 174
Kerr, Lorin, 157, 242–243
Ketterlinus Lithographic Manufacturing Company, 284
kidney diseases
cost-effectiveness analysis of prevention of, 357
deaths from, 34
Kindig, 318
Klarman, Herbert E.
on chronic renal disease, 357
on cost-benefit analyses, 350, 352
on health care costs to business, 85
on influenza, 358
Klipstein, Kenneth H., 179–180, 370–371
Knowles, John H., 134
Koos, E.L., 127
Korea, 358

labor, *see* employees; unions
Labor, Secretary of, OSHA responsibilities of, 214–215
Labor Management Relations Act (Taft-Hartley Act) (U.S., 1947), 211
Lactantius, Saint, 398
Lauer, John, 287
layoffs
environmental hazards and, 209
insurance continuing during, 325
lead poisoning
in battery plants, 166, 259
Franklin's observation of, 12
Hippocrates on, 10
in history of industrial health, 12
measures against, at The New York Times, 275
occupational, 70
Lehr, Lewis W., 391
leprosy, 36
Lerner, Monroe, 28
Lewis, Hallet A., 392
Liberty Mutual Insurance Company, rehabilitation program of, 120–121, 389

Lichtenstein, Grace, 103
life expectancy, 28–29
life insurance, 333–334
    costs of, 91
    first group policy for, 24
life insurance companies, 21
life span, historical, 5
lifestyles, in preventive medicine, 113–114, 129–130, 147–148
liver, angiosarcoma of
    vinyl chloride causing, 71, 373
liver, cancer of, 143
    vinyl chloride causing, 14, 71, 373
liver, cirrhosis of
    alcoholism and, 37
    varying rates of, 129–130
local government
    role in occupational health and safety of, 212
    see also government
Long, Russell B., 317
long-term disability insurance, 328–330
lung cancer
    asbestos and, 70
    radioactivity causing, 14, 174
    smoking and, 39–40, 142, 143
lungs
    carbon in, historical evidence of, 6
    fibrosis of, asbestos causing, 14

McCallum, David V., 368–369
McConnell, William J., 22
McDonnell Douglas Corporation, 390
MacGregor, Gordon, 128
machinery, accidents involving, 73
Mack, James H., 73
Mackintosh, J.M., 131
McNerney, Walter J., 336
Macy Mutual Aid Association, 24
Magnuson, Harold J., 266
major medical insurance, 323, 324, 340
    see also health insurance
malaria, 10
males
    disability rates for, 75, 80
    drug abuse in, 38
    mortality data for, 28–33
malnutrition, 41–42
mammography, 363
management
    access to medical records denied to, 244
    alcoholism control supported by, 299
    in industrial health programs, 234
manganese, 70
Mann, Thomas, 1
Manufacturing Chemists Association
    on costs of pollution, 260, 261
    on occupational injuries, 384
    on off-job injuries, 188–189, 370
    on water pollution, 172, 173
Marx, Karl, 14
Massachusetts Institute of Technology, 17
measles, 36, 43
    immunization against, 107, 359–360
Medicaid, 312–313
medical care, *see* health care
medical insurance, *see* health insurance
medical schools
    history of, 8
    preventive medicine in, 402
Medicare, 311
Melbank Memorial Foundation, 18
Mellon Institute, 18
men, *see* males
meningitis, 34
mental illness, 37, 45
    costs of, 54, 86
    in employed people, 79–81
    in industrial health programs, 297–298
    insurance coverage for, 324
    prevention of, 108–110
mental retardation, 35, 45
    costs of, 54
    prevention of, 145–146
mercury
    pesticides containing, 104
    poisoning from, 14
Merrill, John P., 34

methodology, of cost efficiency analyses, 371–372, 391–394
Metropolitan Influenza-Pneumonia Commission, 23
Metropolitan Life Insurance Company, blood pressure testing by, 195
Middle Ages, life span in, 5
Migrant Health Act (U.S., 1962), 228
milk, pasteurization of, 16
miner's phthisis, 12
Mines, U.S. Bureau of, 16
mining, 12, 14
Mohammed, 10
mongolism (Down's syndrome), 145
monoxide poisoning, 12
mortality
 in U.S., 28–30
 see also death(s)
Morveau, Guyton de, 12
motor vehicle accidents, 43, 146–147
Mountain States Telephone Company, 256–257, 379–381
multiphasic screening (multiple screening), 115–117, 192, 269
 cost-effectiveness of, 362–363, 385–389
 costs of, 259
 health education and, 196
multiple sclerosis, 343

National Advisory Committee on Occupational Safety and Health (U.S.), 215
National Association of Manufacturers
 industrial health and safety surveys of, 251–252, 282
 industrial health assistance from, 294
 on industrial health programs, 370, 379
 multiphasic screening study of, 387
National Board of Fire Underwriters, 20–21
National Cancer Institute (U.S.), 224
National Cash Register Company, 274
National Commission on Product Safety (U.S.), 170, 226

National Commission on State Workmen's Compensation Laws (U.S.), 215, 310–311
National Council for Industrial Safety, 17–18
National Council on Rehabilitation, 119
National Environmental Policy Act (U.S., 1970), 224
National Foundation on Arts and Humanities Act (U.S.), 214
National Foundrymen's Association, 19
National Health Assembly (1948), 316
National Health Council
 on costs of industrial health programs, 370, 372
 examination campaign of 1923 of, 18
 on shortage of industrial health personnel, 241
National Health Forum, 112
National Health Policy Planning and Resources Development Act (U.S., 1974), 60, 401
National Health Service (Britain)
 formation of, 139–140
 misestimate of savings from preventive medicine by, 346
National Health Service Act (Britain, 1946), 139, 140
National Institute for Occupational Safety and Health (NIOSH) (U.S.)
 functions of, 215, 216
 on preplacement examinations, 185
 standards recommended by, 218–219
National Institute of Alcohol Abuse and Alcoholism (U.S.), 228, 390
National Institutes of Health (U.S.), 16
National Insurance Act (Britain, 1911), 24
National Labor Relations Act (Wagner Act) (U.S., 1935), 211
National Organization of Public Health Nursing, 237
National Safety Council
 on cancer prevention, 410
 founding of, 18
 on fuel conservation, 175

476 INDEX

National Safety Council (*Cont.*)
  instructional aids from, 295
  no-injury record of, 188
  on occupational safety and health, 161–162
  on safety in small businesses, 180–181
  standards set by, 293–294
  survey of 1939 of, 264–266
National Tuberculosis Association, 17
Nehru, Jawaharlal, 411
Nelson, Norton, 129
neoplasms, *see* cancer
nervous system diseases, 54
New York Telephone Company
  physical examinations used by, 184–185
New York Times, The
  fire safety at, 259
  headlines from, 164–165
  industrial health program of, 275–276
New York Transit Authority, 390
nickel, 70
Nixon, Richard M., 57, 171, 317
noise
  avoiding hazards of, 163
  concern for, by unions, 207
  costs of, 54, 259, 261
  effects of, 105–106
  fines for, 219
  government responsibilities on, 223–224
  hearing loss caused by, 70
  at The New York Times, 276
  pollution from, 276
Noise Control Act (U.S., 1972), 106
nonwhites
  causes of death for, 29–33
  drug abuse in, 38
  venereal diseases in, 36
Norris Industries, 277
Northern Pacific Railway Beneficial Association, 24
Northwest Industrial Clinic, 286–287
Norton Company, 381
Norway, 15

noxious odors, 103–104, 175
nurses
  employed by industry, 186–187
  in history of industrial health, 19
  in industrial health programs, 237, 240, 241, 266–268
  salaries of, 258
nutrition, *see* diet
Nutrition Foundation, 179–180

obestiy, 41–42, 147
occupational diseases
  incidence of, 69
  in industrial health programs, 163–166
  prevention of, 144
  under workmen's compensation, 309
  see also injuries, occupational
occupational health
  employers' role in, 181–188
  preventive medicine in, 110–111
  see also industrial health; industrial health programs
Occupational Health Institute, 232
Occupational Health Programs Accreditation Commission, 232
Occupational Health/Safety Programs Accreditation Commission, 232
Occupational Safety and Health Act (OSHA) (U.S., 1970), 111, 153, 212–222
  employer responsibilities under, 158
  enforcement of, 251
  industrial health programs in, 208
  small businesses and, 289
  text of, 415–445
  workers' role in, 404–405
  workmen's compensation and, 310
Occupational Safety and Health Administration (OSHA) (U.S.), 216
Occupational Safety and Health Commission (U.S.), 215, 372–373
O'Connor, Robert, 179
odors, noxious, 103–104, 175
office workers, injuries among, 74

INDEX   477

Oil, Chemical, and Atomic Workers International Union, 207
Olin Mathieson Company, 275, 379
oral cancer, 143
overeating, 7, 10, 22
overweight, 41–42, 147

Pap tests, 142, 363
Paracelsus, 10, 12
Pare, Ambroise, 12
Paris, water supply in, 11, 12
Parks, Everett, 309
Parran, Thomas, 134
Partnership for Health Act (U.S., 1967), 60
Pasteur, Louis, 13
PCBs (polychlorinated biphenal), 105, 207, 260
pediatrics, 108
Pennsylvania, University of, Medical School, 17
pension plans, 329, 330
personnel
  in future of industrial health programs, 405–407
  for industrial health programs, 236–245
  for preventive medicine, 131
pesticides, 104–105
  in water pollution, 174
Petrie Clinic, 287
Petty, Sir William, 348
Philadelphia Health Council, 286
Philadelphia Medical Society—Chamber of Commerce Small Plant Program, 286
physical examinations
  effectiveness of, 362–365
  factors inhibiting, 130–134
  insurance coverage for, 336–339
  in preventive medicine, 114–117
physical examinations of employees
  confidentiality of, 206
  cost-effectiveness of, 376–377, 385–389
  costs of, 258–259

  first, 19
  at IBM, 276
  by insurance companies, 21
  periodic, 186, 190–193
  preemployment, 182–185
  savings from, 380
  at Union Carbide, 272
  at Western Electric, 270–271
physical fitness programs, 274–275
physicians
  earnings of, 258
  employed by industry, 186
  in industrial health programs, 236–244, 265–268
  in small businesses' health programs, 283–285
physiotherapy, 340
plague, 6, 32, 36
planning, health, 200
Pliny the Younger, 6, 10
Plutarch, 6
pneumoconiosis, 33, 70, 144
  costs of, 86
  workmen's compensation benefits for, 307
pneumonia
  deaths from, 32–33
  prevention of, 144
poisoning, 105
  deaths from, 44
  food, 7
  industrial, in history, 12, 14
  TNT, 22
  see also lead poisoning
policy statements, 234–235
  on alcoholism control, 299
  of Union Carbide, 272
  of Western Electric, 270
polio, 36, 107
Pollock, Jerome, 134, 204, 309
pollution
  cost-effectiveness of control of, 391
  costs of control of, 92, 260–261, 409
  industrial responsibility for, 171
  total impact of, 102
  see also air pollution; water pollution

polychlorinated biphenal (PCBs), 105, 207, 260
Porterfield, John D., 134
Portland Industrial Clinic (Oregon), 287
Postal Service, U.S., 390
postnatal care, 108
Pott, Percivall, 12
Pott's disease, 6
pregnancy, 33
prenatal care, 108
prepayment medical plans
  history of, 24–25
  see also health maintenance organizations
President's Commission on the Health Needs of the Nation (1951–1952), 316–317
preventive medicine, 97–99, 100–122
  cost-effectiveness of, 347–365
  costs reduced by, 62
  deterrents to, 127–138
  future of, 401–402
  government's role in, 227–228
  health maintenance organizations and, 288–289
  industrial health programs in, 151–152
  insurance in, 334–343
  national health insurance in, 315–319
  potential for, 139–149
  practitioners of, 123–126
  unions' support for, 205
  at Western Electric, 270
prices, *see* costs of health care
proctosigmoidoscopy, 142
product safety, 170
  government's role in, 225–227
professional service review organizations, 60
Prussia, 9
psychiatric care, 337
psychological testing of employees, 19
psychoneuroses, 80
psychoses, 80
psychotherapy, 340
public, *see* community

public assistance programs, 313–314
public health
  cost-benefit analysis of, 349–350
  government's role in, 227–228
  history of, 9–12
  in preventive medicine, 101–106
  see also preventive medicine
Public Health Service, U.S.
  air pollution and, 223
  on costs of industrial health programs, 252
  Division of Industrial Hygiene established in, 16
  small business survey of, 283
  survey of 1939 of, 265
pulmonary emphysema, 33

questionnaires, on employee health, 182–185

Radiation Control for Health and Safety Act (U.S., 1968), 226
radioactive substances
  in air pollution, 174
  in water pollution, 172
radioactivity
  cancer caused by, 14, 71
  monitoring for, 166
Rasmussen, Elmer, 389
records, health
  confidentiality of, 192, 196, 206, 244, 270, 272, 407
  physician in maintenance of, 239–240
recruitment
  of handicapped, 176, 201
  health in, 182–185
rectum, cancer of, 142
Red Dye No. 2, 105, 226
reevaluations of industrial health programs, 246
rehabilitation
  cost-effectiveness of, 389
  in industrial health programs, 176, 201
  insurance in, 339–343
  in preventive medicine, 118–121

under Social Security, 311, 312
workmen's compensation and, 307–309
Rehabilitation Act (U.S., 1973), 211–212, 228
religion, in history of medical care, 7–8, 11
renal disease, see kidney diseases
research, in preventive medicine, 117–118
respiratory diseases
  as cause of disability, 75
  costs of, 53
  among farm workers, 70
  preventive medicine for, 143–144,
  rates of, 32–33
retirement plans, disability and, 329, 330
Reuther, Walter 317
Rhase, 8
rheumatism, 35
  costs of, 54
  detection of, 145
Rice, Dorothy P., 49, 349
Richard II (king, England), 11
Richardson, Elliot L., 408–409
Roberts, Norbert J., 191–192
Robertson, Logan T., 279
Rockefeller Institute for Medical Research (Rockefeller University), 17
Roger of Salerno, 8
Roman Empire, 5, 7, 10
Rome, ancient, 6, 24
Rosen, George, 98, 115, 117, 401–402
rubber industry, 207
rubella, 54, 360

safety
  at Baltimore Gas and Electric, 277
  of buildings, 163
  nonoccupational, employers' role in, 188–197
  in small businesses, 280–281
  at Union Carbide, 272
  see also accident prevention; accidents; industrial health programs
safety committees

companies having, 188
  at The New York Times, 275–276
safety lamps, 12
safety programs
  savings from, 382–384
  see also accident prevention
safety standards
  development of, 69
  under OSHA, 216–219
salary-continuance insurance, 331–333
  costs of, 259
  varieties of, 89–91
Samuels, Sheldon W., 206
Savot, 11
Sawyer, William A., 204–205
Schweitzer, Stuart, 364
sclerosis, 34–35
Scovill Manufacturing Company, 390
screenings
  effectiveness of, 362–365
  employee examinations in, 185, 186, 190–193
  in preventive medicine, 114–117
  for tuberculosis, costs of, 357–358
  for urinary tract infections, 359
  see also multiphasic screenings; physical examinations; physical examinations of employees
Scribner, Belding, 34
scrotal cancer, 12, 14
Seamen's Bank for Savings, 257, 385
second-injury funds, 307
self-employment, disability insurance for, 89
self-insurance, by employers, 322
sewage disposal, 11–12
Shaw, George Bernard, 399
Sheehan, Jack, 222
shipping, environmental concerns in, 199
shock (electrical), 71
Shop and Mills Health and Welfare Thrust, 386
short-term disability insurance, 326–330
Sigerist, Henry, 129
silicosis, 71, 144

Silver, George, 133
skin cancer
　industrial causes of, 14
　sunlight causing, 142
Small Business Administration (U.S.), 296
smallpox, 36
Smith, Adam, 348
smoking, 39–41
　cancer and, 142, 143, 410
　costs of, 360
　disability correlated with, 78–79
　occupational diseases and, 69
　reduction of, 148
　respiratory diseases associated with, 143–144
Snow, John, 13
Social Security, 311–312
Social Security Act (U.S., 1965), 312
Society of Actuaries, 22
Socony-Vacuum Oil Company, 379
Spanish-Americans, 38
Sparks, Richard, 363–364
sphygmomanometers, 21
staffing, *see* personnel
Stamler, Jermiah, 358–359
standards
　under OSHA, 216–219
　set by voluntary agencies, 293–294
states
　environmental control responsibilities of, 222–223
　Medicaid in, 312
　public assistance in, 313–314
　role in occupational health and safety of, 212, 215, 216, 220, 221
　workmen's compensation in, 306, 307, 310–311
statistics
　on accidents, in prevention, 118
　cautions regarding, 28, 44
　collected under OSHA, 216–218
　gathering, on cost-efficiency, 371–372, 374–376, 392–394
　insurance companies' uses of, 23
Stellman, Jeanne M., 209, 404

Stern, Bernard J., 155–156
Sterner, James J., 179
Stevart, Guy, 128
Steward, William H., 406
Stieglitz, Edward J., 97–98
stomach cancer, 143
stress, 147–148
strikes, insurance continuing during, 325
strokes
　preventive medicine for, 141–142
　rates of, 31
suicides
　alcoholism and, 37
　as cause of death, 29
　of kidney disease patients, 34
supervisors, 245–246
surgery
　history of, 8
　insurance coverage for, 323, 324
　unnecessary, 58
survivors benefits
　under life insurance, 333
　under workmen's compensation, 307
Sweden, 157, 208
syphilis, 7, 36, 54

Taft-Hartley Act (Labor Management Relations Act) (U.S., 1047), 211
taxes, disability benefits and, 330
Teeling-Smith, G., 362
temporary disability benefits, 314
tetanus, 394
Textile Workers Union of America, 221
TNT poisoning, 22
tobacco, *see* smoking
toilets, in history, 7, 10
toxic substances
　testing of, 198–199
　in water pollution, 172
　in workplace, 71, 163–164
Toynbee, Arnold, 2, 398
trade secrets, under OSHA, 215
training, in safety and health, for employees, 188
transplantations, 32, 357

treatment
  in alcoholism control units, 299
  for nonoccupational disease and injury, 193
Truman, Harry S., 315, 316
tuberculosis
  costs, 53
  deaths from, 33
  Framingham experiment on, 23
  prevention of, 144
  screening for, costs of, 357–358
typhoid fever, 16, 36

unemployed workers, health insurance coverage for, 326
Ungerleider-Gubner tables for heart size, 22–23
Union Carbide Corporation, 256, 272–273
unions
  alcoholism control supported by, 299
  concern for occupational safety and health of, 110
  in future of industrial health, 404–405
  in history of industrial health, 20
  history of insurance plans of, 26
  industrial health assistance from, 295
  in industrial health programs, 204–209, 220, 235–236
  labor legislation affecting, 211
  number of members of, 67–68
  preventive medicine and, 125
  in small businesses' health programs, 285
United States
  alcoholism program for employees of, 390
  eighteenth century life span in, 5
  environmental control legislation in, 223–224
  health care financed by, 305–319
  health care planning in, 60
  health status in, 28–46
  history of industrial health legislation in, 15–17
  history of medicine in, 8–9
  history of sewage disposal in, 12
  occupational safety and health legislation in, 211–222
  see also government; states
universities and colleges
  industrial health assistance from, 295–296
  in industrial health history, 17
  preventive medicine and, 124–125
Upjohn Company, 383
uranium, cancer and, 70
urinary tract infections
  preventive medicine for, 144–145
  rates of, 33–34
  screening for, 359
uterine cancer
  cost-effectiveness analysis of prevention of, 355
  Pap tests for, 142

vaccination, *see* immunization
Van Buren, Martin, 15
van Steenwick, John, 336
venereal diseases, 36, 107
  control of, 145
ventilation, artificial, 12
Veterans Administration (U.S.), 313
vibration, as occupational hazard, 71
vinyl chloride
  angiosarcoma of liver caused by, 71, 373
  cancer caused by, 14
  FDA actions on, 226
  jobs affected by, 209
violence, as cause of death, 7
viral diseases, 34
vision, impairments of, 39
Visiting Nurse Association of Chicago, 19
voluntary agencies
  in history of industrial health, 17–18
  in OSHA standards, 216
  in preventive medicine, 124
  standards set by, 293–294
von Humboldt, Baron Alexander, 12

wages *lost* through illness and accident, 85
Wagner Act (National Labor Relations Act) (U.S., 1935), 211
waiting periods, for disability insurance, 330
Walsh-Healey Act (U.S., 1936), 211, 213, 214
waste disposal, 104
water pollution, 171–173
　costs of, 261
　in history, 11–13
　in public health, 104
water supply, 10–12
Watt, James, 12
weight, body, studies on, 22
Weisbrod, Burton, 48–49, 348–349
welfare programs, 313–314
Western Electric Company
　industrial health program of, 255–256, 269–272
　safety program of, 383
Westinghouse Corporation, 386
Weyerhaeuser Company, 301
White, Kerr L., 137
Williams, Alan, 351
Williamson, Sherman, 183
women, *see* females
workers
　in future of industrial health, 404–405, 410

see also employees; unions
workforce, 67
working hours, regulation of, 15–16
workmen's compensation insurance, 305–311
　costs and benefits of, 91
　industrial health programs reducing costs of, 378, 379, 382
　occupational health measures and, 111
　private disability plans and, 326
　rehabilitation under, 120–121, 389
workmen's compensation laws
　enactment in U.S. of, 155, 211, 231, 305–306
　first, in Germany, 15
Workmen's Compensation Service Information Bureau, 21
workplace
　accidents at, 43
　mental illness prevention at, 109
　health care system in, 152
　see also industrial health; industrial health programs
World Health Organization, 407–408
world population, effect of disease on, 6–7
Wynder, Ernst, 319, 360

Zapp, John, 403
Zenz, Carl, 386

RETURN PUBLIC HEALTH LIBRARY
TO: 42 Warren Hall

JAN 8 1981